TERTIARY LEVEL BIOLOGY

Tropical Rain Forest Ecology

Second edition

D.J. MABBERLEY

Fellow and Tutor in Plant Science
Wadham College
University Lecturer in Plant Science
University of Oxford

BLACKIE ACADEMIC & PROFESSIONAL

An Imprint of Chapman & Hall

London · Glasgow · New York · Tokyo · Melbourne · Madras

**Published by Blackie Academic & Professional, an imprint of
Chapman & Hall, Wester Cleddens Road, Bishopbriggs, Glasgow
G64 2NZ, UK**

Chapman & Hall, 2-6 Boundary Row, London SE1 8HN, UK

Blackie Academic & Professional, Wester Cleddens Road,
Bishopbriggs, Glasgow G64 2NZ, UK

Chapman & Hall Inc., One Penn Plaza, 41st Floor, New York
NY 10119, USA

Chapman & Hall Japan, Thomson Publishing Japan, Hirakawacho
Nemoto Building, 6F, 1-7-11 Hirakawa-cho, Chiyoda-ku, Tokyo 102,
Japan

DA Book (Aust.) Pty Ltd, 648 Whitehorse Road, Mitcham 3132,
Victoria, Australia

Chapman & Hall India, R. Seshadri, 32 Second Main Road, CIT East,
Madras 600 035, India

First edition 1983
Second edition 1992
Reprinted 1994

© 1992 Blackie & Son Ltd.

Typeset by Thomson Press (India) Limited, New Delhi
Printed in Great Britain by St Edmundsbury Press, Bury St Edmunds,
Suffolk

ISBN 0 7514 0160 9

A catalogue record for this book is available from the British Library
Library of Congress Cataloging-in-Publication Data available

Tropical Rain Forest Ecology

TERTIARY LEVEL BIOLOGY

A series covering selected areas of biology at advanced undergraduate level. While designed specifically for course options at this level within Universities and Polytechnics, the series will be of great value to specialists and research workers in other fields who require knowledge of the essentials of a subject.

Recent titles in the series:

Social Behaviour in Mammals	Poole
Genetics of Microbes (2nd edn.)	Bainbridge
Seabird Ecology	Furness and Monaghan
The Biochemistry of Energy Utilization in Plants	Dennis
The Behavioural Ecology of Ants	Sudd and Franks
Anaerobic Bacteria	Holland, Knapp and Shoesmith
Biology of Fishes	Bone and Marshall
Environmental Microbiology	Grant and Long
Evolutionary Principles	Calow
The Lichen-Forming Fungi	Hawksworth and Hill
Seabird Ecology	Furness and Monaghan
Virology of Flowering Plants	Stevens
An Introduction to Marine Science (2nd edn.)	Meadows and Campbell
Seed Dormancy and Germination	Bradbeer
Plant Growth Regulators	Roberts and Hooley
Plant Molecular Biology (2nd edn.)	Grierson and Covey
Polar Ecology	Stonehouse
The Estuarine Ecosystem (2nd edn.)	McLusky
Soil Biology	Wood
Photosynthesis	Gregory
The Cytoskeleton and Cell Motility	Preston, King and Hyams
Waterfowl Ecology	Owen and Black
Biology of Fresh Waters (2nd edn.)	Maitland

Preface

Since the first edition of this book was written, public awareness of tropical rain forests has become so great that issues involving their exploitation are the stuff of daily newspapers, radio and television. The plight of forest-living peoples has become an international issue; concerns over the greenhouse effect and other climatic changes are often linked to rain forest destruction. At the same time, there has been an unparalleled scientific interest in the workings of the rain forest and an increasing concern by economists as to its potential in balancing the books of many developing countries. The need for an advanced yet concise and up-to-date synthesis of recent studies and a key to the increasingly voluminous literature on rain forests is even greater than it was in 1983.

There are now many highly illustrated popular books on rain forests, as well as new editions of K.A. Longman and J. Jeník *Tropical rain forest and its environment* (2nd edition, 1987) and T.C. Whitmore *Tropical rain forests of the Far East* (2nd edition, 1984, many of the splendid illustrations from which are to be found in his rather less ambitious *Introduction to tropical rain forests*, 1990). Other very welcome regional accounts of rain forest biology in various parts of the tropics have appeared, notable being D.H. Janzen (ed.), *Costa Rican natural history* (1983); Earl of Cranbrook (ed.), *Malaysia* (1988); G.T. Prance and T.E. Lovejoy (eds), *Amazonia* (1984); A. Whitten *et al.* (eds), *The ecology of Sumatra* (2nd edition, 1987) and *The ecology of Sulawesi* (1988). With more popular books such as N. Myers *The primary source: tropical forests and our future* (1984), proceedings of numerous symposia and new journals such as *Evolutionary ecology, Functional ecology, Journal of tropical ecology* and *Trends in ecology and evolution* also available, the reader is now able to discover information over a wide range of topics. However, as a reasonably priced account of the whole subject, the first edition of this book found its niche, and I offer the second one in the hope that it will do likewise.

As in the first edition, this short text deliberately dwells on the trees and other plants that make up the forests in which the animals live. Nevertheless, the forest cannot exist without the animals, and it has been possible in this edition to do more justice to the zoological literature. A great deal of the text has been expanded and rearranged to accommodate the surge of new work which has not only added to our knowledge of rain forests but has often questioned certain dogmas in the subject. Although some of the early parts of the book have come through unscathed, having withstood the test of time, every chapter has been updated. Major revisions were called for in the sections on quaternary studies, fire, nutrient-cycling, gap-formation and its relationship to animal biology, epiphytes, lianes, bryophytes, plant and animal cycles, herbivory, seed dispersal, pollination and those dealing with anthropological issues. As with the first edition, my principal aim has been to show the importance of change as well as diversity in the rain forest and to try to indicate how humans and their habits are involved in this.

The greatest synthesis of rain forest biology remains P.W. Richards *The tropical rain forest* (Cambridge University Press, 1952 with many reprints). All subsequent texbooks have relied heavily on it and, although it has not yet been revised, it is still a most valuable source book and should be consulted by the serious student who wishes to take the subject further. The books listed under 'Further Reading' will repay study, for only a concise rendering of the principal areas of rain-forest ecology can be made in a book as short as the present one. I have indicated, by numbers in parentheses, references to striking or original ideas or facts in these books, and in original research papers, for those who wish to follow up particular points. These references, over two and a half times as many as in the first edition, are listed chapter-by-chapter at the end of the book.

I am painfully aware that most undergraduates know little of tropical plants and animals despite the ballyhoo of 'the media', while a grounding in the basics of plant science and zoology, which could be relied on until fairly recently as having been taught at school, is often lacking too. Unfamiliar tree and other plant names are inevitably a hurdle to many. *Flowering plants of the world*, edited by V.H. Heywood (Oxford University Press, 1978 and republished since) and, I hope, my own *The plant-book* (Cambridge University Press (1987), corrected reprint 1990), should put many of them into context and provide some further information about their distribution, uses and so on. Finally, an unsurpassed general text which looks at plants and their involvements with animals from a tropical standpoint is E.J.H. Corner *The life of plants* (Weidenfeld and Nicolson, 1964, with recent reprints and paper editions), which, unlike most textbooks, is not only a sparkling account, to be read continuously from cover to cover, but is also the nearest that any modern biological work I know comes to literature.

I am indebted to many people for their encouragement and help in the production of this new edition, but would like particularly to acknowledge the captain and crew of M.S. *Profesor Mierzejewski* (Polish Ocean Lines, Gdynia), Mary Peat at the University of Sydney, where I was Visiting Scholar in 1989, and Professor S. Balasubramaniam, University of Peradeniya, where I was Visiting Senior Lecturer, 1989–90, all of whom provided congenial working environments. I also wish to thank Rosemary Wise for the original illustrations, and, as so often, Anne Sing for technical support throughout the book's gestation. Finally I owe a particular debt to Alistair Hay and especially Peter Weyde, of Sydney, for their friendship and encouragement.

D.J.M.
Le Havre-Sydney, Peradeniya, Oxford

Contents

Chapter 1 THE TROPICAL RAIN FOREST   1

    1.1 A tropical origin for ecology?   2
    1.2 Tropical forests   4
    1.3 Some misconceptions   9
    1.4 The tropical rain forest in a wider context   11
        1.4.1 Floristics   11
        1.4.2 Numbers of species   13
    1.5 The global interest   14
        1.5.1 The global carbon cycle   15

Chapter 2 THE CHANGING PHYSICAL SETTING   17

    2.1 Continental drift   17
    2.2 Tropical climate   23
        2.2.1 Precipitation   24
        2.2.2 Storms, droughts and vulcanism   27
        2.2.3 Temperature and radiation   29

Chapter 3 SOILS AND NUTRIENTS   31

    3.1 Tropical soils   31
        3.1.1 Soil processes   31
    3.2 Soil types   32
        3.2.1 Major groups of soils   32
        3.2.2 Volcanic soils   34
        3.2.3 Other soil types   34
    3.3 The relationship between soils and forest type   36
    3.4 Nutrient cycling   41
        3.4.1 Precipitation   43
        3.4.2 Litter   45
        3.4.3 Roots and mycorrhizae   47
        3.4.4 Animals   49
        3.4.5 Overall production   50

Chapter 4   THE CHANGING BIOLOGICAL FRAMEWORK   52

    4.1   Tropical rain forest successions   53
          4.1.1   Problems with older views   53
          4.1.2   The modern interpretation   54
    4.2   The ecology of gaps   55
    4.3   The ecology of pioneers   59
          4.3.1   Pioneers and shade-bearers   59
          4.3.2   Pioneer morphology   62
          4.3.3   Effect on shade-bearers   63
          4.3.4   Features of regeneration   63
          4.3.5   Seed-rain, seedbanks, dormancy and germination   65
    4.4   Features of later succession   70
          4.4.1   Seeds and seedlings   70
          4.4.2   Saplings   72
          4.4.3   Effects on nutrients   73
          4.4.4   Effects beyond the gap   73
    4.5   Animals and succession   74
          4.5.1   Effects of gaps on animals   74
          4.5.2   Effects of animals on gaps   74
          4.5.3   Pollination and dispersal   75
    4.6   Primary successions   76
    4.7   Implications   79

Chapter 5   THE COMPONENTS OF DIVERSITY AND THEIR
            DYNAMICS   80

    5.1   Geographical diversity   82
          5.1.1   Flora   82
          5.1.2   Fauna   83
    5.2   Morphological diversity   92
          5.2.1   Tree form   92
          5.2.2   Other growth forms   105
          5.2.3   'Anomalies'   113
    5.3   Infraspecific variation   115
    5.4   Seasonal variation and other cycles   116
          5.4.1   Plant cycles   116
          5.4.2   Animal cycles   126

Chapter 6   COEXISTENCE AND COEVOLUTION   133

    6.1   Herbivory and resistance to it   133
          6.1.1   Plant resistance   136
          6.1.2   Some mammals   139
          6.1.3   Some invertebrates   142
    6.2   Frugivory and seed dispersal   148
          6.2.1   Birds   152
          6.2.2   Dispersal by several agents   156
          6.2.3   Some marsupials and primates   160
          6.2.4   Fish and ants, wind and water   162

6.2.5 'Anachronisms'   165
6.2.6 Specificity   168
6.3   Florivory and pollination   169
       6.3.1 Insect pollination (entomophily)   169
       6.3.2 Bird pollination (ornithophily)   179
       6.3.3 Bat pollination (chiropterophily)   181
       6.3.4 Other mammals   183
6.4   The milieu for 'mutualism'   183

Chapter 7   SPECIES RICHNESS   187

7.1   Speciation   187
       7.1.1 Refugia   189
7.2   Species diversity   192
       7.2.1 Environmental heterogeneity   194
       7.2.2 Biotic factors   195
       7.2.3 Combinations of factors   198
       7.2.4 An element of chance?   201
7.3   Practical problems   202

Chapter 8   TRADITIONAL RAIN-FOREST USE   205

8.1   The fossil record   206
8.2   Early agriculture   208
8.3   Hunter-gatherers?   210
8.4   Some rain-forest societies   213
8.5   Humans as ecosystem modifiers   215

Chapter 9   THE CHANGING FOREST TODAY   221

9.1   Forest conversion   222
       9.1.1 Farming and gardening   222
       9.1.2 Logging and silviculture   227
       9.1.3 Ranching   234
       9.1.4 Fuel   234
9.2   The prospects   235
       9.2.1 Soils   236
       9.2.2 Succession after clearing   241
       9.2.3 Animals   243
       9.2.4 The 'new' forest   245
       9.2.5 Conservation   249

POSTSCRIPT   264

FURTHER READING   268

REFERENCES   269

INDEX   289

CHAPTER ONE

THE TROPICAL RAIN FOREST

A great deal of Botany is based on herbarium sheets, collecting tubes, microscope slides, wood-blocks, and petri dishes. These are little things which carry only little plants or bits of the big. We start with *Ranunculus* or *Capsella* and go down the scale to *Chlamydomonas, Spirogyra, Fucus* (which is a little out of the way, and getting unmanageable), *Agaricus, Funaria* and *Nephrodium.* We learn of trees from twigs, sporophylls and stained preparations. We are led away by chromosomes, *Neurospora,* subspecies, pollen grains, and ultrastructure. We research in grassland, heath, plantations, sea-shore, desert, and savannah. And, if we have the luck to reach the tropical forest, the immensity is baffling: we collect all the common little things which have been collected there before, because there is no time for the big. So botany has been compounded from its myriad particles, its manageable objects, and its easily presentable aspects: the pursuit is minuter. The cry is in the cities. The principles of plant-life can be demonstrated in dishes, flasks, pots, frames and borders. The graduation is from universities where academic distinction lies. Our grandfathers went to the backwoods to find out: our benches are stacked with costly instruments, that detain us. The axe and the pocket-lens, the flora and the systematic eye are recreations for a holiday. The struggle for existence is common sense, natural selection just a sieve through which the smaller pass. Oh, what a desert the cities are making of our science! Turn with me, for a moment, to the great depots of plant life in the tropical forests, and see the effect. The little flowers and fruits are hoist on big trees: and how do little *Ranunculus* and *Alisma, Helianthus* and *Poa, Funaria* and *Nephrodium* compete with them? (E. J. H. Corner, 'On thinking big', *Phytomorphology,* **17** (1967) 24).

The diverse and rich flora and fauna of the tropics have bewildered, overwhelmed and humbled those trained in the biologically impoverished temperate regions. Such an experience is essential for those who wish to pursue tropical studies, and has been that of all the great traveller-naturalists such as Henry Walter Bates and Alfred Russel Wallace (1, 2) in the last century. Long before them, the pioneering Dutchman, Georg Everard Rumpf, set down the more remarkable of the plants of the tiny island of Ambon in the Moluccas, and his fellow countryman, Hendrik van Rheede tot Draakenstein, supervised the first major work to bring a tropical flora to the notice of Europeans. This was the monumental Flora of part of India in twelve folio volumes, the *Hortus Indicus Malabaricus* (1678–1702). These works, like the chain of botanic gardens and other research institutes established later, were the by-products of European colonial expansion. And, indeed, the first indication of the riches of the tropics had come from similar colonial ambition, for Alexander the Great, having defeated the Persian King Darius in 331 BC, had pushed on over the Khyber Pass to the Punjab of India, so that the Indus became the eastern boundary of his extended Asiatic empire. Though the extent of the enormous enterprise was shortlived, the culture and natural history of India was forever thereafter linked to that of the western world. Findings from his invasion were incorporated in Theophrastus's influential *Enquiry into Plants*: the banana, the mango and *jak* fruit (Figure 5.10), cotton and mangroves, the astonishing trees living in the sea, and the banyan (*Ficus benghalensis*), putting out aerial roots, which undermined the ancients' view of what roots did.

1.1 A tropical origin for ecology?

Today, with an increasingly less parochial view of biology, scientists are turning more and more to the tropics as a means of understanding ecological and evolutionary problems set in a temperate context. Such an approach has led to new insights and concepts readily applied in ecology. But it is true to say that this apparently recent change of heart or emphasis is no new departure peculiar to 'modern' biology, for much of the advance in biological science in the last century derived from tropical experience. In the eighteenth century, it had been widely believed that, though the tropics had a number of peculiar organisms, they were not particularly rich in species, at least in species of plant. Indeed, Linnaeus himself seems to have believed that the tropical flora was rather homogeneous and limited. This attitude may well have reflected the collections of pantropical weeds around settlements, and of the widespread littoral species readily accessible to travellers and brought back to Europe. Access to the canopy of rain forest was rarely obtained and such familiar tropical phenomena as cauliflory, the bearing of flowers on the trunks of trees, so little understood that a mahogany relation with cauliflorous flowers was thought to be a parasite by Linnaeus's pupil Osbeck, who named it *Melia parasitica*, the *Dysoxylum parasiticum* (Meliaceae) of today. More extensive travelling and the penetration of the continents led to a discarding of these ideas and, by the turn of the nineteenth century, attitudes were changing rapidly, partly due, at least, to advances in attempts to put some geographical order into the hordes of plants and animals being brought back to Europe.

Nonetheless, most classical ecological studies have been carried out in the more accessible north temperate zone and it is perhaps surprising to learn that it can be convincingly argued that the scientific study of what we now call ecology began in the tropics of South America (3). Early ideas on plant distribution were brought together following two early nineteenth-century expeditions to the tropics: Alexander von Humboldt climbing the Ecuadorean volcano, Mount Chimborazo (6300 m and, at the time, thought to be the world's highest peak) and noting the correlation between climate and vegetation type as he ascended, and Robert Brown, on Matthew Flinders's voyage to Australia, preparing an early phytogeographical account of that continent and, later, of tropical Africa. These two pioneers were mutual admires and corresponded, von Humboldt going on to produce influential treatises, the biogeographical ideas in which were taken up by zoologists, who established faunal provinces similar to the world schemes of vegetation types built on the work of von Humboldt and Brown. It must be remembered that it was the tropical environment that is alleged to have triggered both Charles Darwin and Wallace to formulate their converging views on the mode of evolution by natural selection, where Darwin stressed the biotic relationships while Wallace stressed the physical environment in essentially ecologically based theories of evolution.

It is true that, as in much of science, some of these ideas were not altogether new, for the seventeenth-century Dutch microscopist, Antony van Leeuwenhoek, developed the idea of the importance of food chains, and, from his observations of the shellfishes in Dutch canals, came to some idea of competition between organisms, while in the following century, Count Buffon was groping towards the concept of what is now termed succession. Nonetheless, these figures did not, in these areas of inquiry, affect the mainstream of scientific thought as it was passed on to the nineteenth century. Only then was the scientific world receptive and the key figure is held to have been Eugen Warming, a Dane who spent three years writing what would now be considered an ecological survey, an account of the forest around the Brazilian village of Lagoa Santa. Rarely before had there ever been systematic collecting in a small area, a fine exception, however, being Wallace's intensive study of beetles over 6 weeks in a square mile of forest near the Sadong River in Borneo, where, in collecting over 1000 species, predominantly longhorns and weevils (34 new to him on 1 day alone), he was probably the first to make such a survey within a limited amount of rain forest (4). Warming returned to Europe to give courses in ecology in the University of Copenhagen and wrote the first textbook specifically devoted to ecology (1895), but because, in 1898, A. F. W.

Schimper published his *Pflanzengeographie auf physiologischer Grundlage*, a book to become a standard text in English as well as German, leaning heavily on Warming's work though without even a footnote to acknowledge it, Warming's pioneering contribution has been largely overlooked.

1.2 Tropical forests

It was Schimper who coined the term *Tropische Regenwald*, tropical rain forest, and contrasted it with monsoon forest, which was more or less leafless during the dry season, especially towards the end. His definition of rain forest was 'evergreen, hygrophilous in character, at least 30 m high, rich in thick-stemmed lianes, and in woody as well as herbaceous epiphytes'. Monsoon forest was lower generally and, though rich in woody lianes and herbaceous epiphytes, poor in woody epiphytes. Lianes are woody 'climbers', which are often carried up into the canopy by maturing trees and are very poorly represented in temperate countries. For example, in the British Isles, the only species are ivy (*Hedera helix*, Araliaceae), old man's beard (*Clematis vitalba*, Ranunculaceae) and honeysuckle (*Lonicera periclymenum*, Caprifoliaceae), whereas 170 species have been recorded from one tiny island in the Panama Canal. Epiphytes are plants living perched on, but not necessarily deriving sustenance from, other plants. In temperate countries, such plants, excluding marine and other aquatic forms and parasites, are generally herbaceous and small, chiefly bryophytes, lichens and algae.

The term 'rain forest' has been so long established in the literature that it is retained in the broad sense here. Although rain forest is, in general, associated with heavy rainfall, it embraces certain riparian forests in seasonal climates. Again, in cool everwet climates, there are temperate rain forests, as along the Pacific coast of North America for example. The humid lowland tropics account for less than a third of the tropical land surface (5), which has deserts, savannas and grasslands as well as montane vegetation types. In attempts to assess the present-day status of forests in the tropics, the term 'tropical moist forest' (6) has become current and much of the work discussed in this book has been carried out in vegetation covered by this term, rather than in rain forest in the narrrow sense. Myers has defined it as 'evergreen or partly evergreen in areas receiving not less than 100 mm of precipitation in any month for two out of three years, with mean annual temperature of more than 24°C and frost free'. It usually lies below 1300 m but, according to Myers, does not reach above 750 m in

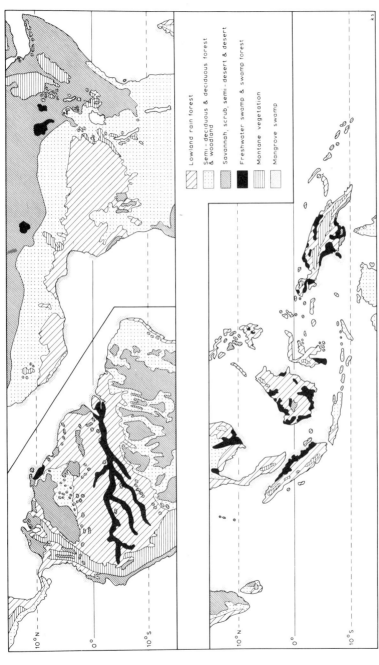

Figure 1.1 Tropical vegetation types. Reproduced with permission from J.R. Flenley. *Equatorial Rain Forest: a Geological History* (Butterworth, 1979). Tropical lowland rain forest represents the major part of tropical moist forests, which in 1976 covered some 935 million ha, the size of the United States. This is 71% of the total area of closed tropical forest and almost 30% of all the world's forests. 472 million ha were in South America, centred on the basins of the Amazon and Orinoco, 187 million ha in Asia, largely Malesia, and 175 million ha in Central Africa with an extension to Sierra Leone. By 1980 (7), the total area of closed broadleaved forest had been reduced to some 860 million ha, of which some 668 million were 'undisturbed', over 153 million logged over and about 38 million were under some form of management.

south-east Asia, and includes cloud forest, riparian forest, swamp and bog forest and the wetter end of the lowland seasonal forest as well as what is regarded as rain forest in the narrow sense. The difficulty of drawing up worldwide definitions led Myers to consider the absence of seasonality as more critical than overall precipitation. For example, much of Latin America has more than 4000 mm of precipitation per annum, a figure rare in Asia and almost unknown in Africa where 1500 mm is often considered the minimum to support rain forest, while in Latin America this is the midpoint for the maintenance of tropical dry forests.

The tall tropical rain forest of Amazonia, 'terra firme' forest, as opposed to various types of inundated ones, occupies 51% of that region. Such forests are not precisely restricted to the tropics in that they extend to 28°N in southern China, though they are largely cleared from there, also into Mexico and southwards down the east slopes of Madagascar, the southern end of the coastal strip of Brazilian rain forest and the eastern coast of Australia, where they are often referred to as 'vine (i.e. liane) forests'. At these extremes, the forest differs somewhat and is often referred to as subtropical rain forest. In turn this grades into subtemperate and temperate rain forests like those, for example, in Japan and New Zealand. In the north island of New Zealand at 34–47°S, the rain forests have conspicuous woody epiphytes and even a strangling *Metrosideros* (Myrtaceae), reminiscent of the strangling figs (see section 5.2.2) of the tropical forests (8). At the edges of its distribution or under stressful climatic conditions, rain forest merges into semi-evergreen rain forest, a belt lying between the evergreen and moist deciduous forests. The more seasonal of the subtropical rain forests are prone to conflagration at the hands of man or from lightning and have been largely degraded to open woodlands and savanna, notably in Myanamar (Burma) and Thailand and much of Africa but such are more rarely encountered in South America. On the coasts, the mangrove swamps become lower in stature the further they grow from the equator and eventually pass into the salt marsh vegetation familiar in temperate regions.

In this biome are several million species of organisms, of which fewer than half a million even have a Latin name (6), and a minute fraction of them are at all understood in terms of their ecology or even distribution. According to Diamond (9), 'No-one doubts, and few people care, that thousands of small terrestrial invertebrates and plants await description, especially in the tropics'. However, even in 'well-known' groups, there are still new discoveries: since 1900, 134 accepted new genera of mammals have been discovered, mostly in the tropics, to give a current total of about 1050.

Of these discoveries, two are being placed in monotypic families; one of these, the bumble-bee bat found in Thailand in 1975 at 2 g and 3 cm long, is the smallest warm-blooded animal known. At present, about three new species of bird are described each year. Even economically important plants have waited until recently to be described: the oil-seeds *Curupira tefeensis* (Olacaceae) and *Acioa edulis* (Chrysobalanaceae), long-collected as flotsam from the Amazon, were not named until 1948 (a new genus) and 1973 respectively.

Based on numbers of insect species associated with tree species in the Neotropics, Erwin has calculated that there might have been as many as 30 million species of insect in the world: this calculation hinges on the specificity of insects, and evidence lately adduced in this area suggests that the total may be more like 50 million. Although covering only 6% of the world's surface, tropical rain forest contains about half the earth's total of animal and plant species, including, for example, some 70–75% of all known arthropods. The most extensive tracts of forest are those in the Americas: the forests of the eastern Andes, the Amazon basin and the Caribbean as a whole account for about a sixth of the total non-coniferous forest in the world.

It is exceedingly difficult to estimate the potential and actual areas covered by the world's tropical rain forest, but it has been calculated from aerial surveys, satellite photographs and other methods that, although some 16 million km² is potentially rain-forest, by the end of 1988 eight million km² (i.e. about half) had been cleared or heavily modified. Africa had lost well over half of its rain forest, and tropical America and Asia had lost at least 40% of theirs. This accounts for about 75% of the total area of closed tropical forest and almost 26% of all the world's forest. In the early 1980s (10), about 100 000 km² were being permanently lost to agriculture each year (i.e. about 1 km² every 5 min) and a similar area was seriously disturbed through logging. By 1987, the rate of clearance in Amazonia alone was believed to be 80 000 km². In short, about 2.5% of the entire biome was being eliminated each year and Myers has predicted that, at these rates, there will be forest left only in Zaire and Amazonia, with smaller amounts in New Guinea and the Guyana Highlands, by early next century. The effect is very uneven, but, as an example of a single country, Sierra Leone in west Africa today has 4% forest cover, when the original was probably 60%.

The richness of some of the forests can be gauged from the statistics from the Malay Peninsula (11, 12), which is estimated to have some 8000 plant species in about 1400 genera (about 28 endemic) in an area half the size of

Britain (which has some 1430 native species in 620 genera, none endemic). Taking mammals and comparing Malaya with Denmark, a similar-sized peninsula, the number of families, genera and species in Denmark is 13/32/45 compared with 32/104/203 in Malaya with 12 bats (83 in Malaya), 14 rodents (54 in Malaya) and eight carnivores (29 in Malaya). Although the rain forests of Fiji may be as poor as those of the richest North American forests, the often-cited species-rich *fynbos* vegetation of southern Africa is comparable in diversity with the poorest rain-forest samples. In *fynbos*, there is about the same total number of plant species as there are trees down to 10 cm diameter at breast height in tropical rain forest, nevertheless, compared with the total number of vascular plant species in the forests, it is much poorer.

On the Rio Palenque in Ecuador, 365 vascular plant species, including 203 woody ones, were found on 0.1 ha. Over a third of the species and half of the plants were epiphytes (13, 14), 13% terrestrial herbs, 10% shrubs and 9% non-epiphytic lianes. In Costa Rica, a 10m × 10m piece of forest was felled and 233 vascular plant species with 32 bryophyte species were recorded (i.e. the equivalent of a sixth of the British flora on half a tennis-court). While the plot was probably not as rich as neighbouring sites (the largest tree had powdery bark and therefore few epiphytes), it demonstrates that with 132 species (57%) of the independent plants less than 1 m tall, the earth's richest shrub community is neither *fynbos* nor Australian heath but the understorey of neotropical rain forest. Of tree species with a diameter at breast height of over 10 cm, over 300 per hectare were recorded in Amazonia, the total number of vascular plant species in western Ecuador being 365 per hectare.

Such forests are marked by certain features besides those in Schimper's definition. They are notable for the strangling habit most familiar in the pantropical *Ficus* (Moraceae), but also in *Clusia* (Guttiferae, Neotropics) and *Wightia* (Scrophulariaceae, Indomalesia); there is a poverty of leaf litter, reflecting a rapid decomposition rate and speedy recycling of nutrients (a tree may disappear completely 5 years after falling, through the action of fungi and termites); the involvement of animals in herbivory, pollination and dispersal is various and more marked than in most other vegetation types, with a consequent range of flower and fruit presentation much greater than in temperate zones; there is a wide range of tree form, both in general *Bauplan* and in relative massiveness of leaves, buds and twigs; and there is a marked parallelism between families in the range of response to the ecological opportunities arising from the formation of gaps in the forest through the fall of trees.

On a world scale it is difficult to gauge floristic richness in rain forests.

Should the generic diversity or family diversity be considered more important than the number of species? A further complication in using the latter criterion is the man-made one, for it has been argued that recent workers on the tropical Asiatic flora have had a wider species concept than, say, many of those working in the Neotropics. Nevertheless, it is generally held that some of the richest forests are those in south-east Asia, the wettest forests of which correspond to the ombrophilous forest (or rain forest in the narrow sense) of earlier classifications. These are remarkable for the abundance and diversity of one particular family that is of enormous commercial importance, the Dipterocarpaceae. Malaya alone has nine genera and 155 (127 endemic) species. Such forests in Sarawak and Brunei in Borneo have 2000 species of tree with a diameter greater than 10 cm. This compares with a figure of 849 in the heath forest and 234 in the peat-swamp forest, two forest types to be considered below, in the same island. Compared with these rich forests, those in the same region with an annual water stress of at least a few weeks' duration are poorer and differ in a number of features, although they are considered here under the umbrella of tropical rain forest (15). These would include the extension down the east coast of Australia, and parts of the Indian and Indo-Chinese rain forests, also the Philippines, Lesser Sunda Islands and parts of New Guinea. Such forest is probably the most extensive of all rain-forest formations in much of central Amazonia and the Caribbean and probably corresponds to the moister of the African rain forests. In it, there is a greater tendency for a marked gregariousness of trees while deciduous trees may make up to a third of the taller tree species. The bark tends to be thicker and rougher in these tree species, and cauliflory and ramiflory, the bearing of the flowers on the branches, is rarer. The canopy height is somewhat lower and there are fewer large lianes, while it is more easily destroyed by fire and thus replaced by grassland.

1.3 Some misconceptions

Possibly springing from boys' adventure books of the nineteenth century and from the more purple passages of over-enthusiastic travelogues of the period, is the idea that tropical rain forest is dark, steamy and impenetrable, overrun with snakes and crawling with poisonous insects and other dangers. Much of this comes from a superficial knowledge of the tangle of secondary forests or forest along river edges. Such places were the first to be encountered by travellers and, indeed, are difficult to penetrate. Further-more, hacking a path through them is likely to disturb animals, some of which at least may react less than pleasantly to the destruction of their habitat. Within the forest proper, however, it is generally rather easy to move about without cutting a path and it is rarely so wet as to make progress difficult. It may be dark here and there but the mosaic pattern of regenerating forest and treefalls and the movement of sunflecks belie the myth of overall dismalness. It is often cooler than cleared land outside, although the absence of a breeze may make human beings insensitive to this.

There are diurnal fluctuations in temperature, so conditions cannot be described as uniform, and spatially there are many differences in physical features between the canopy of the forest and at ground level, near the bases of trees and away from them. Furthermore, as noted above, there is a marked seasonality in many areas and in most regions 'rainy' or 'dry' seasons are readily recognizable and generally rather predictable. The myth of constant growing conditions is often linked with that of constant growth, but few plants grow continuously and few animals breed continuously. So far as the trees are concerned, most have an obvious flushing period and some have bud-scales like those of temperate trees. Another widely-held misconception is that the forest resembles some flower-stuffed greenhouse. Such ideas derive from the magnificent displays in conservatories and botanical garden hothouses in temperate countries on the one hand and from the brightly-coloured plantings seen in tropical cities and their gardens on the other. The massing of cosseted flowering specimens in serried pots and the gaudy pantropical garden flora contrast markedly with the monotonous green within the forest. In general, spectacular flowering occurs only at canopy level, although occasional brilliant cauliflorous trees provide beautiful exceptions.

Perhaps more serious is the belief, still surprisingly widely held, that the forest stands on potentially good agricultural land. In temperate countries this if often so, as in Britain where the limewoods (*Tilia cordata,* Tiliaceae) were probably the first to disappear at the hand of man, for they stood on the best land in terms of agriculture as practised then. In the tropics, good agricultural land is to be found but much forest stands on poor land and its removal may lead to irreparable degradation of the soil. This point will be returned to later, as will be the variable physical features of the tropical rain forest environment.

An increasing unfamiliarity with forest, associated with greater sophistication of urban living, may lead to the forest being considered backward, despicable, even to be feared and, at least, not agreeable. Embedded in some of the more horrific fairy tales of Europe is such a fear of the forest with its darkness, unknowns, ogres and beasts. Today, the urbanized tropical inhabitant may have little time for the 'primitive' and 'backward' forest people and their habitat, their knowledge of how to live in and manipulate the forest in terms of gathering, hunting and shifting cultivation. Rather, they may mimic the early European colonists with disastrous consequences.

A final misconception is one due largely, it appears, to the wishful thinking of biologists themselves, although possibly it represents a higher sentiment, namely the desire to work in primeval, original, virgin, untouched undisturbed or primary forest. What is primary forest? Secondary forest deriving from the revegetation of completely cleared tracts of land can persist for a very long time. Around the great temples of Angkor (Kampuchea), forest cleared 600 years ago and since recolonized still does not resemble fully 'undisturbed' forest patches, and indeed it has been estimated that at least a thousand years may be required (16). Forests long considered primary are now known to have archaeological remains, arguing wholesale clearance in historical time. Networks of canals more than a thousand years old have been seen as a grid pattern by Synthetic Aperture Radar, possibly indicating raised field cultivation of maize and cocoa in Maya times in areas now covered by the Guatemalan rain forest. In the Petén of that country and the Yucatán of Mexico, present-day abundance of *Swietenia macrophylla* (mahogany, Meliaceae), *Manilkara zapota* (chicle, Sapotaceae) and *Brosimum alicastrum* (ramón, Moraceae) is thought to reflect the cultural practices of that civilization, where these species were encouraged to regenerate. There is now plenty of evidence of European archaeological remains, not to speak of aboriginal settlements in Central and South America, and it is estimated that since 1825 (6), the forest of Venezuela increased in area from 21% of its original range to 45% in 1950. Are such forests 'restored' or of a new type? How can they be recognized when there is no absolute yardstick for comparison? There is a long history of settlement in West Africa and Central America and it could be argued that rain forests there represent 'old secondary' rather than 'primary' forests typical of territories like Borneo but even those have long been influenced by man, as naturalized fruit trees and discoveries of medieval Chinese pottery indicate. In short, most forests are probably 'disturbed' but some are more disturbed than others.

1.4   The tropical rain forest in a wider context

1.4.1   *Floristics*

It has become fashionable among biologists trained in temperate regions to consider rain forests as some kind of repository, where the outmoded and effete survive, away from the pressures of the harsh exacting climate elsewhere. Others, with more tropical experience have argued that rain forests are regions of active speciation and that many groups have had their origin there. The controversy rages particularly around the origins of the angiosperms. Stenseth (17) has proposed two models for analysing rates of evolution: speciation and extinction as well as species diversity—comparing the physically somewhat predictable rain forests with the less predictable temperate forests. He concludes that the rain forests ought indeed to be considered the cradle for many taxonomic groups but that, even though the species diversity is high in the tropics, they make a bad museum, since extinction rates are high there also. In short, the richness is dynamic.

Certain groups of plants familiar to the biologist of the temperate zones are also well-represented in the tropics. Compositae, for example, are common there, although many of their number in forest vegetation are trees and shrubs. Euphorbiaceae and Leguminosae have similar patterns while, in the monocotyledons, Gramineae, Liliaceae and Orchidaceae (largely epiphytic in the tropics) are also widespread. Rubiaceae, represented by herbs in temperate regions, are predominantly woody in the tropics. Indeed, many of the families exclusively tropical in distribution are exclusively woody. Many genera of the temperate regions, after which families have been named, turn out to be very atypical, such as *Lythrum, Polygala, Verbena* and *Urtica*, which are mostly herbs in Europe, while their relations in the tropics are largely woody. Conversely, in the tropics there are woody members of widespread and predominantly herbaceous temperate families such as *Bocconia*, a woody poppy from South America, and the 'giant' *Lobelia* species (Campanulaceae) of the African, American and Pacific tropics.

Studies of the tropical flora lead to the blurring of the clear-cut differences between families recognized first from their temperate representatives. Thus Araliaceae and Umbelliferae have now been merged by some and many monocotyledon lily allies brought together in an enlarged Liliaceae including Amaryllidaceae. The differences between the figs (Moraceae), the elms (Ulmaceae) and the nettles (Urticaceae) are slight indeed as are those between Scrophulariaceae and Bignoniaceae, to take another example.

If the flora of a temperate country, Great Britain for example, is analysed we find it has an interesting relation to that of the tropics. The native British angiosperms are considered to belong to some 43 orders, all of which are represented in tropical regions. Furthermore, with the exception of aquatics, parasites and insectivorous plants, they are all represented by woody, sometimes exclusively woody, species there. The only striking exceptions are the Centrospermae, which thrive in open dry habitats and are scarcely woody anywhere except in salt deserts, the Cucurbitales, which are tender forest-scramblers and lianes, represented by *Bryonia* alone, the sea-pinks (Plumbaginaceae), the plantains (Plantaginaceae), the orchids, sedges and grasses, although it must be pointed out that there are dendroid forms of all of these in the tropics or subtropics. Some of the British plants take on an extra interest when they are seen as outposts of large tropical families; such are the black bryony, *Tamus communis* (Dioscoreaceae, yams), ivy in the almost exclusively tropical Araliaceae, the naturalized periwinkles, *Vinca* (Apocynaceae), as well as the white bryony, *Bryonia* (Cucurbitaceae).

1.4.2 *Numbers of species*

It is well known that the tropics harbour an enormous number of plant species and such samples (Table 1.1) as the 23 ha of West Malaysian rain forest with 375 species of tree (in 139 genera in 52 families) with a diameter greater than 91 cm are a staple of textbooks, but in the species-rich *Festuca* turf of the chalklands of Britain up to 40 or 45 species of angiosperms and bryophytes have been recorded in one square metre. With such differences in scale, comparisons are rather pointless, but the comparison of species numbers within large genera in vegetation types is perhaps more interesting. Excluding the hordes of apomictic clones of *Hieracium* and *Taraxacum* in the British flora, the largest genus by far is *Carex* with 72 species, which compares favourably with the 66 found in the whole of Malesia, but such statistics pale into insignificance when the estimated numbers of tropical species of *Piper* (> 1000), *Solanum* (1400) or *Vernonia* (1000) are considered. Nevertheless, such enormous genera as *Astragalus* (2000 spp.) and *Silene* (500 spp.) are largely temperate and the extraordinary diversity of the southern African *Erica, Aspalathus* and *Helichrysum* is outside the tropics. Furthermore, there are some very species-poor forest types in the tropics: in Ghana there are some comprising monospecific stands of *Talbotiella gentii* (Leguminosae, (18)) and, throughout the tropics, mangroves have, at best, 25 tree species, often many fewer, while tropical deserts have a small number of woody species and lack the large ephemeral flora of the temperate deserts. In wet places too, species diversity may be low, as in the papyrus swamps of northern Uganda. These do not appear to be unstable and it is perhaps as instructive to ponder why such species-poor tropical ecosystems are not subject to an enormous grazing pressure from animals as it is to ask why rain forests at their richest are so diverse (19).

Table 1.1 Numbers of individuals and species of trees with a diameter at breast height of 10 cm and over in mixed tropical rain forest

	Plot size, ha (approx.)	No. of trees per ha	No. of species on plot*
Borneo, Andalau For. Res.			
Ridge	2.0	740	199(118)
Valley bottom	2.0	640	219(129)
Sepilok For. Res.			
Ridge forest	2.0	667	198(144)
Malay Peninsula, Bukit Lagong For. Res.			
Hill forest	2.0	559	251
Sungai Menyala For. Res.			
Probably alluvium forest	1.6	489	197
New Guinea			
Hill forest	0.8	652	122
	0.8	691	147
	0.8	526	145
Forest on flat river terrace and gentle foot slope	0.8	430	116
Cameroun, Southern Bakundu For. Res.	1.5	368	109
Nigeria			
Okomu For. Res.	18.5	451	170
Okomu For. Res.	1.5	390	70
Omo For. Res.	1.5	521	42
Guyana			
Moraballi Creek	1.5	432	91
Panama, Barro Colorado Is.	1.5	489	—
Surinam			
Plot 5, Coesewijne Ri.	1.0	—	116(106)
Plot 1, Mapane Cr.	3.0	—	168(108)

*Figures in parentheses represent approximate numbers of species on 0.8 ha gauged from available species/area curves. (From K. Paijmans (ed.) *New Guinea Vegetation*, CSIRO & ANU Press, Canberra, 1976, p. 71).

1.5 The global interest

For biologists, the riches of the tropics are a great attraction. For other people in temperate countries, the rain forest represents a large resource, 50% of the earth's standing timber, and, it has been estimated, some 75% of the potential production of wood products. It is an enormous gene-pool resource and the source of a wide range of products, often unrecognized in the slick retailing of temperate countries, from teabags (from the leaf sheath of a banana originating in the Philippines) to Brazil nuts, the only tropical food crop marketed in Europe and collected from the forest rather than cultivated. There are drugs, gums and other extractives, tropical fruits and other foods, but it is timber that is the largest resource at present exploited with the utility timbers such as teak and ramin (*Gonystylus bancanus* (Thymelaeaceae) from the peat-swamp forests of Borneo and widely used in mouldings), merantis and their allies with a wide range of uses including plywood (*Shorea* spp., Dipterocarpaceae of Malesia) and luxury timbers such as the mahoganies and peroba (*Paratecoma peroba*, Bignoniaceae, Brazil). To tropical peoples, the forest may represent productivity on otherwise useless sites, especially erodable or creep sites or on deep deposits of peat, and to many it is home.

1.5.1   *The global carbon cycle*

Myers (6, 10) has argued that tropical forests contribute to the workings of climate via their impact on rainfall regimes, not only at the local or regional levels, but also globally. He alleges that this therefore has an important effect not only on the agriculture of the 2000 million people living in the wet tropics but that deforestation may also lead to disruption of climatic patterns way beyond the tropics. Despite popular views on impending disasters, meterologists have been less unequivocal in that they cautiously point to certain linkages, but the most significant interaction seems to be that forests of all kinds exchange moisture and energy more intensively than do other types of land cover. The energy is from solar radiation and depends on the surface albedo, which, in turn, depends on the vegetation, which depends on available soil moisture. With the removal of vegetation, there is less evapotranspiration and less moisture to be recycled as rainfall, while forests appear to increase cloudiness in the air above them and such clouds affect both temperature and rainfall in forest regions. In Amazonia, between half (50%) and four-fifths (80%) of the moisture remains within the region, being recycled as rain about every 5.5 days. Removal of the forest, it is suggested, would lead to a self-reinforcing increasing desiccation for the remaining forest cover. Removal of vegetation would also lead to self-promoting albedo enhancement, decreasing rainfall, evapotranspiration and cloud cover. How does this fit in with the greenhouse effect of the build-up of CO_2? Between about 0.9 and 2.7 gigatonnes of CO_2 per annum are derived from burning biomass, notably in tropical forests, although the bulk is from the combustion of fossil fuels (5.2 gigatonnes per annum).

Other gases are even more efficient at absorbing infrared radiation and thus warming the earth. They include nitrous oxide and methane, which may also be derived in significant amounts from burning rain forests, but the overall extent is unknown. The Amazon forests also give off as much as 60 million tonnes of isoprene per annum (i.e. 2–3% of their primary production). This, with nitrous oxide, may affect ozone levels, increasing them in the lower atmosphere and increasing the chance of photochemical smog (20). Although the involvement of tropical forests (21) has been played down, using figures estimating the biomass for undisturbed forests at 176 tonnes per hectare, it is important to note that, at least in the tall diptero-carp forests of Malesia, with emergents to 70 m tall, the above-ground biomass has been measured at 509 tonnes per hectare (22).

CHAPTER TWO

THE CHANGING PHYSICAL SETTING

An extension of the myth of the 'undisturbed' forest is the idea that rain forests are static and unchanging, that for a very long time they have been as they are today. It has been argued that rain forests are museums of period pieces of botany and zoology, even though the arrays of closely related species there might indicate rapidly evolving groups of organisms. That the forests, even without man, have been changing is evident from the geological history and that they are in a constant state of turmoil comes from modern ecological work, which is the subject of Chapter 4. Here, the ancient history of rain forest is discussed.

2.1 Continental drift

It has been established that the land masses of the earth are moving with respect to one another through the forces of sea-floor spreading and detailed maps of the relative position of continents at different times are being drawn up with increasing precision. Even for the Americas, though, there is still room for a broadening of ideas in that much of the land accreted along the western coast of the Americas since the Middle Jurassic comprises various fragments, some from far away in the Pacific (1, 2). At the same time as the continents were fragmenting in the Cretaceous, the angiosperms and their associated animals which have given rise to the modern flowering plant forest were evolving. Long before, there may have been forests in the Devonian, but the best known, through abundant fossils, are the coal-swamp forests of the Upper Carboniferous of Europe and North America, which were then in the equatorial tropics. These trees exhibited a wide range of form seen in modern but unrelated plants, where much of this range is still restricted to tropical regions. There were massive unbranched ferns with big buds and leaves (pachycaul), like modern tree ferns, and the oldest coniferoid gymnosperms had stilt-roots and a combination of anatomical features today seen only in some mangrove trees. These all disappeared with the increasing aridity of the Permian. The next tropical forests were of conifers, which are represented today by, among others, the enormous *Araucaria* species of New Guinea and New Caledonia, and it was these forests that the angiosperms largely replaced in the Cretaceous.

There are many tropical pollen sequences from Cretaceous angiosperms but leaf floras are almost unknown and reconstructions of the takeover have been derived from modern ecology. The most inviting picture has been painted by Ashton (3) where the *Araucaria* forest with ferns in the shade beneath has cycads and Caytoniales on the open fringes, along ridges and in swampy plains. The opportunities in ecological terms of occupying the gap phase, the holes made in the forest by falling trees, may have selected the fast-growing opportunist tree that is almost unknown outside the angiosperms. These protoangiosperms may thus have crowded out the gymnosperm seedlings and, as a group, radiated to become climax trees as well. The angiosperms in this reconstruction are seen as a basically tropical group, initially poorly adapted to colder climates. The closing of the carpel, and hence the distinctness of the angiosperms as a group, is seen as an adaptation to protection from herbivory rather than a protection against desiccation, for which such a device would have been 'pre-adapted' (4).

During the Cretaceous, when the angiosperms were diverging from their seed-fern allies, the apparently temporarily consolidated world continent, Pangaea, was breaking up into several areas of continental crust. The largest piece became Antarctica and Australasia, the smallest India and Madagascar. The sea penetrated the rift between Africa and South America while, by the Upper Cretaceous (Figure 2.1), Madagascar was separating from India. At this time, the chalk was being deposited in enormous volumes over much of Europe as the world underwent its greatest inundation since the Ordovician. It has been estimated that 40% of the continents were covered, i.e. only 18% of the world's surface was land, though estimates of the increase in sea depth vary from 350 m to 650 m. Such features as a trans-Saharan seaway allowed migration of marine faunas, but the evidence from the Pacific region rather supports the view that sea levels there were rather similar to those of today. Much of tropical South America was inundated. In Australia a great inland sea appears to have shrunk in this period, but New Zealand seems to have been scarcely affected as it split off from Australia.

By the beginning of the Tertiary (65 m years ago), the Atlantic was 75% of its present width, while Africa was a few degrees further south than it is

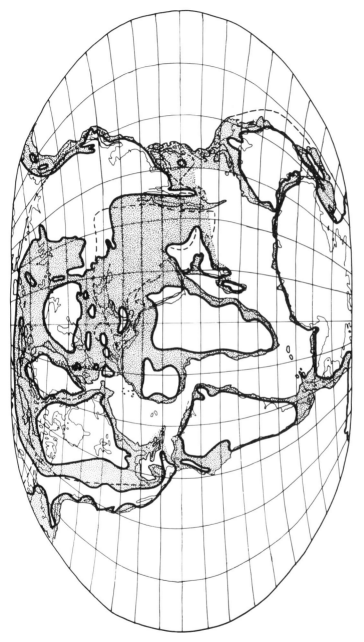

Figure 2.1 World geography during the Upper Cretaceous, 90–80 mi years ago, on a base map reconstructed on a globe 90% of the present earth diameter. Redrawn from H.G. Owen in M.K. Howarth in L.R.M. Cocks (ed). *The Evolving Earth* (British Museum (Natural History) and Cambridge University Press, 1981). The bold lines indicate coastlines in the Cretaceous.

today and India straddled the equator as an island. The sea level had fallen, though the cause is not yet understood, and it was to rise again, but never to the levels of the Cretaceous. The Amazon basin, however, was a site of continental and not marine deposition. The oceans between the southern and northern continents allowed the migration of marine organisms, but the crossing from one to the other by terrestrial ones could only have been by 'island hopping' on temporarily unsubmerged blocks. It has been suggested that in the early Tertiary the sea-grasses spread, leading, in turn, to faunal spread and diversification, as well as affecting sediment patterns. In the Eocene (60–40 m years ago), the well-known London Clay Flora was growing in the Thames Valley and included many plants, which, if they existed now, would be placed in a wide range of temperate and tropical groups, many of them of very restricted distribution. It has been said that, in the Tertiary, tropical and subtropical plants could live further from the equator than they do today, while others have argued that the London Clay Flora has a large drift component of material swept into it from further south. It must be remembered that even modern forests as far south as New Zealand still have a flora of mixed temperate and tropical types, though the presence in the deposit of the characteristic fruits of the taxonomically isolated palm, *Nypa*, now known only from the mangroves of tropical Asia and the western Pacific, is not readily explained away. Moreover, associated with the *Nypa* are certain fossils referred to Rhizophoraceae, something like modern-day species of *Bruguiera* and *Ceriops*, genera now both restricted to the Indian and Pacific Oceans. The London Clay Flora is very different from the pollen records of the tropics at this period and pollens are known from south-east Asia, which, if they were found in modern plants, would be referred to palms, dipterocarps, Moraceae, Fagaceae and Leguminosae, while from India are recorded in the Eocene the gymnosperms, *Podocarpus* and *Araucaria*, as well as palms, bananas and so on, but in general the Tertiary fossil record is still vague.

In the Oligocene (40–30 m years ago), the Fayum area of northern Egypt (5) on the southern shores of Tethys was a seasonal subtropical region with a range of vegetation types with forests including lianes and tall trees (and possibly mangroves) with arboreal anthropoid primates. By the early Miocene (25 m years ago), the continents looked very much as they do today, with India north of the equator and Australia only 6° south of its present position, but the Panamanian isthmus did not exist until 4 m years ago, so that only then could there be a complete mingling of the faunas of north and south America. The equids, mastodons, tapirs and llamas moved south and the sloths and armadilloes moved north, while the juxtaposition of Africa and Eurasia allowed the African elephants, bovids and pigs to migrate into Europe. Throughout this period, northern India and the Malesian archipelago were undergoing a great deal of vulcanism and mountain building, the Himalaya rising at this time, while Australasia (but not Papuasia) was stable. Australasia–Papuasia, plus the eastern part of present-day Malesia, had separated from Antarctica about 53 m years ago, two million or so years after India had collided with Laurasia. Earlier, a portion of Australasia may have split off and become embedded in what is now present-day south Tibet and Burma–Thailand, or may have gone down the trench south of Java (12). By the latter half of the Tertiary, South America and Africa were free of major flooding, as were India and Australia but, at its highest levels, the sea covered half of present-day Sumatra and Java, the Philippines, Sulawesi and all but the highest parts of New Guinea. The Gulf of Carpentaria in northern Australia and much of the east side of continental Asia may well have been dry land.

The Quaternary (Flandrian) is remarkable for its climatic fluctuations, with advancing and retreating icecaps and glaciers. These fluctuations were already beginning to build up in the Pliocene about 3.2 m years ago, and it appears that there have been 17 glacial advances and their interglacials in the last 1.7 m years. Although there is some relation between glacial advance and the enlarging of deserts, there is still much work to be done in this connection, as there is in the interpretation of palynological evidence on the depression of vegetation belts. Part of this derives from the breadth of ecological tolerance of some species known today. For example, in a model study, Hall and Swaine (6) have shown that the living *Sloetiopsis usambarensis* (Moraceae) occupies dry coastal forest in Ghana but, in the Ivory Coast, this plant is found in much wetter, even evergreen, forest. Again, one of the typical lowland evergreen forest trees of Ghana is *Angylocalyx oligophyllus* (Leguminosae) but in Nigeria this tree is widespread in dry semi-deciduous forest. Dangers can be deduced even from temperate floras wherever ecologically different but closely allied (vicariant) species occur: but would their pollen grains indicate their peccadilloes? Clearly, assemblages of plants must be and are being used in reconstructions, but much remains to be done to give a completely satisfactory picture of tropical palaeoecology in the Quaternary.

During the Quaternary, the sea level fell by up to 180 m by present-day standards, and it has been estimated that rainfall was then some 30% less than it is now. There have certainly been fluctuations in the upper limits of forest growth but it would appear that is exceptionally high at present and that we are witnessing an unusually high rainfall régime. Except for swamp habitats, which are unrepresentative of lowland rain forest, little can be learnt from direct palynological evidence because sites are rare, lakes are large pollen catchments or have been subject to erosion of their sediments and so on. Nevertheless, over a hundred pollen diagrams have been drawn for the non-arid tropics and they suggest much smaller extents for tropical rain forest about 10 000 years ago and almost every site studied shows some change at that time (8).

Wood samples from 1100 m in Ecuador indicate that the forest in the last glacial was comparable with present-day Andean forest some 700 m higher up, suggesting a temperature drop of at least 4.5°C for the Amazonian lowlands at that time (9, 10). Studies in southern Amazonia tend to reinforce the idea of extensive changes in vegetation in the Quaternary, for there appear to have been long periods when savanna replaced the forest, which has once again returned. Cores in north-east South America show mangrove deposits, as the water level rose, and savanna incursions during apparently more seasonal conditions in glacial phases. Pollen profiles from central Amazonia indicate reduced precipitation in periods shortly after 4000 BP, around 2100 BP and around 700 BP. Accurate radiocarbon dating and pollen fossils in Africa have indicated that the early post-glacial in many places was wetter than it is today and that desiccation beginning about 4500 years BP forced Neolithic herders to abandon previously habitable Old World deserts. Evidence from subfossil freshwater snails in the Red Sea Hills of the Sudan suggests that the hills were wetter about 2000 years BP as may have been much of northern Africa (11).

In short, until relatively recently in palaeoecological time, there was a more seasonal climate. This ties in with arguments from plant distributions and the location of high levels of endemism, where refugia of rain forest organisms have been postulated in the Quaternary of Africa and South America and it has been argued that these have since contracted and expanded. Interglacials differ in their faunas and floras and even the last one in Britain was different from the present one: *Picea abies* was a native tree, while tropical coral reefs were more widespread, as was the giant clam, now restricted to the western Pacific. Further back, in the mid-Pleistocene of Britain, there occurred a rhinoceros closely allied to one now restricted to Indonesia, and a pig similar to one living in Borneo now: indeed, much of the European mammalian fauna was like that of the tropical Asia of today.

In summary, the reconstruction of the history of rain forest or other vegetation on any site in the tropics is an exceedingly complicated task, though recent work in a Borneo peat swamp (6) where forest deposits overlie grassland deposits again suggests a previously more seasonal climate locally. Nevertheless, certain patterns of animal and plant distribution raise fascinating questions, some of which have been resolved through the work of geologists and palaeoecologists. Possibly the most celebrated is Wallace's Line (12), which divides the Asiatic and Australasian faunas in the eastern part of Indonesia. The marsupials are to the east of the line and have an origin somewhere in Gondwanaland, the placentals largely to the west. There are similar distinctions in other animal groups but it is less clear-cut in plants. The geological history of Malesia shows that both plants and animals could have reached the area, without crossing water, from Laurasia, and from Gondwanaland *via* either Australia or India. The explanation of the enigmatic nature of Sulawesi, which puzzled Wallace, is thought to be that it has part of its origin in Australasia–Papuasia and part in the Asiatic block. Either side of it are the two great rain-forest areas of Malesia, Sumatra–Malaya–Borneo and New Guinea, in which there is no general sign in their geological history of the desiccations found in that of Africa and South America. In New Guinea are represented many of the Austral elements typical of Australasia but, in West Malesia, these are found chiefly in the heath-forest flora. In Malesia, two previously isolated faunas and floras mingle and in the angiosperms are the overlapping north-south plant group affinities elsewhere more separated: Ericaceae/Epacridaceae, Magnoliaceae/Winteraceae, Saxifragaceae/Escalloniaceae, to name a selection (12). Biogeographers may therefore be tempted to argue (and it would be very convenient) that the piece of Australia now in Burma or melted under Java held the 'cradle of the angiosperms'.

Taking individual plant groups and a combined biologist/geologist approach, the primitive relictual genera of the tropical mistletoes, Loranthaceae, are seen to be monotypic or small and to be found on old Gondwana surfaces. From this and reconstructions of Gondwanaland, it is considered that Loranthaceae must have attained a wide distribution by the late Cretaceous at the latest, to explain the presence of relictual genera in New Zealand. It may be argued similarly for certain other 'old southern families': Casuarinaceae, Goodeniaceae, Myrtaceae, Pittosporaceae, Proteaceae and Restionaceae. By contrast, the other mistletoes, Viscaceae, have no southern relictual genera and are perhaps of eastern Asiatic origin (13).

2.2 Tropical climate

The equatorial position of tropical rain forest ensures that more radiation strikes it than forests outside the tropics and that there is no winter period associated with reduced day length. Nevertheless, the tropics do not comprise a region of uniform climate, owing to the position of the continental masses, air-flow and sea-currents leading to wide variations in precipitation, relative humidity, temperature, wind and vulnerability to violent storms and so on. This pattern of macroclimate largely governs the general pattern of forest-type distribution. As an example, the present-day rain forest of West Africa is split in two by the Dahomey Gap or Interval. Here the coastline runs WSW–ENE, the prevalent winds make an acute angle with the coast, and little moisture is brought ashore. To the north and south of the forest area, the dry Saharan and somewhat wetter South African anti-cyclonic systems support a rain forest–savanna mosaic, open woodland, dry savanna and grasslands (14).

2.2.1 *Precipitation*

Tropical climates have been divided into five major categories: rainy tropical, monsoon tropical, wet-and-dry tropical, tropical semi-arid and tropical arid. These are not, of course, sharply separable in the field but rain forest in this book is found only under the régimes of the first two. The others support savanna vegetation, or drier formations, like deserts. The major areas with a rainy tropical climate, i.e. the Amazon basin, parts of Central America, the Congo basin, the steep eastern slopes of Madagascar and much of Malesia, have more than 2000 or even 3000 mm of precipitation per annum, more or less evenly distributed throughout the year, and bear tropical rain forest. However, the semi-deciduous forests of Burma, for example, may not have a different overall precipitation but it is markedly seasonal. Precipitation in rain forest may reach 10 000 mm, as recorded in West Africa, but there is always some seasonal fluctuation.

In Africa (15), the Guineo–Congolian rain forest receives between 1600 and 2000 mm of precipitation per annum and not only is this lower than in other rain forest regions but it is also more seasonal, in that mean monthly rainfall throughout the year is rarely higher than 100 mm. In the main block of this forest in the east, there is one severe and one less-severe dry period. Even in the heart of Zaire, dry periods of up to 30 days or more occur every 12 years or so. In the western block, rain is more concentrated into one season, with a dry season of virtually no rain for 4 months, although the overall precipitation may exceed 4000 mm. Increasing fluctuations are associated with different vegetation types, like the African savannas which have a lower standing crop and are believed to be less productive. Conversely, the very wettest regions may also be less productive, which has been explained as being due to poorer growth associated with high levels of leaching, as in the Rio Negro region of Brazil, while greater cloudiness has also been suggested as limiting growth rates (14).

The major source of moisture is rainfall, of which 25% may be lost through canopy evaporation. Some 40% may trickle down limbs and trunks and be partly absorbed by bark and epiphytes and partly evaporated within the forest, so that it has been estimated that, in Malaysia, only some 10% of the rainfall reaches the ground. Just north of Manaus in Amazonia (16), it was found that interception depends on the intensity of the rain for up to 20 mm per hour, there is 28% interception, 0.1% stem-flow and 71.9% throughfall. At 20–40 mm, however, the figures are 23.7, 0.2 and 76.1% respectively, while at 40–60 mm they are 23.7, 0.4, and 75.9% and at more than 80 mm, they are 18.8, 0.4 and 80.8%. In a detailed study at Pasoh, Malaysia (17), valuable distinctions were made between the effects on a 'per storm' basis, on a monthly and on an annual basis. Stem-flow was found to be 0 to 2.65% of the precipitation on a per storm basis, from 0.32 to 0.92% on a monthly and 0.64% on an annual basis. Throughfall ranged from 0.0 to 99.01% on a per storm basis, 65.27–94.64% on a monthly basis and 77.56% on an annual one. The interception varied from 0.15 to 100% per storm, 5.04 to 34.31% per month, and this represented an annual rate of 21.8%. Other sources of moisture are dew, fog and clouds, which are of great importance in maintaining certain montane forests in tropical regions (18).

Rain gauges may therefore be very inaccurate measures of the water régime. Of the annual precipitation in a forest on the Rio Negro in Amazonia (19), some 3664 mm, about 47%, was lost through transpiration, supporting the contention that 48% of the precipitation in the Amazon basin is derived from the vegetation there. This compares with figures of 12% in other non-forested parts of the world and suggests that tree-dominated landscapes are more efficient than other forms of land use in this respect. Run-off and evaporation from the ground will be greater in forests with punctured canopies and, in large gaps, dew may form. But even intact forest may lose more water through evapotranspiration than it gains from precipitation for, in Malaysia, periods of up to a week of cloudless hot days and cold nights have been recorded, leading to just this state of affairs. Within the forest, then, there will be local variation in water availability.

The effect of rainfall is further complicated by the fact that raindrop size in the tropics is substantially greater than in temperate regions (20). In Nigeria, drop sizes range from 0.09 to 5.46 mm in diameter, with drops 2.0–3.5 mm diameter in most precipitations. Formulae used to calculate the energy with which rain hits the ground, and thence the effect on soil erosion, take account of factors such as intensity and wind speed but not drop size: clearly formulae of use in Britain, where drop size ranges from 0.2 mm in drizzle to about 2.0 mm in heavy downpours are not really applicable to tropical-size drops in regions with sudden heavy storms. In Nigeria, drops at the beginning and end of the season were largest but those falling at night when the temperature was lower were smaller, the biggest drops of all apparently falling at the beginning and end of the storms. Existing formulae suppose that drop size increases with rainfall intensity, but in this study, it is not so simple: up to 11 cm per hour dropsize does increase with rainfall intensity, then up to 18 cm it falls (the larger drops probably disintegrate), but above 18 cm it increases again, the highest intensity recorded in the study being 23.3 cm. With losses of soil of 10–21 tonnes per hectare per year in West Africa with slopes of less than 1%, these losses can rise to 55 tonnes on slopes of 1–2%, while in Madagascar rates of up to 250 tonnes have been recorded.

There are also great variations in relative humidity. In the canopy it may fall to 70% whereas, within the forest, it is likely to remain around 90% in the day and 95% or above at night. In Ghana, during the dry season, when a drying northeasterly wind from the Sahara (the harmattan) prevails (15), it may fall to 53% by mid-afternoon. Longman and Jeník (14) have argued that slow evaporation in the lower reaches of the forest may lead to a slow uptake of ions and thus to slower growth on the forest floor. They have shown that 90% of forest undergrowth species have leaves with drawn-out 'drip-tips' and that the removal of these tips extends the drying period after rain of the blade from 20 to 90 minutes. The slow removal of water without drip-tips may also provide an opportunity for the establishment of an epiphyllous flora; the presence of a film of water may also reflect sunlight, perhaps reducing photosynthetic efficiency (21). Some trees, such as the African *Lophira alata* (Ochnaceae), have such tips when they are juvenile and in the undergrowth, but not when they reach the canopy, and Longman and Jeník have suggested that the xeromorphic features of canopy leaves may be adaptations to moisture stress. In some plants, the drip-tip drainage is supplemented by other mechanisms. *Machaerium arboreum* (Leguminosae), a liane of the Panamanian forest, sheds water, or avoids the wetting of the leaves through thigmonastic and nyctinastic movements, i.e. those promoted by touch and by nightfall respectively, which are frequently found in this family (22). The drip-tips facilitate drainage, as has been verified by excision experiments, but the leaflets of the compound leaves fold together at night, when not photosynthesizing, and respond to light rain during the day by closing through the thigmonastic mechanism.

2.2.2 *Storms, droughts and vulcanism*

Some storms are very violent and are termed cyclones, typhoons or hurricanes. These are much rarer in the tropics than out of them (23, 24) and they develop in specific areas at particular times of the year: in the tropical north Atlantic from June to November, in the north Pacific off Central America from June to October, in the western North Pacific from May to November (though this area is prone to strong storms throughout the year), the Bay of Bengal in May, June, October and November, the south Pacific west of 140°W from December to April and the south Indian Ocean from November to May. No such storms have been recorded from the south Atlantic or the south Pacific east of 140°W. Tropical cyclonic systems carry air-masses north (India, Mexico, Florida) or south (East Africa, Madagascar) and thus extend the tropical climate conditions beyond the geographical tropics. On occasion, they can reach as far as, for example, New Zealand where, in 1982, Cyclone Bernie caused extensive damage. Cyclones seem thus to be important factors in forest composition even beyond the tropics. A model estimating the maximum intensity of tropical cyclones under the warmer conditions expected to result from an increased CO_2 level predicts that a doubling of level will be accompanied by a 40–50% increase in the destructive potential of such storms (25). The importance of such cyclones which, in the western Pacific at least, are becoming more prevalent nearer the equator, will be considered in Chapter 4.

Fire has been with ecosystems a long time, for fusains, which can form seams of bad coal burning like coke, have been interpreted as fossil charcoal from forest fires of the Jurassic. Fires due to falling rocks have long been known from monsoon forests (27), but it has been assumed that rain forest rarely, if ever, burns. However, in the Pacific, at least, cyclical droughts occur and are associated with sea-surface temperature anomalies in the central and east Pacific and their associated atmospheric consequences. These are referred to as the El Niño/Southern Oscillation (ENSO) phenomenon and usually last 6–18 months at intervals of 2–20 years or so, averaging about one in four years. The 1982–83 ENSO was perhaps the strongest this century so far and regions as far apart as Australia, Philippines, southern India and southern Africa suffered droughts. The droughts exacerbated fires and, in Kalimantan (Indonesian Borneo) alone, some 3.5 m hectares of forest were damaged. Of these, 800 000 were 'primary' forest, 550 000 peat-swamp forest, 1.2 m selectively logged forest and 750 000 shifting cultivation. Timber losses were estimated at $US 5 billion. In neighbouring Sabah (part of Malaysian Borneo), it is estimated from satellite imagery that 1 m hectares were also affected. The most extensive fires were in August to October 1982 and February to May 1983, when the smoke from them reached Java, Singapore and Peninsular Malaysia, affecting air navigation. The initial cause seems to have been human-set fires getting out of hand in the drought-stressed selectively logged forest, whence it spread to the primary; the lowering of the water table in the peat-swamp forest made the peat and the trees falling over in it very susceptible too.

Such events in Borneo may occur every 50 to 100 years and there is evidence of droughts and fires in Kutai (Kalimantan) in 1877. Carbon-dated charcoals in soils under mature forest in Venezuela indicate that numerous fires have occurred there over the past 6000 years. The char-coal beds are not associated with human artefacts and the earliest ones antedate human presence in any case (29–31). They seem to correspond roughly with the dry periods recognized by palynologists. Evidence can also come from documentary records, as in New Guinea for example, where it has been found that travellers in 1885 noted that the drought then was accompanied by forest fires so intense that the coastal shipping was stopped for days because of the smog reducing visibility to less than 1900 m. The key to forest inflammability is fuel moisture and so rainless periods have to be accompanied by relative humidity at or below 65%, which is more likely in open rather than closed canopies. Severe burning is likely in selectively logged forest, spreading rapidly up lianes.

Another major disturbing force in some areas is vulcanism. In Panama in 1976 (32), 54 km² of forest was removed by landslides following earth-quakes. From the size of that event and the frequency of earthquakes, it has been calculated that 2% of the vegetation of the region is thus destroyed every century. Such natural 'disasters' are a common phenomenon in Papua New Guinea (31), where the forests in some areas are dominated by species of *Anisoptera* and *Hopea* (Dipterocarpaceae), *Albizia, Intsia* and *Pterocarpus* (Leguminosae) and *Pometia* (Sapindaceae) but these are not regenerating: these are old secondary forests, the results of historical natural disasters. Most of New Guinea has been affected by the collision of the Australian with the Pacific and Indonesian plates, so that in the last 10 000 years, some mountains have been lifted some 1000 m, which is reflected in the large number of earthquakes and volcanic events in the area. Besides earthquakes, landslides may be due to excessive build-up of biomass or to intense rain, though the most spectacular are due to tectonic events. An earthquake of force 8.2 in the Madang area, in 1972, caused a pattern of landslides with the removal in some areas of up to 60% of the vegetation and, in an area of some 240 km², an average of 25% was lost. Such are likely on average every 200 years or so, though between such events lesser ones have been recorded, one in 1985 being 7.4 and also causing extensive landslides. Volcanic activity can also be accompanied by tsunamis (tidal waves), which destroy coastal vegetation.

2.2.3 *Temperature and radiation*

The mean annual temperature in rain forest regions is about 27°C, ranging, on monthly average, from 24–28°C, so that seasonal variations are smaller than diurnal ones, which may be as much as 10°. Maximum temperatures rarely exceed 38°C, which is lower than in continental North America or Europe. It is unusual for the temperature to fall below 20°C, though it may reach lower levels at the base of high mountains or in the bottom of valleys. Within the forest, diurnal fluctuations vary. In Ivory Coast (14) in December the fluctuation was some 10.8° at 46 m but only 4.4° at 1 m above the ground, while in June the figures were 4.0 and 1.7°C respectively. The soil temperature probably never exceeds 30°C in closed forest though may be 50°C or more in the surface layers of exposed soil. At a depth of 75 cm no discernible diurnal fluctuations were measurable in closed forest.

The effective day length for emergent trees on the equator is $12\frac{3}{4}$ hours, though less inside the forest or on steep slopes. At 5° north or south, the annual variation in day length is half an hour; at 10° it is one hour and at 17° it is two. The light that reaches the forest floor is of three types. There are shafts of light passing between leaves, sunflecks, which appear to move as the earth moves with respect to the sun; light through perforations in the canopy; and light reflected from leaf and branch surfaces, with that transmitted through one or more leaves. It has been calculated that the total radiation reaching the forest floor is some 2–3% of the incident light energy. In Malaysia, it has been estimated that half of the annual total is from flecks, 6% from canopy holes and 44% reflected or transmitted. In gaps, this last component is much smaller than in closed forest and the proportion of far-red light and infra-red is thus lower, for in closed forest these are the wavelengths transmitted through photosynthesizing tissue. This change may be registered in a plant through its phytochrome system, indicating that its leaves are no longer in shade. The quality of the light passing through to the forest floor is known to affect germination, stem extension, leaf morphology and juvenility. At La Selva in Costa Rica, 56% of the total radiant energy in full sunlight is that in the range 400–700 nm, i.e. that available for photosynthesis and the red:far-red ratio is 1.28 (33).

In a typical understorey shade environment, there is only 1.7% of the range 300–1100 nm above the canopy and only 0.3% of 400–700 nm as is manifest in the red:far-red ratio being reduced to 0.31.

Many herbs in the ground vegetation of rain forest have brightly-coloured leaves, which make them colourful foliage plants for the dingy houses of man. These colours are seen in certain species of *Begonia*, for example, and are due to anthocyanins, which were traditionally held to enhance transpiration by increasing heat absorption (34). When four angiosperms with such colours were tested in Malaysia, however, no temperature differences could be discerned. In each of them the anthocyanin was confined to a single layer just below the photosynthetic tissue, and it now seems that the layer back-scatters light for photosynthesis. The velvety sheen of the leaves of many such plants is due to reflection from the convexly curved outer epidermal cell walls. Experiments carried out by Haberlandt at the beginning of the century (35) have shown that these act as lenses, focusing the light with the result that there are higher photosynthetic photon-flux densities at particular points in the leaf.

CHAPTER THREE
SOILS AND NUTRIENTS

3.1 Tropical soils

Tropical soils are rather more diverse than is widely believed, for the well-known, highly-weathered and leached soils cover only half of the tropical land surface (1), but these 'latosols' are a common feature of rain-forest country. Pedologists classify soil types according to various systems and give the different layers recognizable in profiles different names. The system set out by Burnham (2) is perhaps most comprehensible and is the basis of what follows.

The upper layers or horizons, from the surface towards the underlying geological material, are litter, sometimes partly decayed plant material and well-decomposed humus. When the last does not pass imperceptibly into the layers below, it is known as *mor*, a humus type that develops in aerobic conditions, while, in anaerobic conditions, a peat may form. A *mull* soil is where the surface layer is incorporated with mineral particles and underlies a thin litter layer. Layers with well-integrated humus and minerals are known as A horizons, beneath which lie the weathered layers or B horizons. These are subdivided or may be separated from the A horizon by layers of deposition of iron or aluminium compounds. The C horizon is the parent material.

3.1.1 *Soil processes*

With warmth and copious rainfall, the humid tropics have strong weathering of the soil so that minerals are continuously leached from the upper layers. The moisture and warmth in a tropical soil promote rapid breakdown of vegetable and other organic matter by decomposers. The weathering allows the development of deep soils, which may reach 15 m or more in profile, though very often this is deeper than the levels to which the decomposing material reaches, so that the biologically significant layers lie over considerable depths of saprolite, or 'rotten rock'. Despite mineral poverty, the physical features of the soils are good, in that they are deep with good granular structure, comprising clay bound into stable aggregates, which are efficient at retaining water after the soil horizon as a whole has drained after rain. As a result, water deficits are minimized in drought periods and the spaces between the aggregates permit rapid drainage, thereby preventing waterlogging and reducing leaching of nutrients from the aggregates (3, 4).

The weathering of all the common minerals, except quartz, to clay and oxides of iron and aluminium leads to the formation of soils that are more clayey than in temperate regions. Furthermore, the great depth of the parent material means that erosion rarely removes or even reaches the unweathered nutrient-rich parent soils. Therefore not only the superficial soils are low in nutrients but eroded slopes and transported materials such as alluvium are too. Leaching means that sodium, potassium, calcium and magnesium are often at very low concentrations in tropical soils, though acidity is also low. The clay of most well-developed, well-drained tropical soils does not swell and shrink on wetting and drying, so that the cracking familiar in temperate-zone clays is rare. This kaolinitic clay has a low cation-exchange capacity, so that its nutrient-holding capacity is mainly a function of humus content and is low where that is low. This is true of subsoils, so that roots tend to be concentrated in the surface soil and few are below, even though it may be physically not difficult to penetrate. The rapid rate of decomposition in well-drained soils leads to a thin litter layer and low levels of organic matter in the soil, although termites may be of local importance in transporting rotting wood to great depths in their galleries and chambers.

3.2 Soil types

While considerable edaphic variation occurs within a small area such as the Malay Peninsula, the average properties in terms of exchangeable cation content, and thus fertility, are commonly found, extreme soil types being rare, so that there appears to be great uniformity. However, the soils can be rather varied in colour and texture and are now grouped into three major categories.

3.2.1 *Major groups of soils*

Ultisols (red-yellow podzols) are distinguished by their B horizon, which has 20% or more clay than in adjacent horizons, the fine clay particles being considered to have been brought in from the upper layers in suspension. This layer may be bright yellow-brown or reddish, but the distinct pale horizon associated with podzols in temperate regions is usually absent. These soils are low in bases and, despite their high contents of weatherable materials, leaching makes them infertile. They cover about 28% of the tropical land-surface and are commoner under seasonal climates than they are in the wetter regions of the Malesian region, though in Africa they are found on the Upper and Middle Pleistocene erosion surfaces.

On the more extensive older surfaces in Africa there are *oxisols*. These comprise the only group of soils exclusive to the tropics and cover over a third of the tropical land-surface but are rare in the Malesian region, probably largely due to extensive recent orogeny. They have an oxic horizon in the subsoil and are often coloured dark red. Their clay content is high but they are less sticky than ultisols and roots may penetrate easily. The clay fraction, dominated by kaolinite and hydrous oxides has very little permanent negative charge and thus little cation retention capacity within the pH range found under natural conditions, particularly the acid régimes obtaining under many rain forests. In Malesia oxisols are leached, but are marginally more fertile than ultisols nearby unless they occur on siliceous materials like alluvium, when they may have high levels of quartz sand and be less fertile. *Alfisols*, which cover about 3.6% of the tropical land-surface, are rather like ultisols but have a higher base concentration and hence greater fertility. Ultisols and oxisols are extremely acid throughout the profile and oxisols are very low indeed in Ca^{2+}. Even in the absence of toxic quantities of Al^{3+}, which restricts root growth elsewhere, there is so little Ca^{2+} that it is severely limited in any case, the net result being that roots tend to be restricted to the surface layers and thus prone to water stress. If acidity of oxisols and ultisols is corrected, the capacity to retain bases is increased and the toxic ions such as Al^{3+}, Fe^{3+} and Mn^{2+} are reduced and microbial activity enhanced.

The main groupings intergrade and many soils are completely intermediate: indeed, as there is no readily discernible correlation with fertility levels it has been argued that there is a good case for abandoning such names.

The term 'laterite' is used where there has been redistribution and concentration of sesquioxides in the soil profile. It is most common in seasonal climates or areas drier than the tropics bearing rain forest and results from compaction and several cycles of wetting and drying, not mere clearing of the vegetation, though repeated clearing with compaction will lead to its formation. Often there is a layer in the subsoil with compact clay mottled bright red and pale yellow or grey. Such clay impedes root penetration and in hardened laterites prevents it, so that these soils are seen to be rather infertile though they are not necessarily more weathered than comparable non-lateritic soils.

3.2.2   Volcanic soils

Where there is volcanic activity as, for example, in Java, there is a new input of fresh mineral material. Magmas low in silica appear as lava, those higher, in the form of ash, which is fragments of pumice. Newly-formed soils over lava resemble brown earths and, in time, become oxisols. The weathering of ash, on the other hand, leads to the formation of *andosols*, which contain allophane, an association of hydrous silica and alumina with a large active surface. This last feature gives the soil a high porosity and a low bulk density, with a 'fluffy' consistency, a high water retention capacity, high levels of cation exchange and a capacity to form stable associations of mineral and fine organic matter, which give the upper layers a characteristic dark colour. In humid regions, allophane eventually turns into kaolinite in a matter of thousands of years so that andosols occur only in late Pleistocene and Quaternary ash. In a world context, they are moderately fertile, but in the humid tropics they are the most fertile of all, though phosphorus and some minor elements such as manganese may be deficient. Once dried out, though, andosols may be irredeemable, for a hard pan of cemented silica or of iron and other oxides may form at a depth of 40–100 cm and this is not readily penetrated by roots. Nevertheless, it is on such soils as these that the intensive agriculture of Java with its high concentration of people is based. Furthermore, the ash from volcanoes may spread very far, enriching soils that are not typical andosols.

3.2.3   Other soil types

Of other soil types represented in the tropics, mention should be made of the *podzols*, which resemble those in northern Europe. They lie over predominantly quartz parent material, low in clay and bases, and have a bleached horizon below the humic layer and a dark or brightly coloured B horizon rich in colloidal organic material, sometimes with sesquioxides. Such soils are commonly found in old beach deposits that are now inland, but also occur on sandstone, quartzite or acidic volcanic deposits, and are covered with heath forest. They have very low fertility and a low capacity to retain water or cations. Soils over limestone formations, including the jagged karst landscapes (Figure 3.1) peculiar to the tropics, may be brownish-red latosols or, in heavy rainfall areas, highly humic soils.

Figure 3.1   Forest and chasmophytes on the rugged karst limestone of the Malay Peninsula. The palm at the bottom left. *Maxburretia rupicola*, is restricted to these limestones.

Over ultramafic ('ultrabasic') rocks with high levels of iron and manganese, low levels of silica and high concentrations of such toxic elements as nickel, chromium and cobalt, the soils resemble the oxisols on basic igneous rocks, but they are as variable as the forests they bear. Of alluvial soils, those marine ones colonized by mangroves are, of course, saline and are alkaline to neutral. With consolidation, acidity rises and there is an increase in organic material. Soil micro-organisms reduce sulphate ions, which react with iron to give pyrite and ferrous sulphide. As rain and fresh water replace the saline, the mangrove becomes peat-swamp forest or, later, freshwater swamp forest with soils that are generally very fertile and have been much exploited for agriculture. Alluvial soils deposited in lakes or by rivers are of varying fertility dependent on the age of the eroding materials and whether they are overlain with peat, in which case fertility may be rather good. In mountainous regions, rainfall greatly affects the forest soil type. On seasonally dry mountains, organic matter content increases with altitude, and shallower soils occur because there is less weathering. On wetter mountains, there are peaty soils, which support 'cloud forest'. These soils are leached, podzolized or waterlogged, with a reduction in microbiological activity, and are thus of low fertility.

3.3   The relationship between soils and forest type

Already it has been pointed out that andosols will support intensive agriculture and that podzols support only heath forest. Can other correlations be found? Under increasingly well-watered, but not swampy, conditions, the density of emergent trees increases and there is a higher incidence of crown epiphytes and of lianes. On fertile sites under such conditions there is a significant increase in the percentage of deciduous species. With decreased nutrient availability and sharply drained soils, there is a decline in stature and density of emergents. In areas of impeded drainage, such as some alluvial regions, the majority of rain forest trees may not thrive so that there are fewer constituent species and even single-dominant stands, notably of palms, such as the *Metroxylon* swamps of New Guinea and the *Nypa* stands in Malesian mangrove forests.

Peat-swamp forests are well known in Sumatra, Malaya and Borneo, very rare in Africa but better represented in the Amazon, some islands in the Caribbean and the north coast of the Guyanas, and have a rather restricted flora. The peat usually has a pH of less than 4.0 and may be up to 20 m deep with a solid layer over depths of semiliquid material with rotting wood. No bird or mammal is known to be restricted in its distribution to such forests.

Freshwater swamp forest differs in being regularly inundated with mineral-rich fresh water with a pH greater than 6.0 and in having a shallow peat layer. It is associated with the great tropical rivers of many parts of the world, as in Indochina, large parts of Africa, and South America, notably along the Amazon and the Rio Negro. There, the floodplains spread widely and the distinction between the seasonally flooded (*várzea*), the permanently inundated (*igapó*) and that which is never inundated (*terra firme*) is of great importance for settlement and agriculture. Such forests are not distinct in terms of the families of plants involved but there is, as in the example of *Metroxylon*, a tendency to greater gregariousness. This is important from a commercial point of view in the Borneo swamp forests, as ramin (*Gonystylus bancanus*) grows there in rather pure stands, making its exploitation worthwhile.

Mangrove swamps are poor in species and consist of largely facultative halophytes. Many have been grown in fresh water and are sometimes found under such conditions in the wild (5), e.g. *Acrostichum aureum* (Adiantaceae) and *Acanthus ilicifolius* (Acanthaceae), but also trees persisting after physical changes, e.g. *Sonneratia caseolaris* (Sonneratiacae) by a Malesian lake that was formerly a lagoon and, on Christmas Island (Indian Ocean) *Bruguiera gymnorrhiza, B. sexangula* (Rhizophoraceae) and *Heritiera littoralis* (Sterculiaceae) near springs on the first terrace now 20–30 m above sea-level since the upheaval of the island. The fauna, so far as, for example, the birds of West Malaysia are concerned, is rather distinctive and comprises largely the species typical of cleared land and gardens, to which such birds may have spread.

Of uninundated forest types, the heath forest is found on the poor siliceous soils of the Malay Peninsula, Indochina, Borneo, and in New Guinea, where it is mostly scattered. It occurs in the Amazon basin, the Rio Negro being black through draining it, and in a small coastal sand area in Gabon. It consists of saplings and poles, all rather tidy and orderly but low, dense and impenetrable with no trace of layering. In Borneo, this is the *kerengas* forest rich in the secondary-consumer plants, the carnivorous pitcher plants, *Nepenthes*, and sundews, *Drosera*. In the Amazon, it is the *caatinga*, which is also found in the Guyanas. It occurs in disjunct pockets but on the Rio Negro covers hundreds to thousands of square kilometres. The trees are markedly sclerophyllous, like the trees of cloud forest mentioned below; on nearby soils with only as much as 5% clay there is distinctly different non-sclerophyllous forest. In the *caatinga*, the trees commonly coppice and the roots may make up 60% of the biomass, whereas on oxisols this figure is nearer 20% (6). The roots form mats in the slowly decomposing litter and may be picked up like a carpet. Such forests have a low resistance to fire. In the Malesian forests, the fauna is largely of animals found in other forest types though some groups, like snakes in Borneo, may be poorly represented.

Within 'typical' rain forest, there are in Malesia, rather abrupt transitions in floristic composition at 100 m, 1500 m, 2400 m and 4000 m (7, 8). On a smaller scale at La Selva, Costa Rica, edaphic features vary with small changes in altitude, a feature that is particularly associated with different drainage régimes. Where soil aeration in Australian rain forests is unfavourable, other factors are relatively unimportant in influencing vegetation type. At a local level, then, distinctive communities are associated with different soil types and topography. In Amazonian Venezuela (4), on tops of hills where the clay is nearest the surface, is the most species-rich mixed forest with no obvious dominant, while nearer streams is the more productive forest dominated by *Eperua purpurea* (Leguminosae), and, along streams is the *igapó*, inundated for up to 6 months at a time. Where the parent material for the soil is coarse sand is the *caatinga*: near streams but above the flooding level is the 'high caatinga' with trees to 20 or 30 m tall, while at even small elevations there is sufficient water stress due to the fast-draining sands to support only the dwarfer '*campina*' vegetation; above 2 m above stream-level, this is reduced still further to the low shrubby '*bana*'. Waters from such soils are black with tannins and other organic materials, while on other soils, these are adsorbed on clay particles and retained.

Of other soil types in Amazonia, the red-coloured 'terra rosa', although restricted, is very important as it is highly fertile with plenty of available calcium: it is derived from soils overlying limestone. Forests on limestone are found in parts of the Caribbean and Malesia but are absent from the African humid tropics. They often have a large number of plant species restricted to them and the caves formed in them have been used by man for a very long time. In the moistest regions, limestone may carry seasonal forests, while on fertile alluvium and basalts in seasonal regions, there may be 'moist' forests. Those in the Malay Peninsula (see Figure 3.1) offer very different topographical habitats from those in the surrounding forest but the basal area of trees is the same on slopes up to 45° and only on steeper ones and rocky hilltops do edaphic conditions seriously affect tree growth (9).

Forests (10) on ultramafic soils are often low and may consist of shrubby vegetation juxtaposed to high rain forest, as in parts of Sulawesi, New Caledonia and, perhaps most strikingly, on Palawan in the Philippines. On

Mount Bloomfield there is a sharp transition at 100 m from 10–20 m tall rain forest to a sclerophyllous shrubland 2–3 m tall of plants occurring at least 500 m below their usual altitude off serpentine. In the Solomon Islands, geologists have been able to map the rocks from aerial photographs of the forests but even as close as New Guinea, the physiognomy of the forest is not different away from ultramafics.

At Gunung Silam, a coastal mountain of ultramafic rock in Sabah (11), the surface soils have high Mg/Ca quotients and a substantial nickel content, but at deeper levels the quotient is much higher and nickel concentrations are sometimes very high. The forest at low altitudes is not readily distinguishable from that on other rocks, which is common in west Malesia. Indeed, many species are to be found growing off the ultramafics as well as on them, but they are capable of accumulating high levels of nickel there. Nevertheless, in some places, there are high levels of endemism, as in New Caledonia (12), where it is thought to be 60%. *Homalium guillauminii* (Flacourtiaceae), a small tree there, accumulates it to the extent that its ashed leaves contain up to 14% nickel on which, among other hosts, grows the common rain-forest moss, *Aerobryopsis longissima* with 5000 μg/g of chromium, which is up to 20 times that in the host.

Sclerophylly is also a feature of wet montane forests on peaty soils despite the cloud cover; indeed, cut shoots of trees from such forests in Jamaica and Malaya (13–15) show resistance to water stress not superior to lowland rain-forest trees, and are far less efficient at stopping water loss than are the sclerophyllous plants of, for example, the Mediterranean. Sugden examined the leaf anatomy of 43 woody species in three montane vegetation types of Isla Margarita, Venezuela, differing in their levels of cloud cover and exposure to wind. Again the xeromorphic leaves of the trees were shown not to be notably drought-resistant but the thick cuticle may reduce mineral leaching and minimize the rate of attack by epiphyllous fungi and may also be associated with the avoidance of overheating during the generally sunny days. Others suggest that here, and indeed in lowland forest, sclerophylly is possibly associated with edaphic conditions, but generally, it remains enigmatic.

When topography, water régime and 'unusual' edaphic features are set aside, are the differences in soils under typical lowland rain forest reflected in the species composition of the canopy (16–19)? In northwest Colombia chemical comparisons were made between the leaves, epiphyllae, epiphytes, litter and soil at six sites. There were significant differences in soil composition, but in the vegetable material, only potassium and caesium levels differed. There was a loose correspondence between soils and geological substrate but apparently little between soils and the vegetation. It has been argued that forest development is much quicker than soil development, so that vegetation mapping can scarcely be predictive of soil type. Nevertheless, in the rain forests of Australia there seem to be good correlations between distribution of forest types and nutrient concentrations, particularly phosphorus, nitrogen, potassium and calcium (1), and soil fertility overrides latitudinal climatic factors. In Brazil, a strong correlation has been reported between height, density and other parameters of forest structure, and soil features. In Gunung Mulu National Park in Sarawak (20), no significant soil correlations could be found in dipterocarp forest or forest over limestone, but in forest of alluvial regions and heath forest there were correlations. The first showed associations with changes in pH and the calcium component, while the second had associations with changes in organic carbon and cation exchange capacity.

In the Korup forest of south-west Cameroun, perhaps the most species-rich in Africa (21), an analysis of floristic variation within the forest over an area of some 6000 ha was made by sampling 80×80 m plots, where altitude, slope and soil conditions were measured and pH, carbon, available phosphorus, nitrate and potassium levels ascertained. The largest soil variations were in available phosphorus (2–29 parts per million) and potassium (38–375 ppm). Besides vegetation changing with altitude, there was a strong correlation of a gradient of change with that in available phosphorus. Low-phosphorus sites tended to have high frequencies of Leguminosae–Caesalpinioideae of the tribes Amherstieae and Detarieae, tribes with ectotrophic mycorrhizae that are known to enhance phosphate uptake such that these trees have a competitive advantage on such soils. Also in Cameroun, the vegetation of the low-lying Douala-Edea Forest Reserve was classified according to six soil variables. The plots were first separated into swamp and non-swamp areas, and these latter separated into those with low and those with high levels of phosphorus. The levels of available soil phosphorus are not as low as at Korup and the associations between those sites with 'low' phosphorus and the caesalpinioids is not so marked, but in the swamp soils, there were marked concentrations of *Protomegabaria stapfiana* (Euphorbiaceae) and *Librevillea klainei* (Leguminosae–Caesalpinioideae), the first associated with the more clayey soils, the second with very sandy ones. At a microhabitat level, the effect of trees (stationary, long-lived organisms) on local depletion of nutrients may be marked. Around dipterocarp roots there may be bleached sands, while allelopathic exudates, including phenolics leached from leaves, may also affect the local soil conditions.

3.4 Nutrient cycling

Because of the lack of absolute synchrony of soil and forest development it has been argued (16) that tropical rain forest often becomes independent of soil, efficiently recycling its nutrients. It has been calculated (2) that the rapid release of nutrients in decomposition is matched by rapid re-utilization and that the annual return of nutrients to the soil is some three to four times that in temperate forests. In the Central Amazon basin (22), the soils and parent material have been greatly weathered so that the present-day soils are clays, low in nutrients, consisting of kaolinite with sands from the ancient Brazilian and Guyana Shields. Jordan (22) has shown that the leaching of nutrients from the soil in the forests there was less than or equal to the input from the atmosphere every year between 1975 and 1980 and concluded that the weathering of the parent material does not contribute in a major way to the nutrient economy of the system. In short, the forest

Table 3.1 Above-ground biomass and nutrient content in a variety of moist tropical forests[a]

Site	Biomass (T/ha)	Nutrient (kg/ha)				
		N	P	K	Ca	Mg
Alfisols and moderately fertile soils						
Panama	316	—	158	3020	3900	403
Venezuela	402	1980	290	1820	3380	310
Ghana	233	1685	112	753	2370	320
Infertile oxisols/ultisols						
Ivory Coast-Banco	510	1400	100	600	1200	530
-Yapo	470	1000	70	350	1900	180
Brazil	406	2430	59	435	432	201
Venezuela	335	1084	40	302	260	69
Colombia-terrace	182	741	27	277	432	133
Low-nutrient sandy soils						
Venezuela-caatinga	185	336	32	321	239	53
-tall bana	180	618	62	669	568	200
-low bana	37	212	28	155	276	43
-open bana	5.5	32	2	29	24	7
Montane soils						
New Guinea	310	683	37	664	1281	185
Puerto Rico	197	814	43	517	894	340
Venezuela	348	876	53	1321	745	215
Jamaica-mull	337	857	41	829	940	193
-mor	209	426	30	272	353	155
Hawaii	176	367	28	380	756	72

[a]Modified from Vitousek, P.M. and Sanford, R.C. (23), p. 141.

apparently maintains itself on the nutrients derived from the atmosphere. Attention is thus focused on the nutrient-conserving mechanisms in cycling. However, many forests are 'leaky', losing more than they receive, and this seems to apply particularly to fertile ones. Moreover many rainfall throughfall measurements are unrepresentative and generalizations formerly made from them are unwarranted.

Patterns of nutrient cycling in tropical forests are diverse (23), much of the variation being related to the length and intensity of the dry seasons and to altitudinal gradients. The aboveground biomass does not differ greatly between sites with different soils (Table 3.1) except that it is lower on the

Table 3.2 Foliar nutrient concentrations in a range of moist tropical forests[a]

Site	Nutrients (%)				
	N	P	K	Ca	Mg
Alfisols and moderately fertile soils					
Panama	—	0.15	1.53	2.29	0.26
Ghana	2.52	0.14	0.85	1.54	0.48
Venezuela	2.54	0.15	1.52	1.50	0.48
New Britain	2.08	0.15	1.67	2.04	0.30
Zaire	2.45	0.12	1.92	0.70	0.88
Infertile oxisol/ultisol					
Venezuela	1.27	0.06	0.46	0.19	0.10
Venezuela	1.78	0.06	0.38	0.11	0.11
Brazil	1.84	0.05	0.50	0.42	0.29
Colombia-terrace	1.93	0.07	0.54	0.50	0.22
Low nutrient sandy soils					
Venezuela-caatinga	1.16	0.07	0.62	0.44	0.15
-caatinga	1.08	0.06	0.58	0.53	0.36
-bana	0.74	0.05	0.64	0.58	0.14
-tall bana	1.03	0.09	0.68	0.46	0.26
-low bana	1.29	0.12	0.72	1.03	0.25
-open bana	0.89	0.04	0.55	0.64	0.22
Brazil-campina	1.11	0.05	0.66	0.37	0.26
Malaysia	0.87	0.02	0.35	0.75	0.20
Montane sites					
Venezuela-cloud forest	1.17	0.08	0.55	0.87	0.26
-montane forest	1.74	0.08	0.66	0.64	0.23
Puerto Rico-lower montane	1.36	0.05	0.48	0.63	0.17
-elfin forest	0.99	0.06	0.51	0.67	0.16
New Guinea-lower montane	1.21	0.08	0.61	1.14	0.25
Hawaii	0.61	0.08	0.61	0.79	0.18
Jamaica-mull	1.78	0.08	1.17	1.0	0.46
-mor	1.11	0.06	0.43	0.80	0.33

[a]Modified from Vitousek, P.M. and Sanford, R.C. (23), p. 143.

poor, sandy soils, nor are there any strong correlations between nutrient status and biomass but nutrient concentrations in individual tissues are more likely to reflect the influence of soil fertility (Table 3.2); concentrations of all the major nutrients are indeed significantly higher on the more fertile soils. The infertile oxisols and ultisols have intermediate nitrogen and low phosphorus and calcium concentrations, the latter extremely low in some cases; sandy soils have low foliar nitrogen and phosphorus but intermediate major cations. *Varzéa* receives mineral inundations and leaves are higher in foliar phosphorus and cation concentrations than in *igapó*; montane forest leaves generally have lower nutrient concentrations than those from lowland fertile forests, even though they are growing on soils, which, if they were in the lowlands, would be considered fertile.

3.4.1 *Precipitation*

Most rain input of nutrients (Table 3.3) is comparable with that in temperate regions. Gas inputs are little known and biological nitrogen-

Table 3.3 Precipitation inputs of elements in moist tropical forests[a]

Site	Nutrients (kg/ha/yr)				
	N	P	K	Ca	Mg
Ghana	14	0.12	17.5	12.7	11.3
Ivory Coast	21.2	2.3	5.5	30.0	7.0
Cameroun	12	1.7	12.0	3.8	1.5
Zaire	6.4		2.0	3.9	1.1
Costa Rica	5	0.2	2.5	1.4	1.0
Costa Rica	1.7[b]	0.17[b]	5.4	3.1	2.6
Panama		1.0	9.5	29.3	4.9
Puerto Rico	14			34.0	26.0
Venezuela-lowland[c]	21[b]	25[b]	24.0	28.0	3.0
-montane	9.9	1.1	2.6	5.6	5.2
Brazil	10	0.3		3.7	3.0
Brazil	6[b]	0.16[b]	3.4	ND[d]	ND[d]
Australia			4.0	3.0	2.5
Malaya			12.5	14.0	3.0
Malaya	13.5		6.4	4.2	0.7
Papua New Guinea-lowland			0.8	0.0	0.3
-montane	6.5	0.5	7.3	3.6	1.3

[a] Modified from Vitousek, P.M. and Sanford, R.C. (23), p. 158.
[b] Inorganic forms of the element only.
[c] Wet and dry fall together
[d] Not detectable.

fixation rates of about 200 kg per ha per annum have been estimated for *varzéa*, about 20 for ultisols and about 2 on infertile oxisols. Losses have been little-documented but nitrous oxide production is probably substantially greater than in temperate regions, though it comprises only a small fraction of the nitrogen circulated in the system as a whole (23). In the Amazonian forest of southern Venezuela, Jordan and his colleagues (24) found that the nutrients reaching the ground in gaps were more concentrated than in closed forest and suggested that they were absorbed by epiphytic algae and lichens and possibly leaf surfaces. Bromeliads, living as epiphytes in the canopy, can absorb canopy leachate through the reservoir at the heart of the plant (25). It has also been shown in Puerto Rico that epiphyllae (small plants growing on leaves) enhance cation absorption from precipitation some 1.7 to 20 times (26).

Because of their ability to absorb nutrients and their position in the canopy, epiphytes can be of great importance in nutrient cycling. A single tree, 13 m tall, of *Clusia alata* (Guttiferae) in cloud forest in west-central Costa Rica had a standing crop (live epiphytes) of dry weight 141.9 kg, of which 20% was on the trunk, 45% on the major branches, 23% on the middle-sized ones and 12% on the outer ones, the whole yielding 3 kg nitrogen, 97 g phosphorus, 460 g calcium, 678 g potassium, 126 g magnesium and 207 g sodium (27).

A further explanation for the presence of drip-tips in rain forest (28) argues that soil structure may be less affected and leaching retarded where the incident raindrop size is small and that the tips act as spouts promoting just this. This might get round the problem of why drip-tips are not found in all ever-wet communities if leaf drainage amelioration is the only function of the tips, but it seems likely that other explanations as well as those so far discussed may be appropriate under different conditions and for different species. Moreover, it should be borne in mind that when leaves are in bud, 'drip-tips' comprise a much larger proportion of the leaf, much as the apical thorn of a thistle-leaf does: in the latter, this apical growth affords protection, but little is known about the role of drip-tips in buds. A further explanation for the scleromorphic leaves of trees in nutrient-poor sites argues that the relatively thick cuticle and wax deposits and relatively low water content make them resistant to nutrient loss by parasites and herbivores (29). There is some evidence that insect attack is less severe. Furthermore, the nutrients invested in leaves are used in photosynthesis for prolonged periods, are more resistant to leaching by rainfall, and a large proportion of leaf nutrients are known to be translocated out of them before abscission. In Malaysia (30) on the other hand, it has been found that the nutrients in precipitation increase through canopy leaching, which was richest in potassium, representing some 24.57 kg K per ha per year—98% of these nutrients reached ground by throughfall and only 2% by stem-flow. Little leaching was recorded from seedlings in Puerto Rico (25) though there were measurable amounts leached from crops such as banana and sugarcane. Once more, an explanation for drip-tips can be provided: their rapid removal of water from the lamina surface may reduce the time in which leaching out of minerals takes place.

Nutrients in stem-flow are generally less than 10% of those in throughfall in mature forests. Throughfall seems to be of little importance in cycling nitrogen, phosphorus or calcium, though it is important for potassium and it would appear that forests on moderately fertile soils lose more potassium and calcium through throughfall than those on infertile sites, a finding consonant with research in temperate forests (23).

3.4.2 *Litter*

Forests on fertile soils (23) return more litter at higher nutrient concentrations than others; though those on oxisols and ultisols return smaller amounts of phosphorus and calcium at significantly higher dry mass/element ratios than the fertile soils, although nitrogen levels are similar. Sandy sites cycle small quantities of nitrogen and phosphorus at low concentrations in the litter. Only the sandy sites and montane forests have high dry mass/nitrogen ratios comparable with those in temperate forests. How much is withdrawn from leaves before litter fall? The evidence suggests that sites with high dry mass/nutrient ratios in litterfall have similar ratios in active leaves. Assuming that calcium is immobile once in leaves, comparisons of nutrient/calcium ratios in active leaves and litterfall show that phosphorus retranslocation appears to be greater on the infertile soils, i.e. the efficient phosphorus utilization on infertile oxisols and ultisols appears to be a consequence of both lower foliar phosphorus concentrations and effective phosphorus retranslocation.

In tropical Australia (31), annual litterfall varied between sites from 728 to 1053 g per m^2 per annum. On 1 ha sites of dipterocarp forest, alluvial forest, heath forest and forest over limestone at Gunung Mulu, Sarawak, 'small' litterfall, i.e. all but wood over 2 cm diameter studied for over a year varied from 8.8 to 12 t per ha per annum, the peak coinciding with the peak in rainfall (32). Litterfall from dipterocarp forest was low in calcium and phosphorus, that from heath forest low in nitrogen, but there were high concentrations of potassium from the dipterocarp forest and, scarcely surprisingly, calcium from the limestone forest. No studies have really dealt with the very significant inputs from large pieces of wood and treefalls. Of the rest of the litterfall, flowers and fruits are high in nutrients and it would be worth investigating seedling establishment after a major input from masting, as decomposition of these parts is rapid.

Litter in *Araucaria hunsteinii* forest in New Guinea represents some 5% of the biomass and is formed at a rate of some 8.7 t per ha per year (33). Although the living leaves represent only 5% of the living biomass, they hold some 23.5% of its nutrients. Palms and palm litter are rich in potassium, tree ferns in nitrogen (in a Hawaiian montane forest they accounted for 28% of the biomass but 70% of the nitrogen (23)), while at Gunung Mulu, *Ficus* leaves had the highest nutrient concentrations and available carbon with low polyphenols (34). Under different canopies in Puerto Rico, different rates of decomposition were found, the fastest being under *Euterpe globosa*, while amongst the six common species tested, there was great variation in the rate of decomposition as well (35). The rates are such that whole trees may completely disappear in 10 years or less, a factor that must be remembered in plot reassessments (36). Decomposition rates vary with climate and soil fertility and so cooler montane forests have slower rates and the litter layer on sandy soils is the most voluminous on lowland sites (23). Leaf decay rates vary greatly in tropical to subtemperate rain forests in New South Wales, those of *Dendrocnide excelsa* (Urticaceae) with over 95% lost in only 4 months and the last of its veins gone by 8 months, whereas those of *Nothofagus moorei* (Fagaceae) have only 40% decayed in one year (37). In northern Thailand, it was found (38) that over 56% of annual litterfall was decomposed in 1 year and there was complete mineralization in 20 months. In the litter at Gunung Mulu (34), 80% of the potassium was lost in the first 2 months; magnesium was similar but calcium losses followed the dry weight losses for the first 6 months; nitrogen and phosphorus were conserved in the litter and phosphorus increased as a percentage by 10% after 10 months; sodium effectively doubled its original concentration in the first 8 months and, after 1 year, was still 30% higher than in the initial concentration.

Conversion from organic nitrogen to biologically available nitrate or ammonia, although little-measured in tropical soils is rapid by comparison with that in temperate ones in general, but lower rates appear to obtain on sandy and montane sites in the tropics (23). With increasing altitude, as measured along an altitudinal transect (100–2600 m) up the northern slope of Volcan Barva in Costa Rica (39), the generally accepted hypothesis that nitrogen mineralization and nitrification are reduced in montane tropical forest was supported. Soils from higher altitudes may be limited by their being very wet (80% at 2600 m compared with 42% at 100 m), for soils from high altitudes, brought to lower ones and dried, had higher rates of mineralization than those at lower altitudes; moreover, under very wet conditions, denitrification is likely to be increased.

Little is known of nutrient outputs from undisturbed vegetation. Measurements were made on the Caura River within the Orinoco watershed, where 47 500 km^2 of undisturbed forest is drained. Every fortnight over 2 years (40) total nitrogen, phosphorus and the forms in which they were discharged were measured. The river is black due to dissolved carbon materials (about 3.96 mg per litre) and the loss of nitrogen (9.98 kg per ha per year) is over eight times that recorded in the River Gambia in west Africa and is very much higher than any comparable temperate system. The atmospheric input including that fixed by lightning is thought to be only 1–2 kg. In a system of steady state, the nitrogen loss through runoff must be the rate of fixation less denitrification, i.e. in this forest nitrification must exceed denitrification by about 10 kg per hectare per annum. Densely planted stands of legumes may accumulate 50–100 kg per hectare per annum but for natural stands more than 10 kg is unusually high.

3.4.3 *Roots and mycorrhizae*

Within the forest itself, the most important mechanisms involved in nutrient cycling are the roots. In the Amazon basin, Jordan and Herrera (41) have found that the poorer the soils, the shallower the mat of roots and surface humus, and that, in the poorest, the mat may be only 15–40 cm deep, with 58% of the feeding roots within it. Fine-root dry weights in tropical forest vary from 3–100 t per ha, the highest higher than in other forest ecosystems but the lowest indicating that not all tropical forests have the dense mats described above (42). Indications from forests in Costa Rica are that fine-root biomass may be inversely related to phosphorus and/or calcium availability, just as nitrogen is in temperate forests. Root/shoot ratios also appear highest on infertile soils, notably sandy ones (23). Turnover of fine absorbing roots has been little-investigated but Sanford, using sequential cores and monthly observations on 'plexiglass' windows, estimated the turnover of fine roots of less than 2 mm diameter. In the upper 10 cm of a Venezuelan oxisol, the turnover was 25% per month, an annual fine-root production of 15.4 tonnes per hectare for this size and depth alone. Assuming that nutrients are not retranslocated from roots before they are sloughed, the amount of nutrients added to the soil would be the mortality multiplied by the concentration, which, in this case, would add 343 kg nitrogen and 11 kg phosphorus per ha per annum, which compares very favourably with the 61 kg and 0.8 kg respectively from fine litterfall at the same site. Of course some nutrients may be translocated but, as only 43% of the root mass is in the upper 10% investigated, the figures may not be gross overestimates.

Direct physical adsorption of labelled cations sprinkled on root mats by Jordan and Herrera (41) was 99.9% and when the tracers were added in the form of labelled leaves, no detectable radioactivity leached through. Work *in vitro* has shown that mycorrhizae are important here as they can exploit larger volumes of soil than can roots alone. Most higher plant species are capable of forming root associations with the Endogonaceae, a family of Zygomycetes, to form what are known as vesicular-arbuscular mycorrhizae. There is, apparently, a complete lack of specificity of the host and this is as true for tropical ectomycorrhizae as it is for temperate ones: they have been recorded from at least 17 genera of trees in six families in tropical forests (43–45). Overall, they are smaller in diameter than in temperate examples but the sheaths are relatively thick and the fungus can form up to 80% by weight of the combined system. It is possible that they might improve water, as well as nutrient, uptake and they are particularly conspicuous in the dipterocarps, and also in the caesalpinioid group of Leguminosae. Over 60% of these do not form nodules with nitrogen-fixing bacteria, and those with the mycorrhiza have no root hairs (46). There is some evidence to suggest that the fungi are antagonistic to certain other, pathogenic, fungi and it is well-known that pines do not develop beyond the nursery stage without their mycorrhiza. Furthermore, some legumes do not nodulate on phosphorus-deficient soil unless their mycorrhiza is there.

On very poor soils, species with obligate mycorrhizal associations are likely to predominate and root hairs become redundant, because these latter overlap too much and compete, whereas mycorrhizal hyphae are better dispersed. Indeed, many obligate mycotrophs, particularly in rain forest, have no root-hairs so that the turnover of mycorrhizal hyphae should also be considered as root-hair turnover is in nutrient-cycling studies. Root hairs (47) are more advantageous when fertility is temporarily elevated as in clearings, volcanic deposits and so on. Indeed, many plants reject vascular-arbuscular mycorrhizae under such conditions or are without it in any case, e.g. *Carica papaya* (Caricaceae), the papaya, *Trema* spp. (Ulmaceae), *Cecropia* spp. (Cecropiaceae), weedy Compositae and some Melastomataceae. Vesicular-arbuscular mycorrhizae extract minerals from the soil solution while some ectomycorrhizae may break down the litter, directly recycling nutrients, but these mycorrhizae are more demanding of their hosts too. Nutrient recovery is increased when roots form the dense mats typical of the poorest soils (43).

In the root mat, there may be other nutrient-conserving mechanisms. Algae might take up dissolved nutrients to be released on their decomposition, while there are nitrogen-fixing micro-organisms, both nodular and free-living. Furthermore, it must not be forgotten that a good deal of litter is not at ground level but is deposited in the crotches of trees, in the masses of epiphytes or the crowns of certain dwarf treelets in the lower reaches of the forest. Such sparsely-branched treelets as species of *Semecarpus* (Anacardiaceae) of Malesia have buds and leafbases surrounded by rotting humus.

3.4.4 *Animals*

Populations of litter-eating macrofauna at Gunung Mulu (34) e.g. isopods, molluscs and earthworms, are lower than in most acidic soils in temperate deciduous forests with approximately 200–300 animals per square metre, with biomasses (alcohol wet weight) from 0.18 g (dipterocarp forest) to 6.84 g per m^2 in limestone forest. Earthworms, which are usually larger but much fewer than in temperate woods (48, 49) seem not to be involved in litter breakdown in Mulu, though observations at Sepilok in Sabah suggest that their rate of soil turnover is comparable with or greater than that of worms in savannas.

Humus-feeding termites that do not tackle intact litter represent a major proportion of termite biomass in both dipterocarp and heath forests at Mulu (34). Although the mesofauna was not completely sampled there, Cryptostigmata (Acari) populations of about 22 800 per m^2 were found in alluvial forest, 93 400 in dipterocarp forest and 5600 in heath forest. Although these animals are not much involved in breakdown in temperate forests, their effect in tropical ones is still unquantified. Termites (50, 51) are undoubtedly the most important animals, reaching numbers of 2000–4000 per m^2 and dominating all other soil animals, with a biomass of 1–10 g per m^2 in Malaysia. These figures decrease with altitude and bad drainage but rise in highly organic soils, despite the presence of fewer species there. Their biomass greatly exceeds that of herbivores and their predators and is about three times that of birds and mammals.

All but the Macrotermitinae use some or all of their faeces in building nests or foraging galleries and, as these are abandoned or decay, nutrients are recycled; the nests may form nutrient-rich patches in the forest for the establishment of seedlings. Except for the Macrotermitinae, termites harbour nitrogen-fixing bacteria but the significance of this in overall nitrogen- fixation levels has not yet been assessed. Available nitrogen is also increased through the digestion of materials by the fungi cultivated by the Macrotermitinae. However, there is no evidence for the selection of high nitrogen-containing food plants except perhaps in *Hospitalitermes* spp., which forage over wide areas for lichens; fresh dead litter is consumed by some but generally rotting and rotten material is favoured by many and there is evidence that fungi improve the nitrogenous quality of litter. Nitrogen is also conserved by termite cannibalism, necrophagy and recycling of faeces. In Zaïre, some 870 termite colonies per ha were found in rain forest. This represents 16 t of biomass which immobilizes some 2.6 t per ha of organic matter, 0.7 t of calcium and 61 kg of nitrogen, while some 6 t of plant matter may have been consumed. At Pasoh, West Malaysia, it is estimated that 38.8 kg of leaf litter per ha is thus disposed of every week, which represents some 32% of the leaf fall. Overall decomposition and nutrient-cycling at Gunung Mulu (34) were not particularly fast, however, contrary to the accepted dogma. This is perhaps explicable in terms of the low concentrations of nutrients in the litter, the high lignin content, the low faunal activity and water-logging of the humus layer or interactions between these.

3.4.5   *Overall production*

The leaf-fall in the Manaus area of Brazil has been estimated at some 7 t per ha per year in a standing crop of some 900 t including roots, about 585 t dry matter. Figures from other parts of the world are: 289 for Ghana, 469 for Ivory Coast, 664 for Malaya and 400 in *terra firme* forest in the Amazon (52) while, within a single area, above-ground biomass at Gunung Mulu was 650 t per hectare dry weight in dipterocarp forest, 470 in heath forest, 380 in limestone forest and 250 in alluvial forest (53). More than three-quarters of the carbon in the system is in the wood, whereas in temperate coniferous woodland, half of it is in the soil.

In tropical forests, except where peat is accumulating, the net production of mature forest is nil. The largest trees may in fact be stagnant and the earlier stages of fast-growing trees at maximum annual net production, but it is exceedingly difficult to assess net production because of the problems of sampling and the role of animals. Nevertheless, it has been estimated that the net production may be the same as that of European forests though the gross production may be twice this. The balance is due to greater losses in respiration, the forests of Europe being leafless for over half the year. The observed greater production of plantations at altitude in tropical regions has been explained as a result of depressed respiration during the cool nights but there are other differences such as soils and aspect, which may have an effect too.

CHAPTER FOUR

THE CHANGING BIOLOGICAL FRAMEWORK

As von Humboldt noticed on the mountains of tropical South America, tropical forests at increasing altitudes look increasingly more like temperate ones. In the lowlands, the forest may be some 25–45 m tall, with 'emergent' trees protruding above the general canopy, whereas in lower montane forest the canopy may be at 15–33 m with fewer conspicuous emergents, and in upper montane to subalpine forests the canopy may drop to 1.5 m and there are no emergents. The 'tropical' features of cauliflory, buttresses, lianes and the common occurrence of pinnate leaves decline in frequency with altitude as they do with latitude. This may reflect, to a certain extent, the taxonomic groups of plants that grow at different altitudes, but such trends can be seen even within families. Conversely, the frequency of microphylls and, especially noticeably, of non-vascular epiphytes, increases with altitude, most remarkably in the bryophyte-festooned cloud forests of tropical regions.

Compared with temperate forests, however, there is a marked lack of readily recognizable strata in the lowland forests. In temperate countries, the tree-, shrub-, herb- and ground-layers are familiar in woods, but these are often a reflection of management techniques: stands may be more or less even-aged, or silvicultural practice may have removed smaller trees and bushes that might otherwise blur the clear stratification. In certain tropical forests, a stratification is clear, as in single-dominant ones such as the *Gilbertiodendron dewevrei* (Leguminosae) forests of Africa, but, in general, the most useful division that can be made in mixed tropical rain forest is between the canopy and the shaded area beneath. This can be imagined as the undulating surface connecting the bases of the crowns of the large trees. Clearly this surface is constantly changing as trees grow up and eventually collapse, and the structure of the forest can only be understood in terms of these dynamics. The framework of trees between and on which other plants,

Figure 4.1 A forest profile some 60 m long and 8 m wide in Andalau Forest Reserve, Brunei, Borneo. Only trees over *c*. 5 m tall figured. Reproduced with permission from P. S. Ashton in *Oxford Forestry Memoirs* **25** (1964), fig. 25.

fungi, bacteria and animals live and die is a result of the careers of individual trees.

4.1 Tropical rain forest successions

4.1.1 *Problems with older views*

In older textbooks, much is made of the progress of succession: an orderly replacement of plant and animal species culminating in an optimal climax vegetation and associated faunal assemblages for the physical parameters of any particular region. In turn, this has led to schemes of prediction of vegetation types from the combined observations of physical parameters such as latitude, precipitation and temperature régimes.

When the nature of the successional process is scrutinized critically, however, as it has been in a classic review by Drury and Nisbet (1), it is seen that much of the accepted dogma on the nature of succession is far too rigid in its approach. Temporal sequences on one site under stable climatic régimes are the essence of secondary successions, whereas those on such sites with changing climatic conditions are primary successions, such as the changes in vegetation as a glacier retreats. Observations on what is essentially a temporal phenomenon have often been made spatially, that is, adjacent sites are seen as representative of consecutive time-stages. The observational basis of much of the older work is therefore called into question. Also attacked is the supposed orderly replacement of apparently altruistic species by their successors, for there are many examples of plants in temperate regions emitting allelopathic substances that hinder their replacement by others. Of tropical examples, some species of *Croton* (Euphorbiaceae) and *Piper* produce chemicals that inhibit seed germination, and it is thought that the Brazil nut, *Bertholletia excelsa* (Lecythidaceae), may produce toxic litter. Monospecific stands of *Dinizia excelsa* (Leguminosae) in Brazil produce a very acid humus within 10 years and this supports a very specialized set of understorey plants. As has already been noted in Chapter 3, there is no requirement for a mature soil profile to herald the arrival of a 'climax' forest, so the correlation between these, again often repeated, is not universal. In short, one or more of the features classically associated with succession may not obtain in any particular vegetational change under study. Succession is seen as the sequential expression of conspicuousness reflecting the maturity rates of individuals of particular species, for so-called 'climax' species may occur, as small plants, in the early stages of such sequences.

4.1.2   *The modern interpretation*

In a prescient paper of 1956, Hewetson concluded that concepts such as 'climax community' and 'association' should be abandoned in tropical ecology and dwelt on the unpredictable nature of storms, droughts and so on, arguing that tropical forest is a continuum in which the parts are in unstable equilibrium. Indeed, in tropical rain forest, change is due to the refilling of gaps in the vegetation, made by the collapse of canopy trees. This process is familiar to European foresters, some of whom refer to the regenerating patches as *chablis*, a term taken up in this context by Oldeman (3). In the Old World tropics, Whitmore, taking his inspiration from the work of E.W. Jones and of A.S. Watt in European vegetation, and the African studies of Aubréville, and Hartshorn, among others, in the New, have independently come to similar conclusions on the importance of gaps in rain-forest structure. These forests, then, are seen as patchworks of regenerating gaps and new tree-falls: the essence of this structure is to be found in the understanding of how and where gaps form, by what they are filled and the biological characteristics of the fillers, both plants and animals. This approach has application in many other vegetation types, too.

The concept of the 'climax' forest as that mosaic of filling gaps, which eventually occupies any particular habitat, embraces time and therefore the unpredictable nature of the presence of any particular species at any particular point. The forest consists of patches, or islands, and potential patches, so that much of island biogeography on a small scale may be applicable to rain forest biology (4). Thus those species that thrive in newly-formed gaps may either immigrate afresh, as to islands in an otherwise disagreeably oversubscribed sea, or possibly may lurk in the seedbank ready for the 'island' when it arrives through the collapse of the vegetation above it. Again, those species that require shade when young and are referred to as shade-bearers, persist in a similar way, though as seedlings or saplings, which, quietly ticking over on the forest floor with little net growth, are as 'dormant' as seeds.

Thus forest cycling may be considered as overlapping processes of secondary succession as contrasted with primary succession, that is, the formation of vegetation on previously unvegetated sites, such as vulcanoseres on lava, while the revegetation of landslips may be considered as a rather intermediate state of affairs in that frequently some soil is left *in situ* and often plants creep in vegetatively from adjacent vegetated sites. Clearly, there may be little to choose between this and a 'large' gap formed by fire, anthropogenic or otherwise, and thence gaps formed by storms and smaller gaps made by the falling of a limb from a canopy tree. Nevertheless, it is often useful to refer to primary succession when isolated naked deposits, such as that represented by Krakatau after its eruption in 1883, are considered. Primary succession is discussed more fully in section 4.6.

4.2   The ecology of gaps

In temperate forests (5), gaps caused by one or several trees may range in size from less than 25 m² to 0.1 ha, while largescale storm damage may lead to devastation over as much as 3000 ha and, in mature forests, gaps may average about 1% per annum of the total area. Even during still nights in tropical rain forest, trees can be heard crashing down at alarmingly frequent intervals. Trees are not immortal and, with age, they become prone to attack by fungi, other pathogens and invertebrates, growing 'stag-headed' and dropping branches or major limbs. There is an increase in falls in the wettest seasons, probably because the anchorage may become more precarious as run-off erodes the soil around the roots. Gusts before storms are apt to increase the likelihood of collapse but, in still weather, trees also fall and it has been suggested, on the camel's back principle, that epiphytic loads may be important in the collapse of major limbs, if not of whole trees. The fall may lead to damage to neighbouring trees, or if they are linked together by lianes, to the fall of such trees as well. Sometimes, a falling tree may be half-felled and hang, trussed up by these lianes, continuing in a small way to produce new leaves, flowers and fruits.

On the other hand, localized squalls can fell relatively huge areas: over 80 ha of *Shorea albida* forest was thus devastated at one time in Sarawak, Borneo. Some regions are more prone to such storms than are others and storm frequency is known to be increasing in some areas. To take Sarawak again, the mean number of gaps and the mean gap size increased steadily between 1947 and 1961. It is estimated that 500 000 lightning strikes affect the world's forests every day (6): the effect may range from no obvious disruption to (rarely) the reduction of a tree to slabs and slivers. In an affected group of trees, mortality processes may take days or weeks or even years, and are dependent on tree species, distance from central point, flash characteristics, subsequent attack by other agents and so on. About 50 000 wild land fires are started in the world each year by lightning, which is an important gap-maker in the peat-swamp forests of West Malaysia and Sarawak, for fires in the accumulated dry epiphytic detritus may thus be initiated. Such fires are also well-known in mangrove forests in both West Malaysia and New Guinea. Gaps thus formed differ from others in that the shade-bearing saplings are often killed as well, so that herbs initiate succession. Gaps are also formed by landslips, notably on steep river banks, where they may be constantly renewed and any succession established may be restarted each time (Figure 4.2).

Meandering rivers in the upper Amazon are so effective in this respect that it has been calculated that 26% of the modern lowland forest has characteristics of recent erosional and depositional activity and that 12% of the Peruvian lowland forest is in successional stages along rivers. In New Guinea and elsewhere, earthquakes are also important in gap formation.

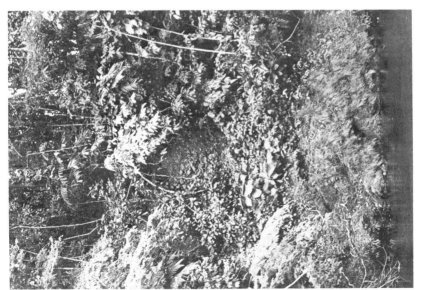

Figure 4.2 Landslips caused by river erosion undergoing colonization in eastern Panama. The early stage shows toppled saplings, bare ground and seedlings while the later is dominated by tree ferns.

Animals may be involved in making or maintaining gaps as at salt-licks or in the perpetuation of grassy glades as by rhinoceros and other herbivores in African montane forests. More sensational have been the results of caterpillar attacks: in 1948 a 31 km strip of peat-swamp forest in Sarawak lost all its *Shorea* trees. From the observation that nests of the termite *Microcerotermes dubius* are invariably associated with patches of dead and dying trees in West Malaysia, and the fact that all such trees in gaps up to 16 × 8 m were infected by termites, it has been cogently argued that these animals, too, are important gap-makers. Possibly a gap is abandoned and the termites move on when exposure increases (7). On a smaller scale, single trees may be killed by the bark-stripping of apes, for instance orang-utans in Sumatra (8).

Gaps, then, may vary from small punctures in the canopy, caused by the loss of a limb, to huge areas of devastation. In a classic study, Poore (9), working on a West Malaysian plot of some 12.24 ha with apparently stable areas of canopy some 45–50 m tall, found that it included some 10% by area of gaps with 75 fallen trees, 90 dead but not decayed trunks and 40–50 standing dead trees, all with a girth greater than 91 cm. The largest gap measured some 20 × 30 m and he calculated from the heights of the tallest trees that the mean gap size was some 400 m^2. The 165 trees that he estimated to have fallen in 12 years would thus form 6.6 ha of gaps, or a little over half of the entire area. Hartshorn has calculated, from gap size and frequency, that the turnover rate in forest at La Selva, Costa Rica, is 118 ± 27 years, and that the forest-cycle in Sumatra is 108–117 years (10). At Los Tuxtlas, Mexico (11), most individuals of *Astrocaryum mexicanum*, an understorey palm, endure treefalls by bending under fallen limbs or boles and, after a year, a bent palm resumes vertical growth, leading to a permanent kink in its stem. Because the age of the palms is reflected in stem elongation, it allows a timescale to be put on gap formation and regeneration, showing that over half of a 5 ha plot under study has been disturbed in the previous 30 years. The overall result of gap-formation is therefore rather similar to that engendered by foresters' selective felling: the devastating effect of machinery and fire used by man is paralleled only by landslips or earthquakes, for then the substrate is greatly modified.

A gap caused by the fall of a tree represents a heterogeneous environment. The dying tree itself will be an almost impenetrable tangle of branches, twigs, leaves and epiphytes at one end and will present bare soil at the other. The rootplate often lies perpendicular to the ground, so that there will be around it areas shaded for part of the day, and soil of different depths and levels of disturbance from ground level to possibly some metres in the air. The bole will lie in the exposed vegetation of the forest floor and may persist for some years. The conditions at the edge of a gap are likely to be less harsh than those in the middle, and the middle of large gaps will be more exposed than the middle of smaller ones. Large gaps made by storms are likely to get larger, through windthrow of exposed trees at the margin.

In short, gaps are variable in size and frequency in both space and time: they are unpredictable.

4.3 The ecology of pioneers

4.3.1 Pioneers and shade-bearers

The variety of ecological opportunities offered by gaps is met by the differing response of different forest species to gap conditions. Species waiting for the collapse of their superiors, shade-tolerant plants generally referred to as *shade-bearers*, or *dryads*, can be contrasted with those that are stimulated to germinate by the new conditions and known conveniently as weed trees, *pioneers*, or *nomads*. There is a spectrum of responses to gap formation and the terms used may seem to make the process of gap colonization simpler than it really is, but so long as it is remembered that not every pioneer species necessarily has all the features associated with the pioneering 'syndrome', this should not occur. Examples along this spectrum can be found in rain forests all over the tropics; in different families of trees, in different continents, these 'syndromes' have evolved in parallel.

Pioneers are not found in the understorey of mature forest but their seeds germinate in gaps open to the sky as the plants cannot survive in shade (12). They also have most or all of the following:

1. small seeds produced copiously and more-or-less continuously, that set when the plant is young and are dispersed by animals or wind;
2. their dormant seeds are usually abundant in forest soil;
3. their height growth is rapid;
4. the seedling carbon-fixation rate is high as is the compensation point;
5. growth is indeterminate with no resting buds;
6. branching is relatively sparse and leaves are short-lived;
7. rooting is superficial;
8. wood is usually pale, of low density and non-siliceous;
9. leaves are susceptible to herbivory and are sometimes with little chemical defence; and
10. they have a wide ecological range, are phenotypically plastic and often

short-lived. They may possibly be subdivisible into classes by height: from pygmy (nanophanerophytes) less than 2 m tall, e.g. *Solanum* spp. (Solanaceae), to small (2–7.9 m (microphanerophytes)), e.g. most *Trema* spp. (Ulmaceae) and many *Macaranga* spp. (Euphorbiaceae) to medium (mesophanerophytes, 8–29 m tall), e.g. *Neolamarckia* spp. (Rubiaceae) or *Cecropia* spp. and *Musanga cecropioides* (Cecropiaceae) to large (megaphanerophytes, more than 30 m), e.g. *Cedrela odorata* and *Swietenia mahagoni* (both America) and the Asiatic *Chukrasia tabularis* (all commercially important Meliaceae), *Chlorophora excelsa* (iroko, Moraceae), *Eucalyptus deglupta* (Myrtaceae), *Lophira alata*

Table 4.1 Examples of the pioneer and climax tree species groups from Africa (AF), the Eastern Tropics (ET) and tropical America (AM) subdivided into height class subgroups[a]

Tree stature	Pioneers (germinate in full sun and require full sun for survival and growth)	Climax (germinate in shade, or rarely in full sun, and seedlings can survive and grow in shade)
Pygmy (Nanophanerophytes) < 2 m tall	probably none, class occupied by shrubs e.g. *Solanum* spp. (pantropical)	*Pycnocoma macrophylla* (AF) many arecoid palms *Coussarea* spp. (AM) *Psychotria deflexa* (AM)
Small (Microphanerophytes) 2–7.9 m tall	*Rauvolfia vomitoria* (AF) most *Trema* (pantropical) many *Macaranga* spp. (AF, ET) *Pipturus* (ET) Some *Piper* spp. (AM)	*Microdesmis puberula* (AF) most Melastomataceae (ET, AM) *Drypetes ivorensis* (AF) *Diospyros buxifolia* (ET)
Medium (Mesophanerophytes) 8–29 m tall	*Musanga cecropioides* (AF) *Neolamarckia* (ET) *Macaranga hypoleuca* (ET) *Cecropia* spp. (AM) most *Sloanea* spp. (AM)	*Turraeanthus africanus* (AF) a few Dipterocarpaceae (ET) Fagaceae (ET) most Myristicaceae (ET, AM) *Minquartia guianensis* (AM)
Large (Megaphanerophytes) ⩾ 30 m tall	*Chlorophora excelsa* (AF) *Terminalia ivorensis* (AF) *Terminalia superba* (AF) *Lophira alata* (AF) *Pericopsis elata* (AF) *Paraserianthes falcataria* (ET) *Eucalyptus deglupta* (ET) *Goupia glabra* (AM) *Laetia procera* (AM) *Cedrela odorata* L. (AM) *Swietenia mahagoni* (AM)	*Khaya ivorensis* (AF) *Entandrophragma* spp. (AF) *Funtumia elastica* (AF) *Aningeria robusta* (AF) nearly all Dipterocarpaceae (ET) *Virola surinamensis* (AM) *Pentaclethra macroloba* (AM) *Couratari* spp. (AM) *Vochysia maxima* (AM) *Eschweilera* spp. (AM)

[a] From Swaine, M.D. and Whitmore, T.C. (12).

(Ochnaceae), *Paraserianthes falcatoria* (Malesia) and the west African afrormosia, *Pericopsis elata* (both important Leguminosae), *Terminalia ivorensis* (idigbo) and *T. superba* (ofram), both important timber Combretaceae from west Africa.

Between shade-bearers there is variation in the amount of light needed for 'release'. At one end, some need a lot and grow fast, tending to have seedlings with rapid mortality below shade and not persisting for long. These are the 'light hardwoods' of Malaysia, e.g. light red merantis, *Shorea* spp. (Dipterocarpaceae) as well as the African *Entandrophragma* spp. (Meliaceae). Regeneration of these is favoured by massive timber extraction: they resemble pioneers except in the point of germination. At the other extreme are those requiring little or no increase in sun. They grow more slowly and have heavy, dark, often siliceous, timber ('heavy hardwoods') and are less likely to regenerate after intensive logging, e.g. the African *Cynometra alexandri* (Leguminosae).

At La Selva, Hartshorn (13) found that the bulk of the tree species, some 150, were shade-intolerant. Factors affecting the presence or absence of any particular species included timing of gaps, the proximity and dispersability of seeds, the gap size, the substrate conditions and density-dependent plant-herbivore relations. The pioneers can have continuous seeding and dormancy but the others do not, the absence of which facility may be counterbalanced by the build-up of a 'bank' of dormant saplings on the forest floor. The pioneers are generally more palatable to insects and other herbivores than are the later gap-fillers, though at La Selva there was some evidence that the common species were less palatable than rarer ones. The leaves are often large and thin, with a high turnover rate, and may be greatly perforated by herbivores: a 20-year-old stand of *Cecropia*, *Dillenia* (Dilleniaceae), *Musanga*, *Macaranga* or *Cestrum* (Solanaceae) can have up to 20% of the lamina removed in this way (14). Leaves well protected by toxins, thick layers of hairs, waxes or sclerophyllous form can be present but (particularly the last) are exceptional. The gradient of palatability associated with the gap-filling in the tropics is not found in such harsher climates as saltwater swamps, semi-deserts or alpine communities where there is little change in palatability with succession.

There are problems in defining 'gaps' based on physical parameters, when compared with the presence of pioneers as indicators of gap conditions. Using the first method in Los Tuxtlas (Mexico), the most northerly neotropical rain forest, the pioneering habitats were underestimated by between 44 and 515%. On average, the colonized area was some

3.4 times larger than the size of the projected canopy opening, the majority of pioneers actually showing a preference for gap edge. 'Gap environment' is the result of a combination of the size of the canopy opening, surface inclination, slope aspect, height of surrounding vegetation and other factors (15).

4.3.2   Pioneer morphology

The form of many pioneers is such that they have many vertical or inclined pithy stems with their large leaves in a single layer. Such a tree is the neotropical *Piper auritum* (Piperaceae). Its leaves wilt in full sun, the leaf temperature dropping by $1-5°C$ with decreased photosynthesis and transpiration, such that the ratio of photosynthesis to transpiration (water-use efficiency) is increased compared with non-wilting leaves (16). Sometimes the stems of pioneers are inhabited by ants. The later trees have a more marked apical dominance, much-layered foliage, and deep roots (often tap roots). Some pioneers have nitrogen-fixing mechanisms, as in the roots by *Albizia* (Leguminosae) or the leaves of Neolamarckia cadamba (*Anthocephalus chinensis* (Rubiaceae)), and some are deciduous.

The wood of many fast-growing tropical pioneer species has thin-walled cells with large lumina and is structurally 'cheap'. With a specific gravity of 0.04 to 0.4, these include most of the world's lightest woods. Again, in harsher environments, the distinction between early and late successional plants is less obvious, so that, for example, wood of early successional legumes in drier environments has a specific gravity greater than 1.0. The tropical pioneers grow at great speeds when young but their life-span tends to be brief by comparison with non-pioneers. In Costa Rica, some pioneer species persist into the canopy and become the large trees of the forest. Similarly, the tallest tree in Africa is the kapok, *Ceiba pentandra* (Bomba-caceae), which is a pioneer and also persists thus. In New Guinea, the long-lived pioneers, species of *Albizia* and *Octomeles* (Datiscaceae), were found to be dying when 84 years old, whereas over- mature *Shorea curtisii* trees in West Malaysia are estimated by ^{14}C dating to have been some 800 ± 100 years old and, in Brazil, a *Bertholletia excelsa*, 14 m in girth, was estimated to be 1400 years old. Pioneers such as *Cecropia* and *Carica papaya*, the papaya, are capable of rapid development without mycorrhizae and in this they resemble some early colonist temperate herbs, which have no known mycorrhizal associations. Of micro-organisms less, in general, is known of their population changes and cycles associated with gap formation, however.

4.3.3   Effect on shade-bearers

It is not known whether shade-bearers acclimatize their leaves or replace them with sun leaves once they are exposed as happens in some temperate-forest herbs. If the first occurs then there must be adjustments in dark respiration, light compensation and saturation points at least (17).

What are the other effects of gap-formation on shade-bearers? The importance of limbfall on sapling populations has perhaps been rather underestimated. On Barro Colorado Island, for instance (18), where the population of the free-standing saplings of a shade-tolerant liane, *Connarus turczaninowii* (Connaraceae) in 1 ha were examined, it was found that in 1 year, up to 46.7% of deaths, the annual mortality rate being 5.3%, was due to limbfalls, the limbs all being less than 10 cm diameter. Moreover, in two successive years, 4.9% and 3.6% of saplings lost 90–100% of their leaf areas from the same cause. Fifty per cent of annual juvenile mortality of *Pentaclethra macroloba* (*P. filamentosa*, Leguminosae) was caused by limb and tree falls and intermediate figures have been recorded for other species elsewhere. As shade-tolerant species persist in the understorey for many years, it is more likely that 'juvenile' examples of such species will be damaged than will those of pioneer species and this damaging force may have increased the selective advantage of certain types of tree-architecture (see Chapter 5), notably the capacity to shoot out afresh, or 're-iterate' after damage.

Examination of sprouting fallen trees over 10 cm diameter at breast height on Barro Colorado Island (19) led to the discovery that 26 out of 88 broken and sprouting trees recorded between 1976 and 1980 were still alive in 1987, and that some of these coppice shoots contributed substantially to the closure of canopy gaps. Their importance is probably greater in small gaps, where they do not compete with pioneers, but it is argued that overall on the island they are more important than pioneers in gap-phase regeneration.

4.3.4   Features of regeneration

Small gaps made artifically in Java refilled with primary forest trees without pioneers. Larger gaps became a tangle of such plants and went through a 'secondary' succession. In large gaps, lianes compete with shade-intolerant and pioneer tree species and only the pioneers may grow fast enough to get away. On Barro Colorado Island, Panama, it has been shown (20) that there were 1974 lianes per ha in what is semi-deciduous forest. In the same area some 22% of the erect seedlings less than 2 m tall were liane seedlings. They are often detrimental to the hosts with which they reach the canopy, for host growth rate is decreased and mortality increased by their presence. Because the pioneer species can outstrip them. However, they cannot completely overcome regenerating forests.

At La Selva, the mortality rates of all liane and tree species with stems over 10 cm diameter at breast height were measured in a 12.4 ha forest. Of the 5623 trees and lianes on the site (21), 23.2% had died after 13 years. Independent of size or buttressing, 36% of these died standing up; 31% had fallen and 7% were buried under tree falls, while 37% had completely rotted, leaving no trace. The annual loss of stems was 2.03%, giving a turnover rate of 34 years for this forest. Study of the 394 canopy openings in 97.3 ha of intact forest, also at La Selva (22), observed from aerial photographs (gaps smaller than 40 m^2 were not taken into account), showed that permanent 'gaps' of early successional vegetation (*tacotales*), with *Cecropia* spp. and balsa, *Ochroma lagopus* (Bombacaceae), comprise 3.4% of the total area and are probably due to poor drainage. The turnover time for the whole area was estimated at 95 years, which compares well with the 118 ± 27 calculated from classical ground surveys of 12 ha (see above).

On Barro Colorado Island (23, 24), there is forest ranging from 100-year-old regrowth to forest at least 300 years old. Thirty gaps (20–705 m^2) were studied over 5–6 years: all were recently formed except two about a year old. It was found that stem densities of both pioneers and shade-bearers increased rapidly and then levelled off or declined by years 3 to 6, but in large gaps the mean rate of growth was higher for pioneers, while early recruits of both types grew faster than did later ones, e.g. 13.5 m height in 2 years in *Trema micrantha*, which only establishes in gaps greater than 376 m^2, *Cecropia insignis* only in those greater than 215, and *Miconia argentea* 102, so that *Miconia* tends to be more common than *Cecropia*. Over 10 years, the differential recruitment and mortality in gaps of shade-tolerant and shade-intolerant species led to a similarity between the gap contents increasing and then falling, the differences being due to great variation in the composition of shade-intolerant species, as has been noted in Malaysia, and the presence of the shade-tolerant species being a reflection of events and conditions before the gap was formed. The origin of differences in composition and density of such stands awaiting the gaps needs to be examined.

However, the importance of gap area may not be apparent in other forests. In heath forest, pioneers do not take over in gaps of different sizes, for example. In a cloud forest at about 1500 m in Costa Rica (25), prone to wind damage due to gusts of up to 100 km per hour in the windy rainy weather of November to March, natural tree and limb falls opened 0.8, 1.4 and 1.0% of a 5.2 ha area in three successive years, with about four gaps per ha per year larger than 4 m². Forty-one per cent of the gaps were formed by uprooted trees, 39% by snapped-off trees and the rest by limb fall, epiphytic masses and dead, standing trees killed by lightning. They were more clumped than would be expected from a random distribution and, in them, saplings of both shade-tolerant and shade-intolerant species were more numerous than in non-gaps, but gap area alone could not be used satisfactorily to predict the course of regeneration.

Over a 10-year period in a 1 ha plot of lowland forest in the north-west of the Amazon (26), 88 of the 786 trees with a diameter at breast height greater than 10 cm died, giving gaps 5–100m, while occasionally larger gaps were formed by windthrow. The gaps were subdivided into microhabitats of trunk zone, open zone between bole and forest edge, crown zone and root pit zone. Nutrient levels did not vary across the zones nor was leaching any greater than in the forest proper. Most of the regeneration was due to the saplings awaiting the gaps. These had an annual survivorship of about 80%, while height growth is only a few centimetres per year and individual leaves may be retained for up to 4 years. Four years after gap formation, regeneration from these saplings accounted for 97% of all trees greater than 1 m tall in single-treefall gaps and 83% in multi-tree gaps, pioneers contributing rather little. Microhabitat did not affect the regeneration patterns in terms of density and establishment or mortality. Growth, however, was influenced by microhabitat, gap age and size as well as plant size, being greater in bigger gaps and in the more open microhabitats of single-treefall gaps.

The stress on gaps *per se* rather than on light levels or canopy closure has recently been attacked (27) and a welcome return to a more dendrocentric approach with efforts to measure actual physical parameters is advocated. However, it is likely that as in studies with temperate herbaceous plants, variation between individuals will be found, such that broad definitions of 'niche' in terms of species will have to be replaced by the sum of the interactions between individual genotypes and their physical and biotic surroundings through time.

4.3.5 *Seed-rain, seedbanks, dormancy and germination*

In mature forest on Barro Colorado Island (28), only 0.09% of the ground comprises pits and mounds caused by uprooted trees (compared with 14–60% in temperate forests), as new ones rapidly fill with soil at a rate of

8 cm per year. Despite the speed, pioneer seeds are concentrated in the underlying mineral soil. Of 742 viable seeds per m^2, most were pioneers, so that large tree-fall gaps are colonized by pioneers since such gaps have more uprooted and therefore soil-disturbing trees. In Asia (29), both *Anthocephalus chinensis* and *Mezoneuron sumatranum* (Leguminosae), a conspicuous climber of clearings, germinate copiously after such disturbance. The relative seed input ('rain') into gaps and understorey has been little-investigated. On Barro Colorado Island (30), in the peak season of wind dispersal of liane and tree seeds, 43 pairs of gaps and adjacent forest in a 50 ha plot were examined. Over 52 000 wind-dispersed seeds and fruits from 14 species of tree and 32 of liane were collected from 1720 pizza-pan traps. Sixty-one per cent of the wind-dispersed species were recorded from the gap sites, where the propagule rain was 328 per m^2, though the rain in the understorey was only 207 per m^2. Only 33% of non wind-dispersed species were in the gap rain: animal-dispersed species are mostly dispersed in the wet season of May to December. The differences between gap and understorey at particular points were not always significant, however, nor were they for any particular species. It was calculated that 105 million propagules were in the rain over 50 ha but, as gaps comprise only a small area at any one time, only 4.33 million will get to them, and thus find suitable conditions for germination and establishment, for none of the plants has any seed dormancy.

There has been some debate as to whether pioneers exist in large seedbanks beneath high forest or whether the seeds are constantly added to forest, should it be in a gap phase or not. Pioneer seeds were first found under mature forests (29) by Symington in West Malaysia in 1933. In artificial gaps created in the Danum Valley in Sabah, soil samples showed fewer than 1000–2000 seeds per m^2 and these samples varied greatly in composition and, though many Melastomataceae and Rubiaceae were generally present at high densities, the number of pioneer trees seemed to be low (31). In 70-year-old forest near Turrialba at 650 m in Costa Rica, the deeply-buried seeds included 21 species in the upper 20 cm and six in the next 20 cm; of the 2236 seeds per m^2, 18% were below 20 cm. This was probably due to soil disturbances by ants and armadilloes, the ants capable of moving subsoil from depths of 30–60 cm. In northern Thailand (32), seeds of pioneering species of *Macaranga, Mallotus* (Euphorbiaceae), *Melastoma* and *Trema*, buried up to 20 cm below the surface of the soil and up to 175 m from the nearest source trees were found to be viable, though in less seasonal West Malaysia (33) few of the pioneer species examined had prolonged seed dormancy. Furthermore, pioneers are notorious for the speed with which they can colonize roadsides and abandoned cultivation. A remarkable example described by Whitmore (34) is that of a tree, *Glochidion tetrapteron* (Euphorbiaceae), known only from two gatherings until, some 50 km from the *locus classicus,* a new road was built where about 12 000 plants appeared.

The presence of pioneer woody species often increases the rate of colonization of a site by other woody species, whose seeds are dispersed by mammals and birds, since the pioneers render the site more attractive in terms of roosts or sites for hoarding seeds (35, 36). Animal vectors avoid gaps on the whole as they are dangerous. Moreover, large seeds are usually dropped by large animals from perches in the canopy, while small seeds are not much more likely to be deposited in gaps because, although birds and bats frequent maturing gaps where they may be attracted to fleshy-fruited pioneers, they tend to retreat to the gap edge to perch. However, turbulence across the broken canopy and convectional currents over heated gaps alter the wind speed and the aerodynamics of seeds so that wind-dispersed seeds tend to fall into gaps.

A comparison of seed-rain and seedbanks was made at 650 m near Turrialba (37), where approximately 75-year-old regrowth was compared with 8- to 11-year-old and 3.3-year-old regrowth at depths of 0–4, 4–10 and 10–20 cm. Samples were taken and the seeds germinated; seed-rain was caught in steam-sterilized soil-traps and then watered daily, the allochthonous seed-rain being estimated from traps in weed-free monocultures nearby, so that the autochthonous was calculable. A forest clearing was made to determine the relative importance of the seedbank, post-deforestation dispersal and vegetative sprouts in regeneration. There were about 6000 to 9500 viable seeds per m^2 in the top 20 cm, though the surface 4 cm of the most mature forest contained fewer than half as many seeds as the two younger stands and, below 10 cm, there were significantly fewer seeds in the youngest stand compared with the mature forest. Fleshy multiple-seeded fruits accounted for 79% of the seeds and 48% of the species in the forest soil, but the other seedbanks were not dominated by any particular propagule type. Eighty-nine per cent of all seeds had germinated in 6 weeks with seeds of longer-lived species tending to take longer, while seed input was at a rate of about nine seeds per day per m^2, i.e. about 3300 per annum in the young vegetation, while in the mature forest it was about 30% of this. Over 3 years, the species composition of the seeds changed from ruderals to more seeds of lianes, shrubs and trees, while biotically dispersed seeds accounted for 93% of the seeds in the first year, though only 51 and 69 in the next two (possibly due to the high rate of production rather than by introduction by animals). Autochthonous seeds comprised at least 74% of the total seed input. Compared with seed-rain, some species were poorly represented in the seedbank (mainly weeds) but others were more strongly represented there, suggesting seed longevity, e.g. *Piper* spp. and *Lepianthes umbellata* had two or three times as many as arrived in 2.5 years. There was very little seed-rain in artificial clearings and all but one of the propagules came from bird or bat droppings. Overall, the richness of the stands resulted from allochthonous seeds, though the bulk of the numbers of seeds were autochthonous.

Although pioneers have generally longer-lived seeds, rather little is known of their longevity, but of seeds of seven species of trees and shrubs at Los Tuxtlas (38), some, such as *Trema micrantha*, remained viable after 2 years, but seed dormancy of such pioneers in the American tropics is apparently rare (13). Some trees there seem to have rather continuous seed production: *Cecropia obtusifolia*, for example, produces viable seed in 10 months out of 12 in Costa Rica. The majority of pioneer species there produce seed annually but some miss a year, or even up to five, between crops. They are mostly dispersed by animals and are epigeal in their germination, with photosynthetic cotyledons. Because the gaps are unpredictable, it is likely that there are only a few possible source trees within dispersal distance and this first-come, first-served principle is reflected in stands of different pioneer species in apparently similar environmental conditions. Such stands suggest that the build-up of a seedbank of diverse pioneers in moist forests is rather unimportant in general. Nevertheless, there are striking exceptions such as the 'Traveller's Palm', *Ravenala madagascariensis* (Musaceae), a pioneer of Madagascar's rain forests, with seeds that may remain dormant for many years and some *Musa* spp., which are related to it, will only germinate after 6 months.

A comparison (39) of germination in seed-free forest soil with that with its complement of seeds in conditions in full light without root disturbance (thus resembling the sites around roots of uprooted trees) was made at Pasoh in West Malaysia. Under mature forest, there were 131 seeds per m^2, like that in many Old World forests but fewer than in many in the neotropics; the density was even lower under 2- to 5-year-old gaps. Pans of the soils showed that the seedbank gave rise to seven times more seedlings than did the seed-rain in the first 9 months. There were fewer pioneer seeds in the soil percentagewise than on Barro Colorado Island for example. This is possibly linked with the observation that pioneers in the Old World tend to have larger seeds than those in the neotropics, and pioneer trees were very few, because the gap size is small, due to trees dying standing and having smaller crowns in any case.

On Barro Colorado Island, there is a unimodal peak in germination within the first months of the 8-month rainy season (40) and unimodal peaks in germination of pioneers, lianes, canopy trees, wind- and animal-dispersed species and those with or without persistent seed reserves, but there was no distinct peak for understorey species. Eighteen per cent of species were dispersed in one rainy season but delayed germination until 4–8 months later; 42% were dispersed during the dry and germinated in the next wet season; 40% were dispersed and germinated in the same wet season, germinating after 2–16 weeks, though these categories in reality grade into one another.

Pioneer seeds are said to be 'orthodox', i.e. have dormancy (perhaps an unfortunate Europocentric concept), and those that do not and cannot have it induced, called 'recalcitrant' as in non-pioneers. The pioneering *Cecropia obtusifolia* and *Piper auritum* germinate in large light gaps in the south-east Mexican forests under photocontrol, triggered when the red/far red ratio of incident light is reduced by the removal of the canopy cover. Experiments with alternating red and far-red treatments indicate that long periods of red, i.e. not just sunflecks, are needed for irreversible stimulation (41). However, within species there is important variation, e.g. seven provenances of *C. obtusifolia* required light to germinate but two others tested germinated in the dark (29). The temperature of bare soil is considerably higher than under forest cover. In Ghana, in a semi-deciduous forest, the maximum at 13:00 hours was 31°C at 5 cm depth, while it was about 24.5°C in forest, and the minimum was about 25°C at 05:00, compared with 24°C under the forest. It may well be that temperatures in the upper 1–2 cm, where most seeds are, are higher still. Temperature-sensitive seeds include those of the sensitive plant, *Mimosa pudica* (Leguminosae) and *Musanga cecropioides*. Balsa, *Ochroma lagopus* (Bombacaceae), seeds germinate best if they have been heated to 40°C, a temperature reached only in large gaps with little shading.

Temperature, in at least some cases, may be indirect in affecting the moisture regime as in some seeds, at least, alternate drying and wetting of the seeds promotes germination as in idigbo (*Terminalia ivorensis*); this may be due to rupturing of the testa, the high temperature in balsa perhaps doing the same. Neither temperature fluctuations nor increases in nitrogen-promoted germination in the absence of light in forest on well-drained oxisols and in adjacent *caatinga* in southern Venezuela (42) and the removal of understorey vegetation also had little or no effect on germination or the survival of seedlings under the canopy at La Selva (43).

In both field and greenhouse, the mortality of seedlings of 18 Panamanian wind-dispersed species was almost always greater (up to 72 times) from pathogens in shade than in sun. This is because higher light and lower humidity are inimical to pathogens, while the faster growth of seedlings in the light gets them past the susceptible period more quickly, though large-seeded animal-dispersed species seem to be more resistant to pathogens than small-seeded animal-dispersed ones or those dispersed by wind (44). Large animal-dispersed seeds will establish mostly beneath the canopy because few such seeds reach a gap, and those that do are eaten by animals.

4.4   Features of later succession

4.4.1   *Seeds and seedlings*

The lack of non-pioneers in seedbanks is due to short viability periods, rapid germination and predation of the large seeds and their not being incorporated into the soil. Some do have long viability periods however, e.g. *Sacoglottis trichogyna* (Humiriaceae) taking $1\frac{1}{2}$ to 2 years to germinate at La Selva and *Endiandra palmerstonii* (Queensland walnut, Lauraceae) loses no viability over $1\frac{1}{2}$ years in Australia (45). Delayed germination is seen in a small number of species in West Malaysia, where viable seeds of species of *Intsia* (Leguminosae) and *Barringtonia* (Lecythidaceae, *sensu lato*), for example, can be found in all months of the year. Generally speaking, the seeds of non-pioneers are larger than those of pioneers. Such seeds allow a build-up of rooting systems for fungal infiltration at the onset of a mycorrhizal association and can withstand the demands of the initial infection. The consequently robust seedlings may be better equipped to persist under a canopy until a gap is formed. Seeds such as those of the durians, *Durio* spp. (Bombacaceae), germinate at once and those of most of the dominant trees in West Malaysian rain forests do so within a few days of dispersal. If they do not, they rot within a few weeks (33). Many of them germinate around the parent, though they are subject to insect and other predators associated with it.

Size can allow seeds to persist until better times obtain, it can provide secondary compounds for defence of seedlings and energy for construction of photosynthetic tissue under conditions where light allows maintenance only just above compensation point; it can also provide energy for growth into higher light intensities and provide nutrients for replacement of lost or damaged tissues in long-lived seedlings. Such are seen in many 'primitive' families, e.g. Lauraceae and Myristicaceae (nutmegs). They have a low surface-to-volume ratio so water uptake in desiccating gaps poses a problem, particularly as they are unlikely to be buried there. Anti-herbivore toxins include amino-acid analogues, alkaloids and phenolics, and can comprise up to 10% of the dry weight, e.g. seeds of *Virola surinamensis* (Myristicaceae) have 15.4% dry-weight soluble tannins, which is the highest concentration of compounds with allegedly defensive function so far recorded. Large seeds may also allow penetration of deep litter and growing up through subsequent litterfall; they may give an advantage in crowded cohorts of seedlings such as after a masting. Predation may select for small seeds or hard-coated large ones and such coats reduce vulnerability to fungal pathogens. In tropical seeds, there is, however, a continuous range of seed size, which is the result of selective compromises between the forces of predation, dispersal, the number of seeds per fruit and, undoubtedly, others.

At Los Tuxtlas, *Cordia megalantha* (Boraginaceae) grows to 35 m, but requires a light gap for release. Seeds germinate at once but there is massive predation of seedlings, though some can survive for a year or more. When seedlings were placed in the shade of the canopy and in small and large gaps and harvested at 0, 61, 162, 237 and 345 days; after day 162, a third of each batch were put in the habitat of the other two. The greatest biomass increase in seedlings (which had been gathered from under the same parent) was in the large gap and plants moved from the shade to there also increased in size. There was rapid acclimatization when opportunities allowed, i.e. they were capable of responding in quite ephemeral gaps.

In West Malaysia, seedlings of *Shorea curtisii* are much attacked by ants, which eat the cotyledons, though it can be argued that such a mass diversion (for the *Shorea* produces many seeds at any one time) may draw the predators from those seedlings that have 'got away', enhancing their survival. Some 54% of the fruits land within 20 m of the mother in this species, and of the whole crop only some 8% were found to be viable, all of these being dispersed less than 25 m. Those that reached the maximum distance of 80 m were found to be lighter because they had been attacked by weevils (12). Such weak dispersal is found in a species with the smallest fruits in its group of the dipterocarps known as some of the merantis, and despite the fact that fruits may be carried up to a kilometre or so in high winds before storms, perhaps it is not surprising to learn that this species often grows in 'family groups'.

In those studies where wind and animal dispersal have been compared in Nigeria and the Solomon Islands, it has been concluded that animal dispersal is the more efficient. Nevertheless, wind dispersal appears to have evolved again and again within families that exhibit both fleshy fruits or seeds, associated with animal dispersal, and dry fruits, often dehiscent and with winged seeds. Such an example is the Meliaceae with the winged seeds of the commercial mahoganies, *Swietenia* and its allies, the bat-dispersed indehiscent fruits of the *lanseh, Lansium domesticum,* a Malesian fruit tree, and the bird-dispersed brightly coloured seeds in dehiscent capsules typical of *Turraea* and some species of *Aglaia.* Similar ranges can be found in Apocynaceae, Bignoniaceae, Bombacaceae, Lecythidaceae, Leguminosae, Malpighiaceae, Sapindaceae, Sterculiaceae and so on. Usually the wind-dispersed examples are large trees of the canopy, but in more deciduous forests these mature nearer the forest floor.

4.4.2 *Saplings*

The behaviour of seedlings and saplings of the non-pioneer species has been studied in thirty natural tree-gaps on Barro Colorado Island (48). The gaps varied from 20 to $705 \, m^2$ in area and in most of them the density of individual stems and species rose sharply in the first 2 to 3 years after gap formation. In the next 2 to 4 years, the density stabilized or declined as competition within the gaps increased. After 6 years, individuals of the pioneer species were found in gaps of all sizes but at highest densities in large ones. The density of seedlings of those non-pioneer species found in gaps and in the understorey of closed forest did not seem to be correlated with gap size, though they grew faster, but not so fast as the pioneers, in gaps. Of pioneers that live through the successional sequence to become canopy species, a study has been made on the same island of the palm, *Socratea durissima* (49). Individuals within each size class were so distributed that the larger they were, the further apart they were, suggesting differential mortality of the smaller size-classes, possibly due to water stress in the dry season. The result is a regular distribution in old age compared with a clumped one when young. The comparison of plots of mature or 'primary' and clearly secondary forest in East Kalimantan, Borneo (50, 51) showed that even though 'secondary' trees were in the canopy of the 'primary' forest, no seedlings of trees were in the plots, but in the secondary forest were over 70 species of the 200 or so found in the primary. By contrast, 24% of the sapling species and 17% of the seedling species in the 35-year-old secondary were still 'secondary species'. In these forests, it has been possible to recognize four stages of consolidation: (i) stabilization of 'secondary' species number, (ii) stabilization of primary species number,

(iii) stabilization of standing biomass, and (iv) formation of a stable and dynamic system, and to calculate that for these and other forests the minimum period for stabilization of 'secondary' species numbers is 60–70 years, and for primary ones more than 150 years, after which gap formation is initiated, the biomass stabilizing after 220–250 years, while individual trees may exist for over 500 years.

4.4.3 *Effects on nutrients*

Studies of the nutrients during successional sequences (14) have shown that immobilization of nutrients occurs rapidly. Stands 10 months old contained the same quantity of nutrients as a mature field of grass; a 6-year-old forest dominated by *Cecropia obtusifolia* and a 14-year-old one dominated by *Musanga cecropioides* had immobilized almost as much phosphorus as stands 50 years old, while the levels of litter produced in *Cecropia* stands 6 and 14 years old were as great as in mature forests. This is largely due to the rapid replacement of pioneer species' leaves. The leaf-area index of 6-year-old *Cecropia* forest is almost the same as that of mature forest even though the height is one-third of it. By 14 years, microclimatic conditions are very similar too.

4.4.4 *Effects beyond the gap*

So far, the contents of the gap have been considered, but the effects of the formation of a gap can be seen in the surrounding vegetation. In dipterocarp forest of Penang Island, Malaysia, there is no seasonal fluctuation in photosynthetic photon-flux density in the open but it is very pronounced in the forest and in gaps, though the explanation for this is still obscure. However, the improved light environment is beneficial to seedlings up to 20 m from the gap (52). There is a good deal of evidence of the effect on tree growth, collected by Forest Departments in the tropics, much of it published in seemingly lifeless reports. In Sarawak, measurements of figs and breadfruit allies (Moraceae) have been made over 20 years in both 'undisturbed' and logged mixed dipterocarp forests (53). In the first 2 years after logging, the annual growth rates in terms of annual girth increment are often up to 10 times greater than in equivalent trees in the undisturbed forest. The rates decline subsequently, though there is variation within populations, some individuals growing more rapidly over successive measurement intervals, with the rest showing little growth in either logged or unlogged forest.

4.5 Animals and succession

4.5.1 *Effects of gaps on animals*

The influence of the cyclical processes of rain forest regeneration on animals is perhaps less understood, though work on birds on Barro Colorado Island (54) has shown that the debris characteristic of the early stages of gap-filling provides a focus for decomposers and a moist habitat exploited by a diverse arthropod fauna, while a wide range of birds is known to forage primarily in such gaps. By mist-netting, it was discovered that there is a greater diversity of bird species in gaps than in closed forest but that the numbers of birds were almost the same while the gap and forest assemblages of birds were quite distinct.

In well-developed forest, there are low numbers of frog and toad species while ticks and even mosquitoes are uncommon (3), but in gaps and secondary forest where the canopy, the herbaceous layer in which so many animals are living, is near the ground, they become more conspicuous. Of insects (55) sampled in gaps and closed understorey on Barro Colorado Island from the end of the dry season to the middle of the wet, certain species of damselflies and robber flies, both 'sit-and-wait' predators, which capture flying insects, are found to perch predominantly in gaps, others in shade, perhaps reflecting their different spectra of prey. There were greater numbers of Formicidae, Coleoptera and Psocoptera in the understorey— most markedly Coleoptera, of which there were 1.2 to 2.1 times as many in gaps as were caught in closed forest. But, 83% of all those captured were less than 3 mm long and only 4% more than 5 mm and it is suspected that larger insects avoid the transparent 'tanglefoot' traps used in the measurements, such that the samples perhaps more properly represent differences in the poor-sighted, weak-flying small insects.

4.5.2 *Effects of animals on gaps*

As many dispersers carry fruits out of gaps into adjacent forest for processing, enlarging of gaps by whatever means could release seeds and seedlings accumulating there (44). By contrast, gap destruction by large animals, such as the elephant, may have catastrophic effects on regeneration there. The Sumatran rhinoceros, *Dicerorhinus sumatrensis*, feeds selectively in small gaps and frequently snaps off saplings while browsing, and pigs in their rooting disturb soils as well as uprooting saplings. In one documented case at Pasoh (56), pigs (*Sus scrofa*) snapped off

Shorea leprosula saplings to make a nest just outside a gap. Smaller animals may also be important: large seeds and seedlings are most vulnerable to predation in gaps where rodents seek shelter in tangles. By contrast, the predation of small seeds and seedlings is perhaps not so disproportionately high in gaps because rodents are not such a common cause of death.

When understorey birds and fruiting plants at La Selva were censused monthly for 1 year in 13 gaps and 13 closed forest sites, 40% of the birds (and 30% of fruiting plant species) were found to be significantly more often in gaps (57), whereas 5% of birds and no plants were significantly more often in closed forest. It was found that the highest densities of both birds and plants were in large gaps. Frugivorous and nectarivorous birds were especially common in gaps, while the plants there tended to produce more fruits over a longer period there than did conspecifics in closed forest. Seeds may be dispersed more often in the vicinity of gaps than elsewhere, fruiting plants affecting the behaviour of the birds and their behaviour affecting the distribution of the plants.

4.5.3 Pollination and dispersal

With succession, the mean density of individuals of tree species declines and their pollinators consequently have to travel further and would be expected, therefore, to be larger, as they tend to be (14). Plants that are wind-pollinated dominate at first, later those visited by animals dominate, but the number of species visited by small bees tends not to decline, though bird-pollinated tree species are rare in mature forest. In each stage, brightly-coloured flowers and bracts tend to be at the top of the canopy and are usually visited by bees, while those at lower levels are white, cream or pale green. Nevertheless, red bird-pollinated flowers tend to be more evenly spread vertically. This range of colour and its distribution remains rather constant as succession progresses, whereas in temperate or semi-arid regions, yellow predominates in the early stages and white becomes dominant later. The flowering period tends to be longer in the early stages in both temperate and tropical successions.

The increasing importance of animals in succession is associated with dispersal too. Species with small gravity- and wind-dispersed seeds tend to be common in the early stages but are soon replaced by animal-dispersed ones, though species with hooked fruits or sticky seeds are also important early on. Winged seeds or fruits are typical of many lianes and very tiny wind-dispersed seeds are characteristic of many epiphytes. The large gravity-dispersed seeds tend to be common among the 'emergent' trees while species with explosive fruits are rather rare at any stage. There seem to be correlations between very tiny, very large and winged seeds with the increasing height of the canopy.

In summary, early successional plants tend to have flowers pollinated by wind or by small animals, fruits with many small seeds (and some forms of inbreeding), while mature canopy trees may tend to have large flowers pollinated by large animals, large fleshy fruits with few large seeds (and some form of outbreeding). Understorey plants in mature forest might be expected to have smaller flowers and pollinators and to have fruits with a greater number of smaller seeds. As with all generalizations in ecology, which makes it the despair of some scientists and the fascination of others, there are numerous exceptions to all of these.

4.6 Primary successions

In Malesia, the island of Jarak, 64 km from the Malayan coast, is believed to have lost its entire flora some 34 000 years ago when an eruption in Sumatra, some 224 km away, covered it in ash. By 1953, there were in the 40 ha area of the island only 93 species of angiosperms, all but 13 of which seem to have been dispersed by animals. Of the rest, eight are thought to have been dispersed by sea, two by man and only three by wind: two orchids with minute seeds and one *Hoya* (Asclepiadaceae) with plumed ones (12).

Krakatau (Krakatoa) is one of a line of about 100 volcanoes running through Sumatra and Java, at the junction of the Indo-Australian and Eurasian plates, the first being deflected below the second. Newly melted plate material is less dense than the surrounding mantle and rises to the surface, where it may be erupted as lava. The eruption of 1883 was due to the generation of pyroclastic flows by the gravitational collapse of the eruption column after several large magmatic explosions (58–60), the major *tsunami* following the entry of the pyroclastic flows into the sea. The eruption column reached 25 km and ships up to 20 km away reported heavy ash fall including pumice to 10 cm diameter: 20 m of deposit was formed on the remnant islands themselves. Four large explosions yielded the flows that entered the sea and the total outpourings were 18–21 km^3; by comparison with prehistoric eruptions of up to 1000 km^3, this was very modest. The loudest explosion was heard 5000 km away and the *tsunami* was 40 m high, travelling at 25 m per second: about 36 000 people died, mostly by drowning. There was a decreased solar constant for several years, resulting in cooler conditions throughout the northern hemisphere due to the volcanic materials put into the stratosphere. Sea-level changes were recorded as far away as the English Channel and New Zealand.

Although the remnant group of islands lies about 40 km from both Java and Sumatra, nine typical littoral plant species and a number of unidentified fruits were recorded from the shores within 3 years. Inland were several ferns, Compositae, grasses and other vascular plants besides quantities of blue-green algae. Nine months after the explosion the only sign of life had been a single spider, but by 1888 there were more spiders, flies, bugs, beetles, butterflies and a large monitor lizard (*Varanus salvator*). By 1897, the littoral vegetation was very well-developed and there was a rich grassland inland; 132 species of birds and insects were recorded. Nine years later, the coastal vegetation closely resembled that of Java and, inland, there were pockets of scrub, though still a predominance of grassland. In 1908, there were stands of typical secondary forest trees such as species of *Macaranga* behind the beach vegetation, with 200 animal species on Krakatau alone, including two land molluscs, 13 non-migrant land birds, many spiders and centipedes. By 1911, there was a python and by 1917 a human (complete with rats). By 1919, the grassland was changing into a *Macaranga-Ficus* woodland and there were two bats, and earth-worms; by 1934, there was even a crocodile.

Ninety-two per cent of the animal species there before 1933 could have been windborne. The colonization by plants with different dispersal mechanisms is set out in Figure 4.3. By the 1980s (61, 62), the number of plant species still seemed to be rising, largely through introduction by animals, but several of the species recorded earlier could no longer be found. Although initially suggesting some turnover of species, the losses are mainly sea-dispersed ephemerals and species of temporary habitats. Furthermore, the forest seemed to be held at a late secondary forest stage, representing a kind of truncated succession in the absence of a source of propagules of later successional species. The inability of propagules of typically late successional species to reach islands, even so close to the mainland, is even more clearly seen in truly oceanic islands, where only groups of plants with small or long-lived propagules can colonize and diversify (4, 63). Similarly, truncated successions have been recorded from the isolated cloud forests of northern Colombia in a model study by Sugden (64, 65) who argues that the differences between these montane floras that are poor in endemics are due partly to chance.

There are differences between the vegetation covers of the four islands now in the Krakatau group. The effects of repeated volcanic activity on Anak Krakatau, which emerged in 1927 and was still active in 1988, are

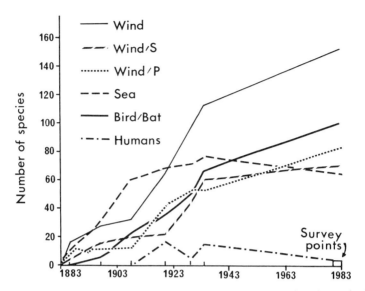

Figure 4.3 Means of dispersal of the Krakatau flora: all vascular (S = seed plants; P = pteridophytes) plants and for all islands (combined) except Anak Krakatau. Redrawn from R.J. Whittaker *et al.* (61).

apparently the cause of much of the differences, notably the predominance of the early-successional *Dysoxylum gaudichaudianum* (Meliaceae) on areas of nearby Sertung, which was destroyed in the eruption, while areas destroyed in 1953 are now predominantly occupied by even-aged stands of *Timonius compressicaulis* (Rubiaceae).

There have been suspicions that some of the organisms on Krakatau survived the eruption there and that many were later brought in by man. A recent study on a newly-created island, which resulted from an eruption in a lake on an island off the north-east coast of New Guinea (66), showed better monitoring. The island of Motmot lies some 3 km from the nearest land and was surveyed several times between its eruption in 1968 and subsequent eruption and devastation of most of the flora in 1973. By 1972 there were 14 vascular plant species, predominantly sedges, most of them probably carried to the island ectozoically by black ducks. Of the fauna, many strand beetles, ants and bugs are thought to have arrived on drifting organic matter, but there were also ants, earwigs and spiders as well as other invertebrates associated with algal crusts around the crater pond. Despite their being carnivorous, the spiders had already colonized by 1969 and it is interesting to note that this is paralleled in other biomes for it is spiders that are some of the earliest animal colonists of spoil tips in South Wales. A neotropical volcano that has been recently re-examined is the Soufrière on St. Vincent in the West Indies, which erupted in 1718, 1812 and 1902. In 1902, all the vegetation was destroyed, its carbonized remains lying underneath a 30 cm deposit of volcanic material. The slopes that have been undisturbed since are clothed in forest but, contrary to expectation, their composition has changed little in the 30 years up to the re-examination (1972), although the vegetation belts there seem to have moved up the mountain about 30 m with the virtual removal of a moss and lichen assemblage found around the crater-rim in 1942 (67).

4.7 Implications

Non-mycorrhizal plants, such as sedges, tend to perpetuate on poor soils as there are no mycorrhizae to allow other species to get established unless fungal propagules are introduced at the same time as the seeds of their hosts (14). In primary successions of the type discussed here, this is unlikely because of the poor dispersability of vesicular-arbuscular fungi. It may well be that such fungi are dispersed only by rodents or by movement of soil by wind or water for, in the tropics, spore production is low. Other formations without such mycorrhizae are certain savanna types and vegetation maintained by fire.

The features of succession in tropical moist habitats contrast with those in drier ones. In these latter, the cycles are usually simpler and frequently involve coppicing of old stumps. In very wet forests in swamps, both freshwater and marine, the system may be very simple, too. For example, in mangroves in the New World, the succession is simply from *Acrostichum* fern stands to mangrove forest. It would seem easier to manipulate such systems than to manipulate the complexity of rain forest and, indeed, disturbance of rain forest may lead to the invasion of other vegetation types from harsher régimes such as drier forest or from higher altitudes. Furthermore, the introduction of early successional trees from such harsher environments could have a serious effect on the native vegetation, much as the introduction of *Eucalyptus* (Myrtaceae) and *Hakea* (Proteaceae) from Australia and of exotic pines has adversely affected the extremely rich ecosystems of the Cape region of Africa.

CHAPTER FIVE

THE COMPONENTS OF DIVERSITY
AND THEIR DYNAMICS

Much of the previous chapter will be of interest to those concerned with
forests as entities: those concerned with biomass, particularly its exploitation;
those interested in a holistic framework for a particular animal species they
may be studying; or those searching for generalizations. But with continua,
arbitrary categories, exceptions galore, probabilities and variation from
place to place, a failure to generate predictions and a general 'woolliness',
there is perhaps something somewhat unsatisfactory intellectually about
the 'community' approach. This is more than a holism/reductionism
dilemma, though that takes a significant role. The fact is that with the
continual change in rain forests and the unpredictability of orogenic,
meteorological and biotic events (to be considered further in the coming
pages), an air of desperation can be sensed in some recent work. We have
mere documentation and there is no 'science', though Simberloff (1) argues
that the fuzziness or 'noise' of the physicist is music to the biologist.

There have been moves to use 'chaos' equations to find some order, in
the belief that the underlying equations describing certain phenomena are
completely deterministic, though the resultant 'trajectories' may be so
complex as to be apparently random. In the meantime, some have
despaired of distinguishing between such 'chaos', experimental error and
true stochasticity; however, it must be pointed out that such are common-
place in meteorology and even the most sublime of physics and mathema-
tics. The important issue is that the very materials of ecological interactions,
as well as the forces acting on them, are in a state of flux. At an apparently
prosaic level, Harper and his colleagues (2) have demonstrated the build-up
of 'adapted' genomes in temperate, herbaceous plants, fitting particular
micro-evolutionary pressures, while selection also favours the capacity to

break these genomes apart to produce new combinations to be selected by new circumstances. This 'Sisyphean' fitness is clearly of profound importance in biological systems but, with the longevity of the rain forest tree, it will be difficult to assess that importance in the tropical rain forest. Nevertheless, consideration must now be given to individual components and their variability and dynamics, or demography, in an attempt to approach this, the most truly biological, problem.

That it was possible in the last chapter to be able to discuss tropical rain forest dynamics in general is a result of the remarkable structural similarities between the rain forests of the three great tropical regions. Nevertheless, floristically, they differ greatly, which would indicate that the components of these remarkably similar-looking forests have evolved in parallel and in isolation.

The rain forests of the Far East are dominated by dipterocarps and their biology, for at their maximum, these may make up 80% of emergent trees and 40% of the understorey, whereas in Africa there is only a handful of species in a different subfamily, and in South America, just one recently described species. (This is placed in its own subfamily but is of somewhat enigmatic relationship and possibly represents the direct descendants of a line that existed before the families allied to Dipterocarpaceae were as recognizably distinct as they seem to be today.)

In all three blocks of rain forest, by contrast, there are emergent legumes: for example *Mora* in the New World, *Koompassia* in Asia and *Cynometra* in Africa. Sometimes closely related genera are found in the same stages of the rain-forest cycle in different regions, such as *Cecropia* in America and *Musanga* in Africa (Moraceae, *sensu lato*), both pioneers, while other closely allied genera may have diverged in their ecology so that *Xylocarpus* is a genus of mangrove, sandy shores and rocky headlands in East Africa and eastwards in the Old World; *Carapa* is a swamp-forest genus in both America and Africa (Meliaceae). Within the same family, distinct but closely allied genera are found on either side of the Pacific—*Cedrela* and *Guarea* in the Americas (the second also in Africa), *Toona* and *Dysoxylum* in Asia. Remarkable amphipacific affinities may occur at the species level, as in Chrysobalanaceae where the single Asiatic species of *Maranthes*, a largely African genus, is also recorded from Panama. Rarely though does the same species occur in two or more continents. Notable, in the same family, is *Chrysobalanus icaco* on either side of the Atlantic and *Ceiba pentandra* (Bombacaceae) in all three tropical regions: though not native in Indomalesia it was introduced there a long time ago.

5.1 Geographical diversity

5.1.1 Flora

A glance at Table 5.1. shows the parallelisms in different types of forest component in the three forest blocks. Remarkable are the epiphytes, for all continents have orchids and ferns but America alone has bromeliads (the one African species of this family is not an epiphyte) and, with the exception of species of the genus *Rhipsalis*, all the cacti. In Indomalesia there are epiphytes in Rubiaceae and Asclepiadaceae.

An analysis of the African flora as a whole by Thorne (3) showed that there were some 500 species of plant found on both sides of the tropical Atlantic, representing about 0.63% of the total. Of these, some 108 are

Table 5.1 Examples of families and genera containing dominant, abundant, conspicuous or subendemic woody plants in the major rain-forest regions with their main groups of epiphytes and secondary forest trees (largely from K.A. Longman and J. Jenik. *Tropical Forest and its Environment*, Longman, London, 1974, pp. 70–2).

Neotropics	Leguminosae	*Andira, Apuleia, Dalbergia, Dinizia, Hymenolobium, Mora*
	Sapotaceae	*Manilkara, Pradosia*
	Meliaceae	*Cedrela, Swietenia*
	Euphorbiaceae	*Hevea*
	Myristicaceae	*Virola*
	Moraceae	*Cecropia, Ficus*
	Lecythidaceae	*Bertholletia*
	Epiphytes	ferns, Orchidaceae, Bromeliaceae, Cactaceae
	Secondary	*Cecropia, Miconia, Vismia*
Africa	Leguminosae	*Albizia, Brachystegia, Cynometra, Gilbertiodendron*
	Sapotaceae	*Afrosersalisia, Chrysophyllum*
	Meliaceae	*Entandrophragma, Khaya*
	Euphorbiaceae	*Macaranga, Uapaca*
	Moraceae	*Chlorophora, Ficus, Musanga*
	Sterculiaceae	*Cola, Triplochiton*
	Ulmaceae	*Celtis*
	Epiphytes	ferns, Orchidaceae
	Secondary	*Harungana, Macaranga, Musanga*
Indomalesia	Dipterocarpaceae	*Dryobalanops, Hopea, Shorea*
	Leguminosae	*Koompassia*
	Meliaceae	*Aglaia, Dysoxylum*
	Moraceae	*Artocarpus, Ficus*
	Anacardiaceae	*Mangifera*
	Dilleniaceae	*Dillenia*
	Thymelaeaceae	*Gonystylus*
	Epiphytes	ferns, Orchidaceae, Asclepiadaceae, Rubiaceae
	Secondary	*Glochidion, Macaranga, Mallotus, Melastoma*

restricted to West Africa and South America and 45 of those are aquatic, or more or less so, many others are from maritime or brackish habitats or are riparian, while yet others are thought to be human introductions. Some are real forest plants and have small seeds (like the orchids) or are thought to be bird-dispersed. Amphipacific affinities, extraordinarily, are much stronger, with higher numbers of families and of genera shared between America and Asia than between America and Africa, though, less surprisingly perhaps, the affinity between the floras of Africa and Asia is strongest of all. By comparison with Africa, the forests of Madagascar, now largely destroyed and covered with secondary *savoka* vegetation with introduced genera like *Melia* and *Psidium* (Myrtaceae), have a greater number of endemic families and many endemic genera, even of palms, a family in which Africa is notably poor (5). Nevertheless, the Guineo-Congolian forests of Africa include some 8000 species, of which 80% are endemic and five generally accepted endemic families : Dioncophyllaceae (three species, including carnivores), Hoplestigmataceae (two species), Huaceae (three species), Medusandraceae (two species) and Scytopetalaceae (22 species), even though most African rain forest is 'mixed moist evergreen' and forests dominated merely by a few species are frequent. The rain forest of eastern Madagascar has some 6100 species of which 16% of the genera and 80% of the species are endemic. Even the forest fragments of tropical Queensland have endemic families: the archaic Austrobaileyaceae and Idiospermaceae (one species each).

5.1.2 *Fauna*

Vertebrates Similar parallelisms and differences are found when the fauna is compared (6–8). Although the forest avifauna of Africa is poorer than the Neotropical, the numbers of mammal species are similar, though there are more bat species in America. There are remarkable ecological counterparts (Figure 5.1) in the two regions, particularly striking in ungulates and rodents. In the Costa Rica forests, there are populations of relatively sedentary seed and fruit eaters, notably agoutis and paras, which have no analogue in Malesia, however. Janzen suggests this may reflect the periodic, rather than continuous, fruiting of the major trees, the dipterocarps discussed below, and that, in turn, it may explain why the number of raptors is low.

There are also differences in the gliding and prehensile-tailed vertebrates in the different continents. Prehensile tails capable of supporting the weight of the whole animal have evolved independently in two suborders of

Figure 5.1 Morphological convergences among African (left) and neotropical (right) rainforest mammals. From top to bottom and left to right: pygmy hippopotamus and capybara; African chevrotain and paca; royal antelope and agouti, yellowback duiker and brocket deer; terrestrial pangolin and giant armadillo. Members of each pair of animals are drawn to the same scale. Reproduced with permission from F. Bourlière in B.J. Meggers, E.S. Ayensu and D. Duckworth, *Tropical Forest Ecosystems in Africa and South America: a Comparative Review* (Smithsonian Inst., Washington, 1973).

reptiles, one order of amphibians and six of mammals: they are used either for support or locomotion. In arboreal frugivores, folivores and omnivores like opossums, monkeys, kinkajous and porcupines they give support, while the animals feed at branch-tips, and aid locomotion on unstable supports and during descent. Those of snakes are usually used as anchors while the rest of the body is thrown out to encircle prey; they are used similarly in arboreal ant-feeders such as pangolins, leaving the forelegs free to tear open ant-nests. In Gabon, chameleons and prehensile-tailed mice only frequent thin-stemmed vegetation near the ground: in short, none of the prehensile-tailed vertebrates there use tails for getting through the forest canopy, while most of those in America do. Gliding occurs in three orders of mammals, nine of reptiles and one of amphibians: all the mammals are nocturnal and the large ones are folivores, the smaller ones are frugivores to insectivores, i.e. the gliders are more restricted ecologically and taxonomically. Although it provides an energetically cheap, rapid transport through the canopy and rapid escape from predators, in Africa it occurs only in one small family of rodents and in America in one group of frogs. Most gliding vertebrates are thus in Asia and species with prehensile tails in America, with few of either in Africa. In western Amazonia there are no gliding vertebrates, in west Africa there are only three mammals, but in Borneo there are 33 species including amphibians and reptiles as well as mammals; of species with prehensile tails, there are 38 in western Amazonia in all three groups, nine in Borneo (reptiles and mammals) and eight in Africa (reptiles and mammals). This could be due to historical factors, but may reflect the selective forces of differing forest structure for, although the forests are similar in terms of biomass, canopy height and so on, the liane densities are different, for America has more lianes than does Asia, but fewer than Africa, and canopy trees are linked only by lianes, such that the higher percentage of them in Africa may have obviated the usefulness of specialized locomotory systems there.

Palms are much more important in American forests and, except for rattans and a large number of understorey species, are probably not as important in this context, while, in Africa, they are few in number. They are difficult to climb and seem to have mechanisms preventing invasion of the crown by lianes, but palm leaves are used as routes to the fruit and as pathways through the canopy for arboreal animals, including prehensile-tailed monkeys: on such a difficult substrate such tails are advantageous. Whereas in America small lianes can be collected by pulling them down, those in Africa are more firmly attached and do not break—even species in the same families, e.g. Apocynaceae and Menispermaceae in the two continents show this, suggesting that the vegetation is indeed more fragile in America. It is suspected that browsing elephants, half of the species in the diet of which are lianes, have selected for tougher lianes, and that this may help to explain why prehensile tails are so advantageous in America, where the largest Cebidae have them, the smallest do not and the intermediate ones have semi-prehensile tails.

A far higher percentage of the non-flying mammal species of tropical rain forest is arboreal than is the case in temperate regions. For example, in Borneo the figure is 45%, in Virginia, USA, 15%. Food is available in the form of buds, leaves, flowers and fruits so that bats, primates and rodents live in the canopy, as do American sloths and African hyraxes and a number of insectivores and carnivores. Although the biomasses of arboreal and ground-living mammals are estimated to be about the same, the latter are represented by fewer individuals, for they are large or medium-sized ungulates, feeding on the scanty foliage, fallen fruits, roots and so on or grazing in glades and along riverbanks, or they are rodents, shrews and a few strictly terrestrial carnivores. There are no true fossorial mammals, possibly due to the wet nature of the soil, little litter and few worms. Although there are herds of buffalo, sounders of pigs and so on, most non-arboreal mammals are solitary, whereas the arboreal animals may be solitary, in family groups or troops, or continuously wandering and coalescing with others.

In mammals in general, species diversity is high, population density low and related species have differences in times of activity, food specilization and preferences for different levels in the forest. In addition, even in a small area (10), local differences in forest structure can have an effect: in northeast Gabon, seven of the nine terrestrial ruminants, a group particularly well-represented in Africa, were individually associated with forest in which undergrowth was of a height causing least obstruction in terms of the shoulder-height of the animal. However, mixed troops of monkeys are found, apparently affording protection, while the members of different species maintain their own food preferences.

The species diversity of mammals is always lower than that of birds, which can be five times as rich (Table 5.2). There are more insectivores among the birds (up to 75% in north-east Gabon, 65% in Peruvian Amazonia, 55% in French Guyana, 52% in Papua New Guinea), the percentage in mammals being 42–19%, for birds are able to feed on small mobile invertebrate prey in a three-dimensional environment independent of plant support. Bats have the same advantage and make up a large percentage of the mammalian insectivore faunas.

Table 5.2   Comparative species richness of vertebrates (excluding amphibians) in seven rain forest sites. From Boulière, F. (11)

Sites	Birds[a]	Mammals[b]	Reptiles[c]
La Selva Reserve, Costa Rica	385 (96)	137 (91)	74 (23)
Cocha Cashu, eastern Peru	536 (22)	99 (25)	—
Barro Colorado, Panama	366 (83)	97 (46)	64 (18)
Makokou (M'Passa), N.E. Gabon	363 (52)	114 (33)	65 (15)
Gogol Forest, Papua New Guinea	162 (?)	27 (?)	34 (?)
Baiyer River, Papua New Guinea	163 (15)	22 (7)	20 (9)
Perinet, Madagascar	73 (2)	38 (9)	43 (26)

[a]Number of long-range migrants indicated in parentheses following the total.
[b]Number of bats indicated in parentheses following the total.
[c]Number of lizards indicated in parentheses following the total.

Fifty to ninety per cent of tropical species of trees and shrubs are visited by frugivorous vertebrate dispersers (12, 13), frugivorous species making up 80% of the avian and mammalian biomass in tropical moist forest in Peru and 50% of the avian biomass in Panama. Of arboreal and flying vertebrates, granivorous squirrels, parrots and pigeons have been excluded (Table 5.3).

The Neotropics seem richest in species of frugivorous bats and birds and are similar to Africa in frugivorous primate diversity but, in terms of areas of forest, the Neotropics and Africa are similar and 1.3 times as great as Asia for birds, with primates 3.2 times higher in Africa than elsewhere and bats 1.8 times denser in south-east Asia. In Africa and America, there are six times as many bird species as bats or primates, whereas there are twelve times as many bird species as bats in south-east Asia. Madagascar is very poor in frugivorous birds but primates and bats occur at African rates, while Borneo and New Guinea are richer in birds than is mainland south-east Asia, though with similar numbers of bats and primates (Borneo).

The size ranges of avian frugivores is similar in all: manikins to guans in America, barbets to hornbills in Africa, flowerpeckers to hornbills in Asia. Some hornbills are much bigger than toucans, though they are often considered counterpart. Some Old World monkeys, e.g. mandrills, chimpanzees and orang-utans (and humans) are much larger than New World primates. In bats, all the New World frugivores are less than 100 g in weight, whereas Old World frugivores can attain 1500 g. Avian foraging is similar with the bulk of avian biomass in the canopy, the canopy frugivores (guans,

Table 5.3 Richness of arboreal and flying frugivorous vertebrates in different parts of the tropics. From Fleming, T.H. *et al.* (12)

| Region | Number of genera (G) and species (S) | | | | | |
| | Birds | | Primates | | Bats | |
(Area × 10⁶ km²)	G	S	G	S	G	S
A. Major regions						
Neotropics						
(5.06)	117[a]	405[a]	11	33	24	96
Africa[b]						
(1.68)	55	149	8	32	12	26
South-east Asia						
(1.87)	30[c]	143[c]	4	11	29	66
B. Old World Islands						
Madagascar						
(0.59)	5	7	3	9	3	3
Borneo						
(0.76)	25	64	3	5	11	17
New Guinea						
(0.81)	37	95	0	0	9	17

[a]Only South America
[b]Only West and West-Central Africa
[c]Mainland only

toucans and cotingids in America, turacos and hornbills in Africa, fruit pigeons and hornbills in Asia) being larger than the understorey ones (manikins and tanagers in America, bulbuls and starlings in Africa, flowerpeckers, broadbills and bulbuls in Asia). There are terrestrial avian frugivores in America (cracids and trumpeters) and in New Guinea and tropical Australia (cassowaries) but there are none in Africa or south-east Asia, where the density of terrestrial frugivorous mammals (primates and artiodactyls) is relatively high.

The frugivorous American primates are all arboreal, while there are many terrestrial primates in the Old World (mandrills and chimpanzees in Africa, macaques in Asia) as well as arboreal primates. By contrast, New World bats include canopy as well as understorey feeders, whereas Old World species are primarily canopy and forest-edge feeders. In short, there are no strict parallels between the different blocks of rain forest and, in this case, Africa is scarcely the 'odd man out'.

In a comparison of forests in different parts of the tropics, those in western Amazonia were found to have nearly twice as many bird species

($n = 124–159$) as similar plots in New Guinea ($n = 83$) do, with Borneo ($n = 108$) and Gabon ($n = 112$) intermediate. New Guinea has 35% frugivores (up to 53% recorded in some instances), whereas Africa and America had 20–27% and Borneo only 15%. Of tropical understorey birds, America seems richest in species with the Malaysian forests being particularly poor; the Malay Peninsula is poor in frugivorous primates too. In African wet forests, the total primate biomass is higher than elsewhere, and absolute frugivore biomass is also higher, while the community biomass of bats in South America is greatest but the biomass may be similar.

The species-rich Neotropics seem to have had two major radiations, providing food sources for pollinators and frugivores: the Amazon-centred canopy trees and lianes on one hand, the Andean-centred epiphytes and understorey plants on the other. The toucans and cotingas have radiated in the Amazon due to the canopy trees, while manikins, tanagers and bats have diversified in response to the understorey. Nocturnal fruit- and insect-eating marsupials in the New World possibly precluded the evolution of small nocturnal primates there, while Africa's and Asia's fruit-eating ruminants and terrestrial primates have pushed out or prevented the evolution of large terrestrial frugivorous birds like neotropical crassows, Australasian cassowaries and the New Caledonian kagu. The radiation of frugivorous pigeons and doves in Australasia may be due to the absence of primate competitors for figs and other fruits. The allegedly greater abundance and apparency of fruits in the Neotropics have led to the evolution of more diverse faunas characterized by reduced dietary overlap, small body size and sedentary ranging patterns: there are no neotropical terrestrial primates nor true brachiators. In south-east Asia, where fruit crops are patchy in space and time, frugivore communities are fewer in species, with more broadly overlapping diets, larger body sizes and more mobile ranging patterns. Africa with its 'depauperate' fleshy-fruited flora is intermediate in this respect. In the absence of the dominance of dipterocarps and less irregular fruiting régimes (see later), New Guinea supports richer frugivore communities than does Borneo, though other features are similar to the rest of Indomalesia in general.

It must also be remembered that some lowland tropical forests become occupied by migrant birds from the temperate zone during inclement seasons there. The forests of Africa harbour hardly any, though the forests of both the Neotropics and south-east Asia support some. Of the avifauna of the forests of the Malay Peninsula, some 370 species, 282 occur in dipterocarp forest, 96 in montane forest and 50 in the mangrove. Only six species, however, are restricted to mangrove. Around 200 species have been recorded in 2 km^2 of dipterocarp forest but only 71 in 1 km^2 of peatswamp forest there. The nectar- and fruit-feeders are skewed towards the canopy, the insectivores towards the understorey while the percentage of food 'niches' there (2% nectar, 18% fruit and seeds, 70% invertebrates, 10% vertebrates) is very similar to that recorded in Gabon, though the percentages for Papua New Guinea are: 9, 41, 39, 11.

A summary of the birds and mammals has been provided by Whitmore in his *Tropical Rain Forests of the Far East*, where six 'communities' are recognized: (i) above-canopy—insectivorous and otherwise carnivorous birds and bats; (ii) top-of-canopy—birds and mammals feeding largely on leaves and fruits but also nectar or insects; (iii) middle-of-canopy flying animals, predominantly insectivorous birds and bats; (iv) middle-of-canopy mammals, which range up and down tree trunks from crown to ground level, largely mixed feeders and a few carnivores; (v) large ground-living herbivores and their carnivores; (vi) small ground- or undergrowth-dwelling animals with varied diets from the forest floor, largely insectivorous or mixed feeders. In short, the top of the canopy is mostly occupied by primary consumers, the other habitats by fewer primaries and more secondaries.

Of other vertebrates, there is a greater abundance of litter lizards and amphibians (14) in Costa Rica and Panama than there is in Borneo and the Philippines: over 17 animals per 100 m^2 with Borneo only 1.5, while the neotropical uplands have 58.7 compared with 10.8 in the Philippines. In Borneo, the principal reptiles are skinks, whereas in the Neotropics frogs are more numerous in the litter. The Neotropics are also richer in fresh-water fishes (15), the Amazon having the richest freshwater fish fauna in the world: 2500–3000 species, 80% of them being catfishes and characins. These are the main frugivores—the characins with well-developed teeth and strong jaws, the catfish with grasping small teeth—and abound in the floodplain forests. Fruit- and seed-eating fishes are poorly represented outside South America. The most diverse fish group in Asia is the carps, but only one species is a fruit-eater, though a few other groups have them and fossils in Africa indicate that formerly there were frugivore fish there too.

Insects   Insects tend to be concentrated within the canopy, though some show bimodality in their vertical distribution (16), occurring conspic-uously at ground level as well. The vertical distribution of these animals may be obscured in regions of deeply dissected topography. Some insects have specific 'insular' habitats occupied by their larvae. Such are the mosquitoes, and biting midges found in the reservoirs of bromeliad plants in the New World or the pitchers of *Nepenthes* in the Old, which, like the cups formed by the bases of teasel leaves (*Dipsacus* spp.) in temperate countries, hold a mass of rotting organic materials, which the sapropha-gous larvae consume. Their predators are associated with them and, in *Nepenthes*, both seem to be immune to the digestive enzymes that convert other animals into absorbable materials for the plant.

There are geographical differences in insect distribution too. Bees, which are important pollinators for many tropical plants (see Chapter 6) are not evenly distributed (9). Unlike many tropical groups, the bees are well represented by species in temperate regions. They are best represented in

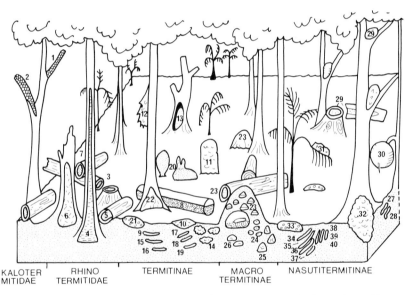

Figure 5.2 Termites in lowland dipterocarp forest in West Malaysia. Kalotermitidae: 1. *Glyptotermes*, 2. *Neotermes*; Rhinotermitidae: 3. *Termitogeton*; 4. *Coptotermes*; 5. *Parrhinotermes*; 6. *Schedorhinotermes*; 7. *Heterotermes*; Termitinae: 8. *Protohamitermes*; 9. *Prohamitermes*; 10. *Labritermes*; 11. *Globitermes*; 12. *Microcerotermes*; 13. *Amitermes*; 14. *Procapritermes*; 15. *Pericapritermes*; 16. *Coxocapritermes*; 17. *Oriencapritermes*; 18. *Syncapritermes*; 19. *Mirocapritermes*; 20. *Dicuspiditermes*; 21. *Homallotermes*; 22. *Termes* (in *Hospitalitermes* nest); Macrotermitinae: 23. *Macrotermes*; 24. *Microtermes*; 25. *Odonto-terms*; 26. *Hypotermes*; Nasutitermitinae: 27. *Hirtitermes*; 28. *Havilanditermes*; 29. *Nasut-itermes*; 30. *Bulbitermes*; 31. *Lacessitermes*; 32. *Hospitalitermes*; 33. *Longipeditermes*; 34. *Leucopitermes*; 35. *Aciculitermes*; 36. *Aciculioiditermes*; 37. *Subulioiditermes*; 38. *Malaysioter-mes*; 39. *Proaciculitermes*; and 40. *Oriensubulitermes*. Redrawn from N.M. Collins in Earl of Cranbrook, *Malaysia*, p. 205.

the tropics of South America, less well in Africa and least in the Indo-malesian region, the palaeotropical bee fauna thus becoming poorer in an easterly direction. There appears to be no close relationship between the number of bee species in an area and the number of species of angiosperms; for example, the Cape region of southern Africa, though rich in plant species, does not have particularly high numbers of bee species.

Of the 2300 species of termite (17), most are tropical and, in rain forest, up to 60 species may co-exist. Their great abundance is rivalled only by ants and their greatest diversity is in well-drained lowland primary forests (Figure 5.2), but few live in the canopy. Ants constitute a third of the insect biomass near Manaus, Brazil (18). By fogging techniques in the Peruvian Amazon, the arboreal fauna was found to comprise some 135 species in 40 genera, the most diversified ant fauna ever recorded. Much of this was due to the coexistence of species, e.g. a single tree had 43 species (of 26 genera), equivalent to all the ant faunas in all habitats in the British Isles. By contrast, the bulk of the 4000 described species of aphid are temperate (19). The short period they can survive without food, their host specificity and low efficiency at locating hosts probably account for this: 90% of plant species are not hosts because these plants are just too rare.

Total animal biomass   There are rather few measurements of the overall biomass of rain-forest animals, though estimates, probably not wrong by more than a factor of two (6), in Amazonia *terra firme* are 210 kg per ha, compared with only 64 for mangrove forest. This is much smaller than the ungulates and other vertebrates of African grasslands, estimated at 100–300 kg per ha, although the bulk of the biomass there as well as in the rain forest is in the soil fauna.

The animals depend ultimately on the vegetable framework of the forest and much of the following discussion will therefore be devoted to a consideration of some of the principal features of the trees and, to a lesser extent, the other plant forms in tropical rain forest.

5.2   Morphological diversity

5.2.1   *Tree form*

H.C. Dawkins concluded that, despite the variation in taxonomic terms of the rain forest of different continents, it was reasonable to argue that a mature forest has a mean basal area, that is to say a total bole cross-sectional area, of 32 m^2 per ha and that where lower figures were found, as they were in Ghana, this could be attributed to major human disturbance (20). Few of the boles, though, have a girth of more than a metre, and none is as tall as the redwoods of California or the big gums (*Eucalyptus*) of Australia.

Figure 5.3   The trunk and crown of *Shorea leprosula*, a common dipterocarp of Borneo. Note the honey-gatherers' ladder.

Emergents and canopy species Most of the forests have conspicuous emergents and some of them are of great commercial value. Such are the dipterocarps of south-east Asia and the Meliaceae of tropical West Africa, but these seldom exceed 50 m. The tallest angiosperm measured was the legume, *Koompassia excelsa*, which was 83.82 m tall in Sarawak. The only other rain-forest tree that has been shown to be taller was a specimen of *Araucaria hunsteinii*, which attained 88.9 m in New Guinea. For comparison, the tallest tree grown in Britain is a noble fir, *Abies grandis* from North America, which had attained 62 m in 1985.

The taller forests are found in areas with an alternation of humid and drier seasons with a total precipitation of some 2000 mm and with well-drained soils. High precipitation areas with little seasonality have lower forests with impeded drainage. In the Amazon, then, the canopy may be at about 30–40 m with the emergent *Mora* and the Brazil nut, *Bertholletia excelsa*, reaching some 50 m, while in Indomalesia the emergent *Dryobalanops aromatica* and other dipterocarps may reach 60 m (Figure 5.3). Contrasting with early-successional trees, the emergents have sympodially-branched hemispherical crowns. Indeed, most of the canopy trees have sympodial crowns too, with shiny scleromorphic leaves with stomata almost confined to the lower surface. The crowns of trees rarely intercept one another and this makes passage for lianes and arboreal animals difficult: the 'crown shyness' also means that sunflecks will pass through the canopy of the trees to juveniles below (Figure 5.4). The gaps may be formed by abrasion of buds or possibly reduction in lateral growth due to lateral shading, though flexible crowns of *Avicennia germinans* (Verbenaceae) in Costa Rica mangrove (21) were more widely spaced than those with stiff crowns, suggesting that abrasion is important in that species at least.

The sheer height of the canopy above the ground has meant that investigations of the biology of the trees and the other plants and animals associated with them are difficult to carry out using conventional methods. Mere collecting of specimens for herbaria may require the felling of a tree, the use of a gun to snipe off branches or the use of tree climbers, human or simian, to collect specimens. Most studies, though, require access to the canopy and a whole range of towers, ladders, platforms, ropewebs, walkways, hoists, balloons and even helicopters have been designed or employed in rain-forest studies (22).

Many trees, including dipterocarps, generally flower only when they have reached the canopy. Although in cultivation these may do so when 5 years old, they set no fruit. Precocious flowering is recorded in many tropical plants including *Citrus*, *Swietenia* and *Melia*, where it may occur in

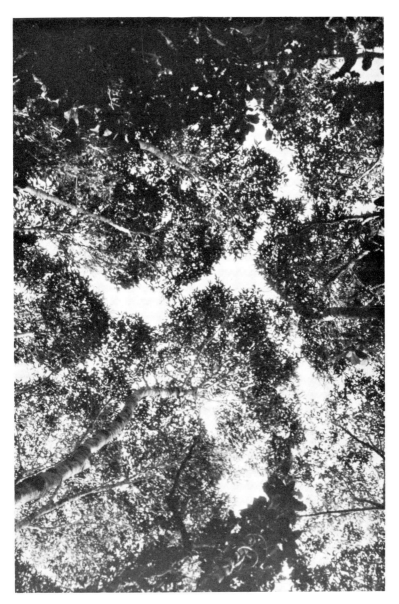

Figure 5.4   View of 'crown shyness' in *Nothofagus* forest in New Caledonia.

the seedling stage. Manipulation of *Triplochiton* (obeche) seedlings with hormones has resulted in precocious flowering and accelerated breeding programmes.

Architecture   As was outlined in Chapter 1, there are features other than height that mark out tropical rain forest. A consideration of the morphological, as contrasted with the taxonomic, variation in the trees themselves has led to a system of categorizing plant *architecture*, a field pioneered by Hallé and Oldeman. By considering the behaviour and fate of meristems, 20 or so *models*, or construction blueprints, for tree structures have been described and are most easily recognized in juvenile trees (Figure 5.5). Of these, only three or four are commonly found in trees of temperate forests where there may be even fewer models represented, but an array of other models may be found in the herbaceous flora of temperate vegetation. The models are found in a wide range of sizes so that it may seem rather difficult to associate any particular model with an 'ecological strategy'. Nevertheless, taking tropical trees, it is sometimes possible to see trends. Although these models are referred to as distinct, there are intermediates and it is preferable to think of the models as particularly frequently expressed parts of an architectural continuum (23).

    In the simplest model (Holttum) the tree is unbranched and the single meristem is destroyed in producing a massive terminal inflorescence. This is found in some palms, such as the talipot, *Corypha umbraculifera*, which after 50–70 years of vegetative growth produces an inflorescence 5–6 m tall and 10 m across and then fruits, flooding the surroundings with seeds. In dicotyledonous trees, this is rarer, but occurs, for example, in species of *Spathelia* (Rutaceae) in the American tropics. In *Corypha*, about 15% of the total dry matter is converted into seeds and, although the time scale is different, it behaves like the annual or biennial herbs familiar in the vegetable garden (lettuce, carrot or parsnip) in this respect. It is argued that the production of masses of seed at rather irregular intervals in any particular place may have the effect of swamping the seed 'predators' with plenty. The increase in predator population lags behind, and so by the time it has increased the uneaten seeds would have germinated and thus escaped. Other unbranched trees, like most palms, have lateral inflorescences that do not herald the death of the tree. This form of architecture (Corner) is known in 39 families of flowering plants as well as tree ferns and several fossil groups (24) and occurs commonly in treelets of the forest floor. Compared with the first model, it is rare in herbs and, in the temperate zones, possibly the most familiar are the plantains, *Plantago* spp. (Plantaginaceae). Both

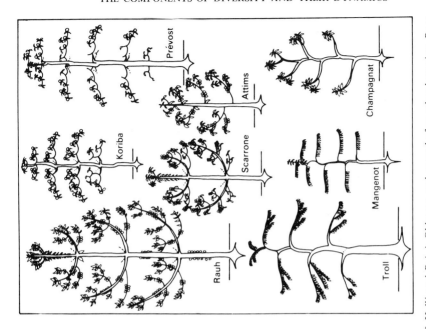

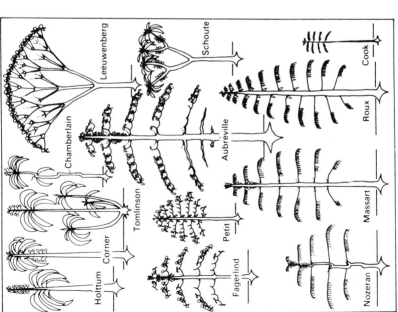

Figure 5.5 'Architectural models' of tropical trees as proposed by F. Hallé and R.A.A Oldeman and named after other botanists. Root systems stylized; shoot systems in one plane, showing position and frequency of branching, branch orientation and inflorescence position. Reproduced with permission from P.B. Tomlinson in P.B. Tomlinson and M.H. Zimmermann (eds.), *Tropical Trees as Living Systems.* Fig. 7.1 (Cambridge University Press. 1978).

models have a variation (Tomlinson), where suckers are formed, repeating the main axis, giving rise to clumps or groups of plants linked by stolons. This is the most familiar model of architecture in perennial herbs in temperate countries and is the stuff of herbaceous borders. Many palms do this, as do cultivated bananas and their relation the 'Traveller's Palm' of Madagascar, *Ravenala madagascariensis* (Musaceae), a species in secondary *savoka*, where it is a pioneer. It is possible that such clumps promote a relatively closed nutrient system (23).

None of these simple models, which, in trees, include massive buds and leaves (pachycaul), is found in temperate trees; unlike one where the innovations arise from aerial parts of the plant after the death of the meristem in flowering. This model (Leeuwenberg, Figure 5.6) is represented by the sumach, *Rhus typhina* (Anacardiaceae), a fast-growing treelet of gardens, native to North America. After each flowering, successive tiers of branches and leaves are formed so that several orders of branching are achieved. With branching, the new twigs are successively smaller, as are the leaves. Although in cultivation this can be reversed by taking cuttings and starting the plant again, as in the frangipani, *Plumeria alba* (Apocynaceae), it leads in the forest to a rather short-lived tree. It is the common construction of a number of *Solanum* and other species like *Manihot esculenta* (cassava) and *Ricinus communis* (castor oil) in the early stages of succession. It is not always associated with such an 'r-strategy', to use a zoological parallel, i.e. with rapid growth, short life-cycle and the production of many offspring, for this model is to be found in the slow-growing pachycaul giant groundsels (*Senecio* spp., Compositae) of the afro-alpine belt of the Central and East African mountains (25).

As can be seen from Figure 5.5 there is a wide range of form in branched trees, some with marked stratification into a tiered form, 'pagoda trees'. These, like most of the models, occur in a wide range of families; in other words, models are *grades* of construction, attained in parallel in divers groups of plants. The pagoda form is sometimes associated with pollination or dispersal by bats, but it represents one of a number of models that can present a multilayered canopy to incoming radiation. Horn (26) has shown that such a system is more drought-resistant than a monolayer system. The explanation lies in the fact that for every unit of incident radiation there are more layers and therefore a greater leaf area for the dissipation through transpiration of the energy not used in photosynthesis. Therefore the mean water-loss per unit of photosynthetic area is less. These layers allow light through, and therefore other plants can become established below them. In this way, shade-bearers can become established under the multilayers so

Figure 5.6   *Polyscias kikuyuensis* (Araliaceae), a fastgrowing 'ivy-tree' with Leeuwenberg's Model, in the montane forest of northwest Kenya.

that there is a correlation between multilayers and early-successional trees.

The monolayers cast a dense shade and nothing, not even their own offspring, may develop beneath them in extreme cases. Such trees in Europe or North America are the beeches (*Fagus*) and it is estimated that some 20–30% of all trees have this model (Troll), which seems to be able to exhibit great plasticity (23). Oaks, compared with their allies, the beeches, have upward-pointing twigs with spirally arranged leaves (Rauh) rather than weeping branches with leaves presented distichously. Oaks often have a rich undergrowth of herbs and shrubs: the forest floor beneath some beeches is inimicable to almost all plants, save a few mycotrophs. In temperate trees, such contrasts are found between the allied willows and poplars and within genera in both the temperate and tropical regions. The two types of architecture therefore represent just the end of the range of tree form found in the tropics. It is interesting to note that Joseph Needham in his monumental *Science and Civilization in China* (VI:1 Botany (1986), p.128) points out that the importance of the distinction between these two modes of growth was appreciated in Ancient China as well as by Theophrastus.

That the various models have been arrived at in parallel can be seen from the study of a single genus, as in *Psychotria* (Rubiaceae), where several models are found and there are useful taxonomic characters at the infrageneric level (27). The selective forces leading to this range are not obvious but the differences in architecture in the two species of the mangrove genus *Lumnitzera* (Combretaceae) are apparently linked to their pollination systems. *Lumnitzera racemosa* (in Old World tropics) has white flowers, pollinated by butterflies, which penetrate the canopy, whereas *L. littorea* (in tropical Asia) has red, somewhat zygomorphic ones, pollinated by birds, which do not. The first has Attims's Model with flowers presented throughout the crown; the second has Scarrone's, where they are presented only peripherally (28).

Reiteration   Juveniles have thin branches that can be lost, but mature trees do not lose major axes so readily, since the apical dominance of the leader is then also lost. (Some tree families, e.g. Annonaceae and Lauraceae, seem not to undergo this 'metamorphosis'). As they grow up, however, trees may suffer smashed limbs from a variety of causes—insect and fungal attack and grazing from animals, such as orang-utans for example, which may break off branches to get at their food. After such damage, some trees may never recover (some palms are an example), but generally speaking buds grow out and repeat the basic model of the species, a phenomenon known as reiteration. This is also manifest in coppice shoots or root suckers, just as in the suckering forms of the simpler trees. Some trees, such as dipterocarps (29), are poor at reiteration in the crown; while young they can do so, often through formation of accessory buds, but later the capacity declines. Apparently this is like other phenomena associated with meristem-ageing, e.g. leaf-retention in beech saplings, retained if the tree is grown as a hedge, for some dipterocarps can reiterate through coppicing.

Eugeissona minor (30), a stilt-rooted palm of Borneo, and a source of walking-sticks, has a reiterating system perched above the ground, trapping litter which is exploited by animals as well as other plants. The complex is able to fragment so that vegetative reproduction occurs as the old central parts die, a demise accelerated by termites. Pieces with only a few stilt roots fall away under their own weight to re-root.

Most trees differ from the simple models set out in detail above in that not all the branch units or 'articles' are equivalent to one another. In the sumach, as in cassava or castor oil, they are, but in most trees there is a 'division of labour' between the branches (specializing in photosynthesis) and the trunk (in support), as in rubber, avocado pear, coffee, cocoa or mahogany. Some trees have meristems that give rise to axes that are mixed, that is they give rise to trunk and branch portions, their orientation changing with development. This is the form in beech and elm and many tropical legumes. With age, all such trees reach a maximum, whereafter they merely replace pieces lost, reiterating the model in an essentially herbaceous way only. The result is the range of crown form characteristic of different species: the thick, dense crowns of *Mangifera* (mango, Anacardiaceae) compared with the light crowns of many legumes for example.

Other variable features   Variation is found in other features. Bark may be black, as in some *Diospyros* species (ebonies), to white through red-brown, but, at the forest edge, barks are often bleached pale grey. They may be smooth or fissured or scaly, sometimes with spectacular strips or patches falling away. The smooth barks are slow-growing. The trunks may be fluted or fenestrated as in some Rubiaceae, while the fluting at ground level is often continued into buttresses, which spread out from the bole. A negative correlation between bark thickness and buttressing has been found in trees of lower montane forest on Dominica in the West Indies (31).

Buttresses may be long and sinuous or may be some metres tall, concave or convex (Figure 5.7). They certainly increase the soil surface area covered by the tree and may, therefore, make it more stable. Indeed, there is a certain degree of correlation: emergents with deep taproots having no such buttresses and *vice versa*. In *Triplochiton scleroxylon* (obeche, Sterculiaceae), there is a significant correlation between tree bole diameter and buttress area but not between weak prevailing winds and directional buttress formation, nor between upper bole fluting and buttress size. However, there is a correlation between bole lean and asymmetric crown-formation and directional buttress formation in which larger buttresses are on the tension side, with a tendency for soils under large-buttressed trees to be shallower

Figure 5.7   Buttresses of *Koompassia excelsa* (Leguminosae) in northern Borneo. Note relative size of man.

(32, 33). In *Tachigali versicolor* (Leguminosae) on Barro Colorado Island, there was no significant correlation between orientation and buttress formation or tree-lean, though buttresses in the direction of the lean tended to be larger, suggesting that buttresses can indeed form tensile members. As buttresses develop above lateral roots, the distribution of these will provide the range on which selection can act. In *Quararibea asterolepis* (Bombacaceae), on the same island (34), buttresses tend to point in the direction of the prevailing winds, but also in the direction of any slope, suggesting a mechanism reducing the possibility of toppling. Buttress height increases faster than tree height as the crown reaches canopy height but, in this species, there seems to be no relationship between buttresses and crown asymmetry. *Pterocarpus officinalis* (Leguminosae) in western Puerto Rico has buttresses independent of compass direction but associated with crown radius. However, the tree grows in deep mud and the buttresses therefore form a broad stable platform (35). Interestingly, breadfruit, *Artocarpus altilis* (Moraceae), develops buttresses on poorly-drained soils but on better-drained ones these are not prominent. There seems, therefore, to be a number of explanations for the presence of buttresses: as with drip-tips and sclerophylly in leaves, there need not, of course, be just one 'reason' for their presence.

Stilt roots, that is roots arising from the trunk above ground level and eventually reaching the ground, where they reiterate, are characteristic of many monocotyledonous trees such as *Pandanus* (Pandanaceae) and palms. The majority of genera with stilt roots is to be found in these families and in Euphorbiaceae, Guttiferae, Moraceae, Myristicaceae, Rhizophoraceae and Sapotaceae. The development of root primordia above ground and the survival of unprotected root apices is enhanced by high humidity and temperature. Such roots stabilize trunks on soft media (36). In some palms, the base of the trunk may wither, so that the tree is supported on a polypodal arrangement of the stilt roots. In *Socratea exorrhiza*, a palm of the Peruvian Amazon (37), such roots may allow the plant to 'move' away from behind or under obstacles. Stilt-rooted plants are not restricted to mangroves or other swamps, though they are conspicuous there: the pioneering *Musanga cecropioides* (*Cecropiaceae*) in West Africa (38) is found to germinate on the soil and root mass turned up when a tree is blown down, or on the surface of the fallen trunk itself, stilt roots straddling the germination point such that erosion or decay of the germination site leaves the original root apex above ground. *Musanga* thus exploits ephemeral germination sites with nutrients up to 3 metres above the ground (cf. figs discussed in section 5.2.2). Some species do not produce the stilt roots if grown in dry conditions. Similarly, pneumatophores or 'breathing roots', characteristic of certain swamp species (Figure 5.8), may not develop under dry conditions. This is a well-known phenomenon in the temperate swamp-cypress, *Taxodium distichum*, which has 'knee-roots', commonly seen in mangrove swamps. The two closely allied *Xylocarpus* species (Meliaceae), which grow together in such swamps in Malesia, can be distinguished from

Figure 5.8 'Knee-roots' or pneumatophores of *Lophopetalum multinervium* (Celastraceae) in
the swamp forests of northern Borneo.

one another by only one having pneumatophores, so that their function is
not readily explained.

Aerial roots occur in other conditions, as anyone who grows the 'Swiss
cheese plant', the Mexican *Monstera deliciosa* (Araceae), knows, for this is a
scrambling and climbing aroid, which roots in pockets of humus in tree
crowns. Roots may be found growing into the humus collecting in the
persistent leaf bases of tree-ferns or other trees. In certain cases, the trees

themselves produce adventitious roots running beneath the mats of accumulated debris and epiphytes. Such root systems short-circuit the usual nutrient cycle and are found in temperate as well as tropical trees (39, 40). Thus through the presence of the entrapping epiphytes, which are often considered detrimental to their hosts, the trees' nutrient status is improved. Even more remarkable, perhaps, are certain 'apogeotropic' roots, which grow out of the soil and up tree stems, usually of different species. They may reach the subcanopy branches, as in the case of a *Swartzia* (Leguminosae) root reaching 13.4 m up a *Caryocar* sp. (Caryocaraceae). In a 100 m^2 plot in Amazonian Venezuela, 12 tree species in five families were found to produce such roots, which grow at rates up to 5.6 cm in 3 days and absorb nutrients flowing down the trunks. These fine roots were found on all stems greater than 4 cm diameter at breast height in the plot, but those of *Eperua purpurea* (Leguminosae) bore only roots of that species. Artificial plastic trunks with an input of nutrients (cattle manure) attracted the most and best-growing roots, the principal attractant possibly being calcium, a gradient of which possibly generates the observed apogeotropism.

5.2.2   Other growth forms

Stranglers and other epiphytes   The most spectacular roots are probably those of the strangling figs, plants that begin as epiphytes and eventually, by producing roots down the host trunk to the ground and by the production of a crown, shading out the host's crown, sometimes completely engulf and exterminate the host, leaving the figs as free-standing trees. Such stranglers occur in unrelated genera besides the figs: in *Spondias* (Anacardiaceae, Philippines); *Fagraea* (Loganiaceae) and *Timonius* (Rubiaceae) both in Papuasia; *Clusia* (Guttiferae) and *Coussapoa* (*Cecropiaceae*) in the New World; but also *Metrosideros* (Myrtaceae) in New Caledonia and New Zealand; *Wightia* (Scrophulariaceae), and possibly *Schefflera* (Araliaceae). Such occur in at least 20 families of dicotyledons, those epiphytes later forming contact with the ground being known as hemi-epiphytes (primary), those lianes secondarily losing contact with the ground being secondary ones (41). The humus in which the primary ones become established may be richer than the ground, e.g. that in the crowns of the palm, *Copernicia tectorum*, in Venezuela is derived primarily from faeces and nests, especially those of ants and termites, and is five times richer in nitrogen, 10 times richer in phosphorus, 1.5 times in calcium and six times richer in organic matter, than the ground. The crowns are infested with fig roots, some from fig trees whose roots have climbed up the palm stems, perhaps as their crowns grew up; similar roots from several different species have been found in the crowns of *Livistona endauensis*, a rare palm of the Malay Peninsula (42).

In cloud forest at 1500 m in Costa Rica (43), many seedlings seen in gaps start as epiphytes in the crowns of trees that later fall, and many others are found to be concentrated on 'nurse logs', some of which started upright, while some epiphytes, like some of the woody lobelioids peculiar to the Hawaiian forests, may have some of the effects of the stranglers. Indeed, the division between stranglers, epiphytes, climbers and scramblers becomes confused. Furthermore, some epiphytes may have the effect of parasites without physically penetrating the living tissues of their hosts: the small fern, *Pyrrosia nummulariifolia*, is a pest of coffee in that it webs the twigs and is believed to cause their rotting. The hemi-parasites in the Loranthaceae and Viscaceae allied to mistletoe, *Viscum album*, are a common feature of tropical forests and may bear hyperparasites of the same family, even to the second degree.

According to recent estimates (44, 45), 10% (well over 20 000) of all vascular plants are epiphytes, which are represented in 19% of families, the largest being Orchidaceae with about 14 000 species in 440 genera. No family is exclusively epiphytic and it is held that the habit has evolved in parallel at least as many times as the number of families, in some of which it appears in addition to have evolved more than once from terrestrial living, e.g. Araceae. The greatest diversity is in the Neotropics to which important epiphytic families such as Bromeliaceae, Cactaceae, Cyclanthaceae and Marcgraviaceae are restricted (or almost so). Ferns occupy similar 'niches' in the Old World but the large water- or litter-impounding habit is not really found in the Old World, though frequent in the New World. Almost three-quarters of epiphyte genera have dust-like seeds or spores; others often have wings or plumes. All these are wind-dispersed, though some groups have species with fleshy animal-dispersed propagules. Nevertheless, the high frequency of wind-dispersal is in contrast to the high level of animal dispersal in the trees. Small seeds are more readily caught in the crevices where they germinate and with a large surface-to-volume ratio rapidly absorb water. In maturity, the plants are subject to alternating dry and wet periods and the storage of water in succulent tissue is a nearly universal feature of vascular epiphytes, e.g. roots in *Medinilla* (Melastomataceae), sometimes grown as pot-plants in Europe, stems in orchids (pseudobulbs), petioles in *Philodendron* (Araceae) and peduncles in *Begonia* spp. (Begoniaceae), but most commonly in the hypodermis of the abaxial surface of the leaf, found in nearly all epiphytic Gesneriaceae, Ericaceae, Marcgraviaceae, Guttiferae, orchids, *Peperomia* spp. (Piperaceae) and very many others including nearly all epiphytic bromeliads. It is known that epiphytic leaves left in a dry room retain turgidity in the photosynthetic tissues while the hypodermis shrinks.

The hypodermis may be important in epiphytes with crassulacean acid metabolism. In these, the stomata are closed during the day and open at night when CO_2 is incorporated into organic acids, such that photosynthesis can occur without water loss through transpiration from open stomata. Cooling of the leaf through transpiration is thus not possible and it may be that overheating is prevented through the presence of the water-filled hypodermis acting as a heat buffer.

Water stress in primary hemi-epiphytes is associated with water-storing swollen stem bases in certain *Ficus* spp., *Didymopanax pittieri* and *Oreopanax liebmannii* (both Araliaceae) and well-watered seedlings of the figs do not develop the tubers. The dangling aerial roots of *Ficus benghalensis*, which so surprised Alexander the Great's troops, have no root-hairs but thick cortex and pericycle with a well-developed periderm with chloroplasts and lenticels, later thickening to become pillar-like and thus resemble stems. The fusion of fig roots begins by the coalescence of root hairs, followed by compression of the cortices of adjacent roots in *Ficus globosa* and, at the periphery of the contact zone, numerous parenchymal cells are produced, eventually fusing to give a parenchymatous connexion, the cambia of the roots being joined when some of the parenchymal cells 'redifferentiate' into cambial cells; fusion may also be initiated by cortical cells in older roots or in species without root hairs. Fusion can occur not only between roots of plants of the same species but also possibly between different ones.

Some vascular epiphytes have unorthodox modes of nutrient supply, perhaps the most remarkable being the ant plants, species of *Hydnophytum, Myrmecodia* and allied genera (Rubiaceae) of the Malesian forests. Through labelling experiments (46) it has been established that ants bring nutrients into the chambered tuber of these plants and that these nutrients are absorbed by the host. In *Hydnophytum*, there is an aperture produced by the plant, whereas, in the New World, ants chew an entry into their host bromeliad, *Tillandsia medusae*. Another bromeliad, *Catopsis berteroniana*, certainly absorbs rotted-prey nutrients through trichomes but apparently no digestive secretions are known (47). Furthermore, it is now known that some epiphytic orchids have fungi in the cortex of their roots and that these are linked thus to the rotting host. Curiously, then, the claim of a 'parasitic' nature of orchids, long-advocated by foresters and denied by botanists, has some substance.

Other prominent epiphytes include the more familiar bryophytes, which are to be found growing on tree bases, especially buttresses and stilt roots. Notable are the Neckeraceae, which are represented in all continents and up to 100 species of them have been recorded in a single habitat, as in Vietnam (48). Bryophytes are also found on the drier upper branches, twigs and leaves of trees, where some lichens also occur. Long-lived leathery leaves of dicotyledons are sometimes heavily infested, though 'unwettable' ones are not; ferns and monocotyledon leaves, notably Zingiberaceae are often similarly clothed, while even the delicate filmy ferns are not immune in that some tiny liverworts in the genera *Aphanolejeunea* and *Cololejeunia* as well as some Hookerioid mosses are most often found on them. The habitat is impermanent, however, and the species growing in the 'phyllosphere' must be thought of as 'nomadic'.

Epiphylls in Peninsular Malaysia (51) include green algae, usually *Trentepohlia* spp., growing from the edges of the leaves or cushion-forming *Phycopeltis* spp. Some crustose lichens penetrate leaf tissue and all occlude to reduce photosynthesis. Lichens may appear within 6 months, liverworts and *Phycopeltis* after about 2 years. Leaves of herbs like species of *Pentaphragma* (Pentaphragmataceae) can become half-covered in 3 years. In South America, the lichens tend to be overcome by the bryophytes and these, in turn, may be invaded by seedlings of vascular plants. The epiphyllae derive nutrients from precipitation, droppings and leachates in a phyllosphere, which also includes small animals such as tardigrades, rotifers and Acarina, and nitrogen-fixing blue-green algae. Obligate bryophyte epiphyllae have special attaching mechanisms and adhesive exudates. In the liverwort *Radula flaccida*, the rhizoids that secrete the exudate later penetrate the host tissue, causing the death of some epidermal cells. Heavy covers of epiphyllae increase water loss from the leaves and labelled phosphate passes from the host to the liverwort. They are perhaps best considered as 'hemi-parasitic' (50).

Liverworts are the most important group of epiphyllae: some genera are pantropical, others restricted to a single continent, but Africa has no endemic genera. Up to 25 species of liverwort have been recorded on a single leaf, though the norm is about six. There are specialized epiphytic communities on pandans and on palms and, at higher altitudes, tree-ferns, the different surface textures leading to different vertical arrangements of species. Although mosses are not very conspicuous in lowland tropical rain forest, they and liverworts also occur, of course, on dead and decomposing wood, disturbed soil and on rocks. At higher altitudes, they make up very significant proportions of the above-ground biomass (49).

Lianes, mycotrophs, saprophytes and more parasites Although some climbers cannot be absolutely distinguished from epiphytes in the broad sense, lianes are very numerous and may make up 40% of the total flora, in terms of species numbers, in some forests. In the Rio Negro basin of Venezuela (52), lianes were found to make up 15.7 t per ha, about 4.5% of the above-ground biomass, though also representing 19% of the total forest leaf area. In the Lambir National Park in Sarawak, Borneo (53), there were 348 lianes with a diameter at breast height of more than 2 cm per ha in the valleys and 164 on hilltops. They infested about 50% of the trees in both sites and connected, on average, 1.4 trees each, while on Barro Colorado Island (54), lianes were found to be at a density of 1597 per ha and to be distributed amongst 43% of the canopy trees, connecting, on average, 1.56 of them. In the understorey, they comprised 22% of the upright plants up to 2 m tall and, of these, 15–90% (dependent on species) were ramets and not seedlings. The importance of vegetative propagation is shown by a study (55) in Veracruz, Mexico, where it was found that *Ipomoea phyllomega* (Convolvulaceae) has a stolon system, which can colonize gaps from adjacent forest and *vice-versa*, though *Marsdenia laxiflora* (Asclepiadaceae) can only reproduce thus in closed-canopy conditions. In Gabon, *Tetracera alnifolia* (Dilleniaceae) also reproduces vegetatively under closed canopy (56), though no seedlings are found there, even though there are abundant seedlings of *Dalhousiea africana* and *Griffonia physocarpa* (Leguminosae). On the whole, lianes are most abundant in the trees at the edges of gaps and they provide a major route for animals to reach the canopy; they also reduce host growth rates. In the deciduous forests of Costa Rica, they also affect the fecundity of *Bursera simaruba* (Burseraceae), it being negatively correlated with fruit production: experimental reduction of liane cover (57) increased fruit production, so that lianes may be considered 'structural parasites'.

Some lianes are tendril climbers with modified leaves, leaflets or stipules; others are twiners; some have adhesive-root or adhesive-tendril systems; yet others are sprawlers, scramblers or are hooked but not directly attached but are hoisted into the canopy by their hosts and are able to extend internodally when apparently mature, thereby compensating for any growth by their hosts that might otherwise cause them to fracture. Their falling when their host falls may constitute an effective dispersal event, in that the new ramets growing up in the canopy will be some distance from the parental germination site. Some lianes undergo 'phase-change', i.e. they have markedly different juvenile and mature forms, most familiar perhaps being ivy in the temperate zone. It occurs in some Bignoniaceae, some *Ficus* spp. and in Marcgraviaceae. In *Triphyophyllum peltatum* (Dioncophyl-

laceae), an African carnivorous liane up to 40 m long, the shoots are trimor-
phic (58), in that there are sterile branches with ephemeral circinate leaves
bearing stalked or sessile glands, as well as long shoots with leaves with two
small hooks and short shoots where the leaves are large but unarmed. The
glandular leaves are produced just before the height of the rainy season,
suggesting synchrony with the insect maximum: the digestive glands with
hydrolytic enzymes are the most elaborate of all known angiosperm glands.
Other carnivorous lianes include the pitcher-plants, *Nepenthes* spp.
(Nepenthaceae), which are found from the spray zone to 3520 m in Irian
Jaya. Essentially Indomalesian, *Nepenthes* is represented by 28 of its 70 or
so species in Borneo, with one in Sri Lanka, one in Seychelles and two in
Madagascar, though up to 25 species may be found in one place on one
mountain in Borneo. They are secondary hemi-epiphytes and are almost
absent from dipterocarp forest, but are frequent in montane and heath-
and peatswamp-forests.

　　A remarkable root climber is the orchid, *Galeola altissima* of Java, which
is a mycotroph and alleged to grow to 40 m, though usually less. Other,
herbaceous, mycotrophs in the forest include species of several pantropical
genera of rather bizarre appearance. They are often conspicuous on the
forest floor for being white, bright yellow or pale bule. The fruiting bodies of
saprophytic fungi are not frequently encountered as they are very seasonal.
There is a problem in deciding to which of a mixture of tree species any
sporing caps may belong. For example, mycorrhiza (59), though in
Peninsular Malaysia, some *Russula* spp. may be associated with dipterocarps and species of *Boletus* and *Cortinarius* may also be involved. It is
interesting to note that in the Solomon Islands, unlike in nearby New
Guinea with which these islands share so much floristically, there are no
dipterocarps nor Fagaceae and there are no species of *Boletus* , *Russula* or
Amanita , a genus associated with Fagaceae, either. In Peninsular Malaysia,
there are few parasitic species and these are mainly facultative, lignicolous
and polyporoid. In central Pahang, both *Ganoderma mirabile* and *Phellinus
setulosus* attack dipterocarps, *Koompassia* spp. and various other unidentified canopy trees, their brackets being up to 80 cm wide and produced up to
40 m on the trees. A species of *Fistulina* rots the aerial roots of strangling
figs, while species of *Phylloporia* attack herbaceous plants, producing
brown brackets on green stems, petioles and midribs. In very humid forest,
bushes may be entangled in threads of toughened rhizomorphs or
aggregated hyphae coated grey-brown or black in agglutinated hyphae:
such 'horse-hair blights' are mostly species of *Marasmius* and the sporing

bodies are produced directly on the threads. There are even fungi in saline mud of the upper reaches of mangrove.

The fact that there is rapid decomposition of organic material in rain forests presupposes that there is a wide array of fungi capable of breaking down the various constituents, not only of this diversity of species in the forest but their individual constituent parts. It is notable that many fungi grow on wood only in its last stages of decay and it seems that higher fungi never grow intermingled, suggesting some kind of 'first come, first served' principle. Of vascular parasites, the most celebrated are the Malesian *Rafflesia* spp., some of the most remarkable of all angiosperms, living entirely within the tissue of certain vines except when in flower. The flowers may reach 1 m in diameter, though the plants are rare and some species are now threatened with extinction. There are root-parasites in the Balanophoraceae, again with brightly-coloured but morphologically much reduced flowering shoots, throughout the tropics. The only 'herbaceous' gymnosperm is a shrubby parasite, which is parasitic on its close allies in the Podocarpaceae of New Caledonia.

Herbs   By comparison with the richness of the canopy, the herbaceous flora of the forest floor is often rather species-poor and the most commonly encountered families are the relatively 'advanced' dicotyledonous ones like Rubiaceae, Gesneriaceae, Acanthaceae or monocotyledons, notably Zingiberaceae, Araceae, Marantaceae and Commelinaceae. Nevertheless, endemism levels in herbs are much greater than in tree groups, e.g. 80% of native Gesneriaceae and 90% of Begoniaceae are restricted to Peninsular Malaysia (49). The grasses there tend to have rather broad leaves and are considered rather primitive in the family as a whole. The coloured leaves of some forest herbs have been alluded to in Chapter 2, but there is great variation, even in a small area such as the forests of Peninsular Malaysia, where they may be blue, e.g. *Begonia pavonina*, black (some *Argostemma* spp., Rubiaceae), with white or reddish spots or stripes (*Sonerila* spp., Melastomataceae) or with veins outlined in white or gold as in *Anoectochilus* spp. (Orchidaceae); marbled leaves are found in *Cyrtandra pendula* (Gesneriaceae), velvety leaves that change colour from blue-green to gold depending on the angle of the light in *Begonia thaipingensis*, while many other herbs have leaves red or purple abaxially. The blueness of *Selaginella willldenowii* is due to the reflective properties of cuticle and outer epidermal wall differentially reflecting blue light; grown in full light, these layers are developed such that they do not look blue. In *Argostemma* spp., the compact palisade cells are completely filled with large chloroplasts, which absorb all the light so that the leaves appear black. Whatever the selective advantage of such colours, pigmented red or purple colouration, for example, varies greatly within populations. In addition, many of the dicotyledons have 'bumps' on either side of the main vein, which may possibly act like drip-tips in the drainage of the leaf.

The form of rain forest herbs varies more than does that of temperate ones (60). Some, like Zingiberaceae, have rhizomes, often long stoloniferous giving scattered clumps, but rather like temperate herbs. Others form patches from the stems becoming decumbent and rooting, as in *Cyrtandra* (Gesneriaceae), many Urticaceae and species with lateral inflorescences. Those with terminal inflorescences, as many Acanthaceae, shoot from buds after flowering: these may either be basal or grow out nearer the top of the plant, which is not found in temperate herbs to any great extent. Others are like the unbranched treelets with lateral inflorescences, e.g. species of *Sonerila* and *Neckia* (Ochnaceae). On vertical faces, the rosette of leaves may be asymmetric, the lower leaves being the larger, up to 70 cm in *Cyrtandra mirabilis*, leaves that could not be held on the flat as they are very thin. The Gesneriaceae also include some plants that resemble a stalked leaf bearing flowers from the base of the lamina. Such organs are intermediate in structure between leaves and stems, terms indeed that are applicable to the majority of, but not all, vascular plants.

Some herbs are hapaxanthic, i.e. they flower once and then die, such as *Strobilanthes* (Acanthaceae), and are often associated with more seasonal forests, as in India, or as *Mimulopsis* and *Isoglossa* of the same family are in Africa. In seasonal forests, there are true storage organs as in temperate corms, bulbs and tubers. There may be tubers or swollen storage trunks as in *Impatiens mirabilis* (Balsaminaceae) of the seasonal limestone forests of northern Malaya. In the seasonal forests of Africa, a species of the aroid, *Amorphophallus*, flowers and then produces leaves from its corm. In the scarcely seasonal Sarawak, however, *Amorphophallus* corms produce flowers in one year, leaves in another, though flowering seems to take place every few years, rather than every other year. It is perhaps unique amongst rain forest plants in having a well-marked annual dormancy.

Rheophytes   A further category of plant forms, embracing both herbs and woody species, is that of the rheophytes, which are plants confined to the beds of fast-flowing rivers throughout the world. Two totally tropical families, the Podostemaceae and Hydrostachyaceae, which are small thalloid angiosperms growing on rocks, make up some half of the world

Figure 5.9   A cauliflorous durian (*Durio* sp., Bombacaceae) in Sarawak, Borneo.

total but, in all, some 60 different families of angiosperms include rheophytes. They usually have narrow leaves or leaflets and some tropical ones have seeds dispersed by fish (61).

5.2.3   'Anomalies'

Characteristics, which when seen in temperate floras are often referred to as 'anomalous', are in fact more typically tropical features of the forest, including cauliflory, as was mentioned in Chapter 1. This is the production of flowers from twigs (Figure 5.10), as they grow into trunks, or from long-dormant buds. While some preventitious ones remain from differentiating shoots, endogenous adventitious buds develop from dedifferentiated parenchyma of trunk 'bark' in *Artocarpus integrifolia* (Moraceae) and the cannon-ball tree, *Couroupita guianensis* (Lecythidaceae). Such endogenous vegetative buds are also found in temperate trees (62). There is frequently variation within species such that young trees may have flowers borne in the leaf axils or just behind them, older trees producing them from the major branches or from the bole. Sometimes the flowers and fruits are produced at ground level, while some figs (the so-called geocarpic ones) produce their inflorescences on long ground-level branches.

Characteristic of some tree species is the production of flowers on the ends of very long inflorescences hanging like bell-ropes in the forest. Within species of the genus *Chisocheton* (Meliaceae) such ropes may be 7 m long (63) while others of the 51 species in the genus are cauliflorous, ramiflorous or even produce their flowers on the leaves. The inflorescences may be

Figure 5.10   The cauliflorous *jak* fruit (*Artocarpus heterophyllus*, Moraceae) under cultivation in Tanzania.

axillary or internodal. They may consist of determinate inflorescences or short shoots bearing such inflorescences. This genus is further remarkable for its leaves, which in some species are imparipinnate or paripinnate like the majority of species in the family, while in others only the juvenile foliage is like this but that of mature trees has crozier-like buds (pseudogemmulae) at the apex of the leaves. With succeeding flushes of the tree, more leaflets are produced from the pseudogemmula, the older ones falling off. Eventually this 'leaf' is dropped, when it can be seen that it has secondary depositions of xylem, like a stem. In short, these plants defy the pigeon-holing of temperate-trained morphologists. Indeed, these branch-like leaves behave, in the epiphyllous species, rather like the leaf-like branches of some Euphorbiaceae, such as species of *Phyllanthus* with their opposite leaves and deciduous branches.

5.3   Intraspecific variation

The variation within species of the plants that make up rain forest can be exposed only by intensive taxonomic studies and, although knowledge of the basis of this variation in tropical plants is still in a preliminary stage, there is now good documentation of the types of variation pattern to be seen in rain-forest species. Again, there are no clear-cut categories, but within any group intensively described, there are often to be found: (i) species that are morphologically rather homogeneous; (ii) species that may be more or less divisible into ecogeographically distinguishable races, usually labelled subspecies; and frequently (iii) others that exhibit a form of variation absolutely intractable to the taxonomist and known as ochlo-species (64). In these latter, at any particular locality there may appear to be two distinct entities, differing in some clearly recognizable features. At a second locality, there may be two other such, but the four together begin to merge in characteristics, and so on. Such forms may represent speciation in progress and have often been labelled 'provenances' by foresters, though variously given specific, varietal or form rank or an informal classification by taxonomists (63). It cannot be expected that a rigid hierarchical system, demanded by nomenclature (or nomenclaturists), should encompass organisms absolutely, if we believe in evolution at all.

Analysis of morphological (or, indeed, biochemical) variation within individual populations can be used to assess gene flow (65, 66). A population of the dipterocarp *Shorea leprosula* at Pasoh, West Malaysia, had leaf-morphological heterogeneity independent of edaphic influence, but a regression analysis of the variation with inter-tree distance showed significant linear correlation, suggesting that trees close to one another are more similar than they are to those further away, and it has been inferred that this demonstrates that gene flow is restricted and that its extent may be as little as 100 m. The findings are supported by isozyme work and pollination studies. On the other hand, studies of 16 different species on Barro Colorado Island, although showing that the taxa maintain large amounts of allozyme variation apparently typical of outcrossed, long-lived species with high life-time fecundities, show very little genetic heterogeneity between populations up to 1–2 km apart. This would suggest that gene-flow in this species is extensive on the island.

5.4   Seasonal variation and other cycles

5.4.1   Plant cycles

Sex   The coming and passing of the seasons are such an accepted part of temperate living that the phenomena of flowering and fruiting seasons, leaf-fall and perennation seem the norm to temperate-trained biologists. In the Asiatic tropics, regular fruiting seasons are well known in such fruit trees as durians (Durio zibethinus, Bombacaceae), lanseh (Lansium domesticum, Meliaceae), mangosteens (Garcinia mangostana, Guttiferae) and so on, while the flowering of the kechapi, Sandoricum koetjape (Meliaceae), was formerly regarded as the cue for sowing rice in Malaya. This regular fruiting contrasts with the state of affairs in cultivated bananas (which are of complex ancestry and are triploid), pineapples and papaya, which produce fruit almost all the year round in tropical countries.

Indeed, the regularly-fruiting fruit trees are exceptional. In lowland rain forest in West Malaysia (67), the phenology of the trees was recorded over a 4-year period from a suspended transect walkway. Here, no strong seasonality was observable, though there appeared to be more species in flower after dry spells at whatever time of the year these occurred. Some 85% of the species there did not flower or produce new leaves at regular intervals and, within species, there was a lack of synchrony. There seemed to be no clear relationship between the flowering trees and their dispersal agents, suggesting that these comprise a wide spectrum of generalists, depending on there being a considerable proportion of the tree species in flower at any one time.

Ficus fistulosa in Hong Kong (68) shows strong inter-tree synchroniz-ation of phenology with four population-wide fig crops a year, though any one tree is likely to have only three. In more aseasonal Singapore there is less synchrony, and four to seven crops a year with some trees apparently never bearing figs at all. Of course, the concurrence of different cycle lengths within the same species would mean the coincidence of flowering of any one particular tree with that of different sets of individuals through time.

Much of the periodicity of flowering is intimately associated with tree architecture. Of plants growing in Malaysia (69), those with ever-growing shoots, ever-flowering laterally include the exotics, coconut and *Muntingia calabura* (Flacourtiaceae) and papaya as well as the native *Trema* spp. Those with ever-growing shoots with intermittent lateral flowering, i.e. the dormant inflorescence buds are triggered by external stimuli, include *Shorea platyclados*. Ever-growing shoots with terminal inflorescences are found in *Dillenia suffruticosa* and the exotic frangipani, *Plumeria rubra* (Apocynaceae). Ever-growing shoots with the capacity to flower only in the basal leaf-axils include those of the exotic guava, *Psidium guayava* (Myrtaceae). Flushing shoots with terminal inflorescences on the new growth are found in *Lagerstroemia speciosa* (Lythraceae) and the rain-tree, *Albizia* (*Samanea*) *saman* (Leguminosae), though sometimes the flush becomes dormant without flowering and resumes later, as in *Peltophorum pterocarpum* (Leguminosae), which changes its leaves every 5–7 months and flowers every 5–7 or 11–12 months. Flushing shoots with basal flowering on the new growth are found in *Diospyros latifolia* (Ebenaceae). Some flushing shoots have irregular flowering positions—terminal or axillary or both—and are found in e.g. *Shorea parvifolia*, but this species also needs a further stimulus as flushing is more frequent than flowering.

In more seasonal forests, most flowering occurs at the end of the dry season so that water stress is invoked as a stimulus to flowering. On Barro Colorado Island, Panama, herbs in clearings, which are most sensitive to climatic change, flower at the beginning of the dry season while herbs in the forest proper flower during the rainy season and probably respond to intense rain following drought. Similarly, trees which usually flower in the wet season will flower in the dry if there is a period of heavy rain then (70).

There is some confusion between the control of anthesis and the control of floral induction, anthesis in many tropical trees being apparently kept in check by water stress, that in coffee (*Coffea arabica*, Rubiaceae) being triggered by the restoration of a good water régime (71–73). Leaf shedding may do the same and can thus lead to synchronous flowering. This can be compared with bud dormancy in temperate trees, where the flower buds are formed the previous season, or the flushing species above, where the rainy season promotes a flush terminating in a flowering episode. *Hybanthus prunifolius* (Violaceae), a shrub in the understorey of Barro Colorado Island, flowers asynchronously in the dry season a few days after the first rain of 12 mm or more. This can be induced by watering after a sufficient drought and it can be prevented by watering during the dry season, rather like the familiar 'summer ripening' of bulbs from Iran and Central Asia. Flowering like this is also known in *Rinorea sylvatica* (Violaceae), *Turnera panamensis* (Turneraceae), *Tabebuia guayacan* (Bignoniaceae), *Randia armata* (Rubiaceae) and possibly others there. Moreover, disturbance caused by cyclones may promote flowering as was observed after Cyclone Winifred hit the forests of Queensland in 1986. Widespread defoliation together with hot dry weather that followed the cyclone, exposed the understorey layers to water stress and increases in temperature and insolation. Some plants are known to respond to cold periods by flowering, the most celebrated example being the pigeon orchid, *Dendrobium crumenatum*, again of Malaysia, which flowers 9 days after such a cold snap.

Where slight seasonal differences in photoperiod exist, some plants seem to be able to respond by flowering like many temperate plants, e.g. *Hildegardia barteri* (Sterculiaceae) in West Africa. Locally 'bad' years for fruits may be due to heavy rains, which may affect or even kill the flowers and their pollinator. Some trees, on the other hand, flower very rarely and little is known of the phenology of the late-successional shade-bearers in this respect, but *Homalium grandiflorum* (Flacourtiaceae) in Malesia, for example, flowers only once in 10–15 years.

Another factor to bear in mind is age, for, rather like the North American herb, *Arisaema triphyllum* (Araceae), which is male when its corm is small and increasingly female as the corm grows, *Attalea funifera* (Palmae), the source of piassava fibre for brooms and thatch, in eastern Bahia, Brazil, is male when it is in the understorey and becomes more female as it grows up into the canopy (74). In *Zamia skinneri*, a long-lived dioecious cycad, at La Selva, reproduction is also a function of size, with females becoming fertile at a larger size than males, while larger males produce more cones and release pollen over a longer period. In the population studied, males dominated through time but more cycads (both male and female) were found to be fertile in secondary forest compared with primary (75).

Mass flowering   In a 5-year study of fruit trees in a lowland rain forest, again in Malaysia (76), two major peaks of flowering were observed (Figure 5.11), although some species flowered annually as a population, the intensity varying year by year. Some, like the dioecious *Xerospermum noronhianum* (*X. intermedium*, Sapindaceae) opened flowers over rather

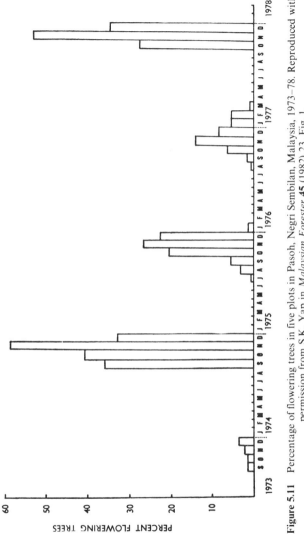

Figure 5.11   Percentage of flowering trees in five plots in Pasoh, Negri Sembilan, Malaysia, 1973–78. Reproduced with permission from S.K. Yap in *Malaysian Forester* **45** (1982) 23, Fig. 1.

long periods on any one tree, thus promoting outcrossing over a long season. Most trees were of the periodic leaf-exchanging type so that flowering lasts longer because those flowering at the bases of new flushes bloom before those flowering irregularly on new growths or those with flowers terminating the flush, so that flushing in March is associated with flowering from February to July (69).

One of the two peaks coincided with a mass flowering of the dipterocarps. Owing to the ecological (and commercial) importance of these latter trees, many ideas have been put forward as to the stimulus for flowering, which takes place heavily every 2–5 or sometimes 9–11 years on average. Ng (77) has made a convincing case for the stimulus in West Malaysia being a sharp increase in mean daily sunshine in January or February of the masting year. In years of such gregarious flowering, the percentage of trees producing good seed is highest, while it is low in years of sporadic flowering and lowest of all when the number of trees flowering is least. These flowering pulses do not seem to be 'all-or-nothing' phenomena like the familiar phytochrome system, for a stronger stimulus will cause a more massive flowering response. In Borneo, such flowering may occur throughout an area as large as Sabah, or may be confined to particular valleys. Because these flowerings do not take place in successive years, it would appear that a build-up of photosynthate must take place first. Indeed there is apparently a correlation with El Niño events, during which prolonged depression of minimum night temperature is associated with the onset of inflorescence development (78, 79). In *Shorea* sect. *Mutica*, it appears that induction is caused by a drop of roughly 2°C or more in minimum night-time temperature for three or more nights. Such periods are indicative of the invasion of the aseasonal tropics by a dry air-mass and some of these periods may derive from the El Niño phenomenon. In Sri Lanka, where some *Shorea* spp. flower annually, night temperature may also be critical, as *S. trapezifolia* flowers 1 month after night temperature first exceeds 23°C. Such stimuli in these more seasonal areas would result in flowering at the beginning of the dry season and ripening of the fruits at the beginning of the wet, suggesting that, although the dipterocarps are most species-rich in the aseasonal tropics today, they may have originated in the more seasonal regions. There is genetic variation in responding to the stimulus as there seems to be in the length of the flowering to fruiting periods of divers rain forest trees (80). This period may vary from year to year but, of 86 species examined in West Malaysia, 90% had completed it in 6 months and all within a year.

One of the many possible evolutionary explanations for the gregarious flowerings had been put forward by Janzen (81), who argues that the predation of the seeds may be heavy, the predators, particularly insects such as weevils, not being very specific about their food plants. He adds that such pressure may have been greater in the past in that dipterocarp seeds are now sought by pigs and man but possibly formerly by elephants and rhinoceros, now largely removed from the forests. In seasonal regions, even the annually fruiting species have unusually heavy fruitings and the intensity of seed predation is high except in those years. Predator satiation in dipterocarps is also achieved in the aseasonal tropics in the absence of an *annual* climatic cue, which may help to explain their success there even if, as Ashton maintains, they are invaders from elsewhere. Gregarious flowering within a species would swamp the predators, like the hapaxanthic *Corypha*, mentioned above, as well as the hapaxanthic branched dicotyledons such as *Tachigali versicolor* (Leguminosae) in America and *Cerberiopsis candelabra* (Apocynaceae) in New Caledonia and, in these species, allow offspring to grow up in the favourable sites vacated by their mothers. With the invasion of a 'core' of masting trees such as dipterocarps, the flowering of individuals of other species coinciding with mass flowering of dipterocarps might promote greater survival of the seeds, so that a response similar to those in dipterocarps might be selected for. As a result, selection would favour a low animal density, generalists and nomadic behaviour. However, it must be remembered that, although most dipterocarps are wind-dispersed, most rain-forest trees rely on animal-dispersal, so that, for them, the 'hitching their wagons' to dipterocarp biology is something of a compromise.

The spectacularly long cycles of the reproductive behaviour of the bamboos, up to a century or more, is rather rare in the equatorial belt, though a similar explanation is sought for those. Janzen suggests that the 'counting' of so many years might be difficult in the tropics and the essential synchrony lost: in the more seasonal regions, where at least 200 species of bamboo flower gregariously, those plants that flower early or late are likely to be exterminated by predators, thus reinforcing the mass-flowering habit.

Flowering is not always so apparently regulated (69) and may vary not only between but within individuals of a species, though usually fruiting of the previous crop has to be completed before the next flowering episode, e.g. in West Malaysia, *Posoqueria grandiflora* (Rubiaceae) fruits take 32 months to mature and flowering is every 36 months. Others may flower more than once a year, e.g. twice in *Symphonia globulifera* (Guttiferae) to five times in *Guarea rhopalocarpa* (Meliaceae), in America, where episodes may vary from 4–20 weeks. This can be viewed as selection for 'bet-hedging' because of the varying availability of pollinators or dispersal agents, or a little but frequent outlay for reproduction.

Besides variation in a general sense, there is much intersexual variation, in that in males the number of matings seem paramount but in females the quality is important, so males of some species flower earlier and, as they offer pollen as well as nectar and offer more flowers, are more attractive to pollinators; in some cases, females lack any rewards and mimic males. Males are found to flower earlier in the day than females and, in those examples investigated, they flower over longer periods and may flower more frequently. Such may also prevail in monoecious species, such that particular trees may appear dioecious in some years. As an example, the dioecious *Jacaratia dolichaula* (Caricaceae) at La Selva has a tree sex-ratio of 1.08 males to 1 female, but an operational sex-ratio of 1.4–10 to 1 during the flowering period as males flower for longer; females have no nectar and mimic the commoner males, their stigma lobes resembling the male perianth, and the female petals mimicking the male buds, such that a single flower looks like one male surrounded by five buds (82).

Within a single family (83), there may be a wide range of behaviour. *Muntingia calabura* (Flacourtiaceae) flowers throughout the year but the flowers last only 1 day, while *Casearia praecox* of the same family has all individuals in a population flowering at once, again just for 1 day. In outcrossing plants, the extended period would lead to more mates and a wider range of new gene combinations for any one plant's genes; pollinators would move from tree to tree, the few flowers on each reinforcing this and the system would avoid the problem of bad weather discouraging pollination, a risk in those having just 1 day for this. However, plants that are rare in space and time may be more attractive to pollinators if they mass-flower, so that plants with supra-annual flowering cycles tend to be mass-flowerers. Mass-flowering may also satiate flower herbivores, which destroy without pollinating. Continuous flowering may lead to a continuous input of seeds into the seed bank and therefore be of advantage to pioneers and, indeed, it is a feature of them. It may be that long-distance canopy-cues for pollinators are favoured in non-pioneering mass-flowerers. It is clear that the different syndromes of flowering have evolved in parallel time and again, that the range in some families is governed by the limits of the family's architecture and that in some groups the behaviour patterns are more deeply canalized than in others.

Vegetative growth   In some rain-forest herbs, growth is very slow. In West Malaysia, *Didymocarpus platypus* (Gesneriaceae) produces four leaves a year and these live for 22 months, while *Pentaphragma horsfieldii* produces two per year and they each last 33 months (49). At La Selva, the number of leaves produced over a 2-year period varied greatly between species in the understorey (84): a species of *Renealmia* (Zingiberaceae) producing up to 12 leaves, though *Protium pittieri* (Burseraceae) and *Myrica carnea* (Myricaceae) produced only two each. 'Evergreen-ness' is attained by shoot growth and leaf development occurring throughout the life of a plant as in the red meranti *Dryobalanops aromatica* (a dipterocarp), palms and most pioneers. Steady rates are seen for example, in the papaya, producing 13–15 new leaves per month, the dipterocarp *Shorea ovalis* about 1–1.5 (85). Vegetative growth in tropical plants is rarely continuous, as was pointed out above. The main peak of leaf growth, of flushing, in West Malaysia occurs after the driest time of the year and a second peak begins before, and extends into the wettest period of the year. Most trees produce one recognizable flush per year. *Terminalia catappa* (Combretaceae) has two flushes per annum though there is some infraspecific variation in this. In the more seasonal forests of Barro Colorado Island, *Tabebuia rosea* (Bignoniaceae) and *Quararibea asterolepis* (Bombacaceae) lose their leaves twice a year, and some *Ficus* spp. lose them several times (69). *Breynia cernua* in Java has a cycle of $5\frac{1}{2}$ months, *Delonix regia* some 9 months and *Heritiera macrophylla* 2 years and 8 months. 'Evergreen-ness' may also be attained by the flush life-span being longer than the interflush periods—this is the system in conifers and is often seen in understorey trees, e.g. *Harpullia confusa* (Sapindaceae) in Malaysia, where it has 10 flushes in 2 years, though mature trees tend to flush less often than juveniles. Alternatively, new leaves are flushed as the old ones fall and this is the case in many upper canopy species in Malaysia. Nevertheless, some trees may become completely leafless, as in *Toona ciliata* (Meliaceae) and *Piptadeniastrum africanum* (Leguminosae), both large trees, and, indeed, deciduousness of this kind occurs in a higher percentage of trees in the emergents than lower in the forest. It is sometimes associated with flowering, making the flowers rather conspicuous.

The categories of 'evergreen' and 'deciduous' are not 'watertight' and trees moved to other habitats can change; they can also change with ontogeny (85), e.g. chengal, *Neobalanocarpus heimii*, another dipterocarp, has up to seven consecutive conifer-type flushes when young but is a 'leaf-exchanger' at maturity, or even deciduous, as when grown in Thailand. Teak, *Tectona grandis* (Verbenaceae) grows intermittently in native habitats but continuously when transferred to aseasonal ones as a juvenile. Sometimes new leaves are brightly coloured or speckled or may hang limp, as in the legumes *Brownea* of the New World and *Saraca* and *Amherstia* of the Old. These extend once the xylem in the stems and leaves has thickened. Withering leaves are sometimes startlingly coloured and may be diagnostic for allied plants, e.g. of two common forms of *Sandoricum koetjape* grown in West Malaysia, the leaves of one wither red, the other yellow (86). Age-related changes in stomatal behaviour lead to there being less control with age and this could explain leaf-fall in dry spells in tropical forests (87). Unless the leaves have secondary thickening (e.g. species of *Chisocheton* , Meliaceae), they will be dropped as the shoot-apex grows in any case. Dropping leaves may get rid of epiphyllae, and parasites; indeed, it is known that leaf disease triggers leaf-fall in rubber (69).

The rhythmic nature of flushing in rubber, a native of South America has been closely investigated. It produces about six extensions a year and branching begins after about nine of these (23). Each branch behaves like the axis except that its branches tend to be on the lower side. This independent endogenous rhythm can be overcome by stress; for example, if most of the leaf laminas are removed, continuous growth follows. Tea (*Camellia sinensis*, Theaceae) has up to four flushes a year, a matter of some considerable commercial importance, as the commodity is prepared from the young shoots. For comparison, a deciduous temperate tree like sessile oak, *Quercus petraea*, has a rhythmic apical activity of 10–15 days, followed by 8–15 days of extension; in other words a 20–30 day cycle (but with winter dormancy, which needs a cold period to be broken, superimposed on this). Even in seasonal forests, leaf deciduousness is not universal, for it is seen to be related to nutrient status—a poorer status is associated with less deciduousness. Species may then behave differently under different conditions.

Plant demography   A feature of trees associated with seasonality in temperate regions is the formation of annual rings in the secondary xylem, that is the juxtaposition of small, thick-walled, xylem elements at the end of a season with the large thin-walled elements at the beginning of the next. Not surprisingly, perhaps, these either do not occur in tropical trees or their periodicity is not fully understood. Tree ages of tropical trees are often difficult to determine because rings, if formed, are often discontinuous, though particular increments can be aged by gauging ^{14}C, which have changed since the 404 above-ground nuclear explosions of the 1960s almost doubled the atmospheric level and this has since declined (88). In temperate trees, a set-back such as defoliation may lead to the laying down of two rings in a season, but the rigid pattern of one ring per season, of such great value in dendrochronology and the correcting of radiocarbon dating, is a general and unique feature of such trees.

Girth increment is often used as a measure of growth rates but it is greatest at different times of year in different species. At La Selva (89), the maximum in *Pentaclethra macroloba* (Leguminosae) is in the driest months when it is not in flower or fruit but *Carapa guianensis* (Meliaceae) is at a minimum then, when it is flushing. In primary forest in Sarawak (90), the mean annual growth rate in girth for *Artocarpus* spp. (Moraceae) was 1.4–1.9 mm per year over 21 years, though some trees showed rates of 6.6 mm and some no growth at all. Trees appear to grow faster when near fast-growing conspecifics or other fast-growing neighbours, or when they are far from neighbours. The results are further complicated by genetic, edaphic, topographic and local factors such as tree-falls and gaps (see section 4.4.4), for there are low mortality rates in 'primary' forest and high ones in logged forest. In different zones of gaps at Los Tuxtlas in Vera Cruz, Mexico (91), for example, there were 72 different species of seedling in the root zone of tree falls but only 33 in the crown zone, with most of the surviving seedlings of *Cecropia obtusifolia* being in the crown zone and most *Heliocarpus appendiculatus* (Tiliaceae) in the root zone even though *C. obtusifolia* started with higher concentrations in both; the entire population of *H. appendiculatus* disappeared from the crown zone in 6 months. Although many deaths were through unknown causes, injury from limb falls and herbivory were important, plants in the crown zone suffered the most. Although gaps are unnecessary for the germination and establishment of *Dipteryx panamensis* (Leguminosae), an emergent at La Selva (92), growth is enhanced by gap formation though there is often only one emergent per ha there. From 7–60 months after germination, there is a 97% seedling mortality, a proportion of this being due to litter fall.

The apparent dichotomy of lifestyle, pioneer and non-pioneer (93), stresses the importance of the early stages but is less concerned with the demography of later ones. So pioneers like *Cecropia obtusifolia* have cohorts of age classes associated with gap ages, while shade-tolerant ones like *Astrocaryum mexicanum* (Palmae) have a homogeneous distribution of age classes, because mosaic environmental variation is likely to affect all age classes of *A. mexicanum* but primarily the dispersal and survival of seeds of *C. obtusifolia* . Seed occurrence in *A. mexicanum* is highly correlated with the distribution and fecundity of palms with most falling under the trees and not accumulating in the seedbank. The seeds of *C. obtusifolia* are dispersed by various frugivores to gaps of all ages, but remain dormant for periods dependent on gap age. Over 12 years, the population of *A. mexicanum* seemed fairly constant but actually those in closed forest were increasing at 1.2% per annum and those in gaps declining at a similar rate.

Overall, the annual mortality of trees in a forest is between 1% and 2% (94), with a range of projected half-lives of 35–69 years, apparently independent of tree size over 10 cm diameter at breast height, but the highest death rates are in suppressed trees. At La Selva (95), life-spans of up to 442 years (*Carapa guianensis*) have been estimated; understorey species have shorter spans but slower growth; shade-tolerant understorey species have the same maximum growth rate but twice the life-span; canopy and sub-canopy shade-tolerant trees require release and have long life-spans and high maximum growth-rates, while short-lived pioneers have the maximum rates. Of course, these categories intergrade. In general, there is a very high mortality rate in the young stages of trees and a high survivorship during their reproductive stages. In sexually and vegetatively-reproducing palms, compared with adults and ramets, seeds and seedlings have high mortalities. Fecundity may increase with age as in *Pentaclethra macroloba* (Leguminosae) in Central America, while *Podococcus barteri*, a West African palm, peaks at intermediate ages as does *Araucaria cunninghamii*, a conifer of eastern Australia; *Astrocaryum mexicanum* increases and then becomes fairly constant (96), such reflecting a constraint imposed by its architecture.

5.4.2   Animal cycles

A number, if not the majority, of rain-forest mammals breed seasonally and, in primates, there may possibly be a correlation between conception and the coolest time of the year (6). Animals in general, even insects, tend to have single offspring or broods when compared with their temperate allies. The small size but high frequency of broods may be an adaptation to a sustained yet small production of their food. Small terrestrial anurans in the Ivory Coast (97) have very fluctuating populations, with early maturity, high reproductive effort and continuous breeding cycles, all reflecting generally favourable conditions for amphibian life, and both the unpredictable filling and drying of pools where they breed, and the high predator pressure.

The cycles that affect plants sometimes affect animals, which are sometimes indirectly affected by them through the plant response. Mast fruiting occurs in both tropical Fagaceae as well as dipterocarps (28), though their masting does not seem to coincide in Malesia. The lack of synchrony appears to largely explain the remarkable migrations of hordes of wild pigs (*Sus barbatus*) in Borneo (the only wild pig species there) and probably formerly elsewhere. One of the most spectacular migrations was in 1983 when many seeds were available (98): information on migrations has beei. collected from tribespeople and school-meal records (this local produce being bought when available). The build-up of fat by pigs feeding on the oily seeds allows reproduction and mating, the gestation time being the same as the fruit-ripening, i.e. about 3–4 months. They exhaust one area and move on to the next. The flowerings of the dipterocarps were worked out from local information and, again, school records (they buy the fruits too), while the levels of exports of 'illipe' nuts used in chocolate manufacture, etc., are recorded for custom duty purposes. 1982 was a good season for diptero-carps, followed by big crops of non-dipterocarp fruits in 1983, which sustained the population. In the Upper Baram River, there were possibly 800 000 pigs, and may be many more, but about 8% were caught and killed for food by local people. The pigs then circled back to the fruiting oak forests in the upper reaches once more, a circuit lasting about a year. The cycle is dependent on the initial build-up of dipterocarp fruits and support by both oaks and other fruiting trees.

Invertebrates　The general flowerings appear to be an Asiatic pheno-menon and are not recorded elsewhere, but they present a resource for insects required by plants for pollination. The populations of *Apis* and *Trigona* bees 'explode' (99), while carpenter bees (*Xylocopa* spp.), which normally forage along rivers and in secondary vegetation, move into the forest, as at Pasoh, West Malaysia. The flower buds of trees of *Shorea* sect. *Mutica* form breeding grounds for minute flower thrips (Thysanoptera), which complete their development in the spent flowers on the forest floor before returning to the canopy. It is probable that elsewhere other small, rapidly reproducing flower-feeders, such as beetles, flies and bugs may use and be used by other dipterocarps. Sequential flowering in tree species pollinated by *Xylocopa* start with *Xanthophyllum discolor* (Xantho-phyllaceae), a gap plant that acts as a reserve for the bees.

The effect on animals is to maintain a low biomass with sudden bursts of not only herbivores but also their predators and so on. In south-east Peru (100), biomass was found to be greater in wet seasons with consistent and distinct seasonal patterns in both Coleoptera and Collembola, while among psyllids (jumping plant lice) in Panama, some show continuous generations, though most are seasonal, some strongly so and this appears to be linked to the flushing of the leaves on their host plants.

The activity seasons of tropical insects in general, when compared with those of temperate ones, tend to be longer with seasonal peaks less well-defined on average (102) but there is a wide range of seasonality patterns, e.g.

aseasonal parts of Panama, some Neuroptera (dobsonflies) as active adults for only 2 months of the year. In others, be four or even six to eight generations a year. On Barro o Island, only 39% of 426 species of Homoptera occur as active throughout the year and only 7.5% are really non-seasonal; this in is broadly repeated in many insect groups in many parts of the ti pics e.g. aphids, Hymenoptera, Lepidoptera, Psocoptera and even *Drosophila*. Seasonal inundations, as in the Amazon, may be associated with the vertical migration of insects or their survival underwater with reproduction usually, but not always, taking place on the 'dry' forest floor. Large irregular, non-seasonal fluctuations can also occur.

Changes in day length are associated with the end of diapause in many temperate insects and this has also been found in some tropical ones. Humidity and rainfall are also associated with emergence but whether these are the actual proximal causes is often unknown, though the onset of rain can be a direct trigger in some cases at least, e.g. the beetle, *Stenotarsus rotundus*, in Panama is in diapause during most of the rainy season and throughout the dry, the state being ended by photoperiod and humidity but the beetle remains inactive until the first rains of the rainy season, when it mates and disperses. There may be other, biotic, causes, too, including food availability, e.g. *Jadera obscura* and *J. aeola* (Hemiptera) are seedbugs, which, in Panama, break diapause when they sense the appearance of the Sapindaceae seeds on which they feed. How can such a variety of patterns coexist? The active periods coincide with adequate food, tolerable physical conditions and tolerable levels of predators, parasites and pathogens, such that the season may therefore be a complex compromise. Food may become less palatable as e.g. leaves mature or other herbivores induce chemical change in them. What is needed is detailed life-history studies of individual species with experiments on changing their water régime and so on as has been done for many plant species.

So far as the parasites of the insects are concerned, it has been possible (103) to model insect-parasitoid populations to demonstrate that discrete, rather than continuous populations of hosts are favoured by parasitoid lifecycles 0.5 or 1.5 times that of the host, whereas parasitoids with generation times approximately the same as or twice as long as those of the hosts are more likely to promote continuous generations. Since parasitoid generation times of the order of half those of their hosts are widespread in tropical systems, it is argued that generation cycles often seen in tropical insect populations can be explained by their interactions with natural enemies, compared with the one generation of each per year in temperate regions, e.g.

coffee leaf miners in East Africa, coconut moths in India. A similar moth in Fiji (*Levuana iridescens*) was controlled by the introduction of its parasitoid fly, *Bessa remota*, when generation cycles as opposed to continuous populations appeared.

'Outbreak' insects include *Zunacetha annulata* (104) a lepidopteran, found from Texas to Paraguay. On Barro Colorado Island, its larvae feed on *Hybanthus prunifolius*, which they severely defoliated in 1971 and 1973. The outbreak was ended by a fungus or other disease that killed eggs, larvae and pupae of the insect, which seems to be a seasonal immigrant on the island. Seasonal immigrants may also include pollinators as in the mangrove swamps of the Sundarbans of India, where the nectar produced by the flowers of *Xylocarpus* spp. (Meliaceae) is an important resource for *Apis dorsata* bees, which move to the forests from March to July, a migration of some hundreds of kilometres (105).

Soil invertebrates seem to increase with the fungal mycelia that burgeon in the wet period after dry spells. In southern Peninsular Malaysia, for example, forest agaric fungi have two general 'fruiting' seasons in most years, in March to May and August to October or November (106). The return of wet weather after relatively dry periods of several weeks brings the sporing bodies up but they last only a few days, when different species and genera appear in succession, so that for most of the year the mycological richness of rain forests is concealed. On Barro Colorado Island (107), some groups of litter arthropod populations increased in the dry season, though the majority did so in the wet season, while others, e.g. Hemiptera, Coleoptera and Formicidae, showed no fluctuations or, at least, the fluctuations were independent of seasonal patterns.

Vertebrates   Very few species of bird and probably no individuals (108) breed all year round, most species probably beginning to breed when food supplies increase after a lean period. This may differ from year to year and between insectivores, frugivores or nectarivorous species. Despite this, the post-breeding moult occurs at the same time every year. It has been argued that as adult survival is greater than in temperate birds, yet reproductive rate very much lower, and moult is an energy-demanding process, there may be as high a selection pressure for timing the moult in the tropical belt as there is for breeding seasonality in temperate birds.

Seasonal variation in the avian community structure decreases with increasing vegetation complexity in Panama, where the diversity and abundance of insectivores generally varies more seasonally than do those of frugivores; there is less change in the understorey than higher in the canopy, where the fauna is poorer (109). In Sarawak, insects are available throughout almost all the year with a slight dip just before the north-east monsoon, whereas fruit is patchily distributed and erratic in season. Breeding for both insectivorous and frugivorous birds is from December to June when insects are most abundant: the explanation for the frugivore breeding then is perhaps because nestlings are fed a protein-rich diet including insects (110). Annual recruitment is very low— about 10%. Some bird species there have a 9-month cycle: such occurs in *Stachyris erythroptera* (the red-winged tree-babbler) and *Arachnotera longirostris* (the little spider-hunter) but how such cycles are maintained is not understood for it is suggested that the protein condition of birds governs breeding as when the levels are high at the beginning of the monsoon, when insects are increasing.

At the Brown River in south-eastern New Guinea (111) the density of birds remained remarkably constant for the majority of 96 regularly observed bird species. Only 19 showed periodical density changes: these included some rails and pittas, which declined in the dry season, probably because their feeding was hindered by the hardness of the ground and some others probably affected by human activity. Nevertheless a swift species regularly moved roosts while populations of six lorikeet, two fruit dove, and a fruit-pigeon species fluctuated with flower and fruit abundance. Canopy insectivores and mixed feeders tended to breed in the dry season, while ground-living insectivores and seed eaters bred in the wet season, with frugivores breeding throughout the year, though breeding success was generally low. Some species moved their vertical foraging regions in different seasons there, *Tanysiptera galatea* (common paradise kingfisher) feeding low down in the forest in the wet season, higher in the dry, probably because soil arthropods were fewer and the soil harder in the dry season. *Ptilinopus magnificus* (the wompoo fruit-dove), alone in its genus, forages throughout the vertical structure of the forest, shifting to the higher levels when fruits are abundant in the canopy in the dry season but returning to the lower reaches to eat less palatable food less attractive to competitors when canopy fruits are scarce in the wet season (112). For fruit pigeons in Queensland, by contrast, fruit seems to be available all year round because of asynchronous fruiting of the trees, the most significant of which appear to be Lauraceae. Fruit abundance is in the dry season when the pigeons breed, though there are differences between years, 1970 and 1972 being 'good' years, 1971 very poor (113).

Long-distance movements by vertebrate frugivores seem to be more common in the Old World than in the New (114). Besides the nomadic behaviour in pigs discussed above, it is also found in fruit pigeons and fruit doves in New Guinea and tropical Australia, hornbills, parrots, starlings and broadbills in Borneo as well as in immature orang-utans there. Nine of the 35 frugivorous bird species eating fruits of 20 species of canopy tree in New Guinea are nomadic, whereas only 8% of the resident bird species on Barro Colorado Island are. There is some seasonality in the canopy fauna there in that there is an influx of small omnivorous tanagers in early dry to the early wet season and the presence of temperate zone migrants from late-wet through the dry seasons (115).

An example of an intensive study of a single species concerns a 10-year project on a bulbul, *Andropadus latirostris*, in north-east Gabon, where it breeds during the short dry season (January to March), when there is sunny weather, low rainfall and a slight drop in fruit and insect abundance preceded by a peak in food production, during which the female increases in weight before laying, and followed by a second period of high food availability, corresponding to the increasing demands of the young (116). The building of the nest, the laying and the incubating of the eggs as well as the feeding of the young take place when conditions are optimal for the thermal comfort of the eggs and fledglings. Extraordinarily, during the 10-year period, there were 2 years, 1971–2 and 1972–3, when there was no breeding at all; it resumed in 1973–4 with a very high fledgling success, as the predation rate was much lower than the norm of 46%. The success was maintained for 1974–5 but thereafter fell as predation increased. The cause of the fluctuation seems not to have been food shortage or rainfall, while other birds continued to breed during those years. It has been suggested that the pause may lead to a predator-satiation phase but how such could evolve and be maintained is unclear.

Frugivorous mammals rely to a large extent on figs, which are available (different species) through the year. Indeed mammal breeding cycles are throughout the year (117) in many species, though many insectivorous bats have pronounced seasonality but may vary from place to place even within a species. In *Tylonycteris platypus*, a flat-headed bat in Peninsular Malaysia, the maximum spread of births, however, was found to be only 20 days, even though mating is over an extended period: sperm is stored in the female and the synchrony comes from ovulation, which may be associated with high daytime temperatures in the roost.

In addition, there are, of course, daily cycles of behaviour contributing to the everchanging framework of the forest. Many birds in southern New Guinea (112) come down from the canopy as temperatures increase during the day, while nocturnal mammals in Peninsular Malaysia (117) are active during the early part of the night with diurnal mammals early-rising and 'peaking' by mid-morning. However, leaf monkeys there have a midday rest between eating bouts perhaps because of the slow rate of digestion of foliage compared with fruits. Similarly *Cercopithecus cephus* monkeys in Gabon (118) have two peaks of locomotion: the first half of the morning and the second half of the afternoon; their phytophagy has the same rhythm but they feed on animal prey and rest in the middle of the day. Dense and low vegetation is chosen for resting but more open, taller forest is chosen at night and for fruit gathering; insect hunting and social interactions take place at levels lower than those used for locomotion and fruit gathering. Rainy seasons can over-ride certain daily rhythms as in the agile gibbon (*Hylobates agilis*), in the Malay Peninsula, where in heavy rain, gibbons tend to retire to their sleeping trees earlier than at other times of year (119).

The diurnal, seasonal, annual and longer cycles of growth, behaviour and demography of so many millions of species, occurring within the dynamic matrix of the forest framework itself, all set in a changing meterological and geological context, provide the living milieu in which individual plants and animals interact. It is to these interactions, already alluded to *en passant* — predation, dispersal and pollination — that our attention must now be directed.

CHAPTER SIX

COEXISTENCE AND COEVOLUTION

At its simplest, the relationship between animals and plants, with the exception of carnivorous plants, which are not restricted to the tropics, is one of eater and edible. We have seen how animal cycles are often associated with plant ones and how the activity of animals affects gap formation and mortality of plants. It is now time to examine in more detail the intricate relationships between plants and animals, beginning with 'predation' of plants by animals, or herbivory, its extent and plant resistance to it, then the interactions that are more clearly of advantage to plants, namely dispersal of propagules (i.e. genetic material at the diploid level) and pollination of flowers (i.e. dispersal of genetic material at the haploid level). Both would appear to have evolved from herbivory.

6.1 Herbivory and resistance to it

Some of the earliest known fossil vascular plants,those of the Rhynie Chert of the Devonian (1) 370 m years ago, show damage, which might be attributable to biting arthropods. Cockroaches and dragonflies appeared in the Upper Carboniferous (300 m years ago), at the same time as the giant clubmosses, ferns and seed ferns; beetles are known from the Permian, and flies and wasps from the end of the Jurassic and Early Cretaceous when the angiosperms were diverging from their seed-fern ancestors. According to Southwood (2), phytophagy arose from both saprophagy, the consumption of decaying or at least dying plants, and from feeding on fallen propagules including spores and pollen grains, which led to living on the strobili themselves, a route followed by the extinct insect orders, Dictyoneurida and Diaphanopterida. The earliest known herbivorous vertebrate is the late Carboniferous pelycosaur, *Edaphosaurus*: it and other early herbivores were large (2–4 m in length) and it has been suggested that size was an adaptation to low-energy food, as their ancestors were carnivorous and smaller. It is generally held, then, that there were some 70 million years between the origin of the land plants and any significant herbivory by either invertebrates or vertebrates. Compared with animal tissues, those of land plants are low in fat and protein (and this is often built quite differently from animal protein), while cellulose is almost universal and the phenolic toughening compound, lignin, is found in all woody ones. Lignin, the mechanical pre-requisite for erect growth of any size, and thus the origin of trees and forests, is an effective herbivory deterrent. Tannin, another phenolic derivative, is common in the outer parts of land plants and is instrumental in the inactivation of proteolytic enzymes. The inedibility of the early land plants to animals and, apparently, fungi led to the great Coal Measures of the Carboniferous and thus to the fuel of the Industrial Revolution and thence the technology for the destruction of those forests' successors.

Indeed, in large part, the problems of digestion have been associated with the rise of the organisms in the other kingdoms in response to these unused resources: bacteria as well as fungi, and, perhaps, their associations with both animals and plants, from rumen bacteria to mycorrhiza. This then allowed not only the efficient re-cycling of nutrients, enhancing plant growth, but also the opening of a 'Pandora's Box' in terms of the exploitation of plants by animals. Perhaps because of the symbiotic requirement for such exploitation, phytophagy is restricted within the

Table 6.1   The distribution of land-plant herbivory in the extant orders of animals. From Southwood, T.R.E. (2)

Group	Approximate total 'terrestrial' orders	Nectar/ pollen	Propagules only or mainly	Vegetative parts (= 'grazing')
		Orders associated with plants		
Nematoda	13	—	—	2
Arthropoda	36	8	—	9
Mollusca	1	—	—	1
Amphibia	3	—	—	—
Reptilia	3	—	—	—
Aves–Ratites	4	—	—	3
–others	21	2	7	4
Mammalia	16	1	2	7
	97	11	9	27

animal kingdom as shown in Table 6.1. In addition, there are problems concerning particular nutrients. Folivores (leaf-eaters) may satisfy their sodium requirement only by consuming soil or water plants, while animals like litter-feeding cockroaches in Amazonia would need to consume 30–40 times their energy requirement of litter to satisfy their phosphorus needs: in fact they eat dead animals as well.

Today, herbivory in the crowns of mature trees in lowland rain forest in geographically distinct regions in the New and Old Worlds leads to 13.8–14.6% tissue loss (3) in the leafy branches of trees, but consistently less in epiphytes, while in Ghana it has been shown that those plants with a rapid flush of leaves are consistently less severely attacked than those with a slower leaf production (4). Long-term observations of leaves in Australian forests (5), using labelled leaves and shoots, contrasted with the usual technique of measuring leaf damage on one occasion, gave much higher herbivory levels, particularly in temperate forests, since such measuring does not account for leaves totally eaten; 'long-term' (3 months) study showed similar trends in tropical forests, where up to 21% losses were recorded (6). Closeness of conspecific plants can increase the rate and there are many examples (see Chapter 7) of increased attack of seedlings associated with closeness to the mother tree. This is not an exclusively tropical pheno-menon: seedlings of Chinese Elm, *Ulmus parvifolia*, under the canopy of the mother tree at Albuquerque, New Mexico, were found to suffer some 580 times as much attack from the elm leaf beetle, *Pyrrhalta luteola*, as those beyond it (7).

Damage due to phloem feeders can also be substantial (8); leaf feeders can be miners or raspers; there can be root feeders e.g. some insect larvae, nematodes, stem-miners and gall-formers where the allocation patterns of plants are over-ridden. Moreover, it is not just live tissues that are consumed as bark feeding is widespread among mammals (9), e.g. orang-utans, elephants, rhinoceros and squirrels, besides many invertebrates. The stomach contents of the small squirrel, *Sundasciurus lowii*, on Siberut Island west of Sumatra, contain an average 4% bark in which 90% of its food trees are represented. This animal does not strip bark like other squirrels, but merely removes flakes of the large smooth-barked trees with low levels of hydrolysable tannins. The concentrations of fats, protein and carbohydrate and the calorific value of the bark of the different species seem to have no bearing on the choice of food.

Of non-flowering plants, it has been argued that ferns have a lower associated insect fauna, though it has long been known that bracken has a large number of associated arthropods. In Costa Rica (10), three species of different genera of ferns were examined and have been found to carry a fauna as rich as that on angiosperms; damage was just as great (10). Of fungi (11), spores, hyphae and spore bodies are eaten by a wide range of invertebrates and fleshy spore bodies soon become infested with maggots; snails rasp them; beetles lay their eggs in them; carrion- and dung flies take the foul-smelling slime of phalloids, while tortoises, rats, pigs, deer, macaques and probably other large mammals, in the Malay Peninsula at least, also eat them.

6.1.1   Plant resistance

Herbivores are, of course, also affected by the carnivory régime, weather, and other physical features, but, in the groups of plants that have survived the animal onslaught, there has been evolution of defensive features, both physical and chemical. Bark loss is generally repaired by the formation of new bark, the inevitable result of secondary thickening in dicotyledons as the phloem dies. In some plants, thin roots are produced beneath the bark, thickening and coalescing, as in *Ormosia nobilis* (Leguminosae), while many palms (monocotyledons) have harder wood on the outside (12).

Hormone analogues and other anti-herbivore compounds have evolved, followed by their being overcome by certain groups of herbivores, some even using the toxins in their own defence (13–18). For example, some Asclepiadaceae synthesize cardiac glycosides as defence but some danaid butterflies convert them to their own defences against carnivores, mainly birds. In turn, there are non-toxic mimics of those butterflies. In many plants, secondary compounds are concentrated in the trichomes and, for example, a ketone is thus concentrated in the hairs of tomato plants, but that same ketone has been found in the defence secretions of termites and caterpillars. Again, the male adults of certain danaids ingest material from withered Asclepiadaceae and Apocynaceae before mating and without this their pheromone systems are unable to function adequately to ensure successful mating; the dihydropyrrolizidines in their odour is similar to the heterocyclic moieties of the plants' pyrrolizidine alkaloids.

Rubber latex is a physical defence but also contains insecticides; such occurs in unrelated genera in a range of families, such as Apocynaceae, Caricaceae, Guttiferae, Moraceae and Sapotaceae, while, in other families, there are defensive resins and gums. Latex is avoided variously by insects, e.g. the butterfly, *Lycorea cleobaea*, avoids papaya latex as the larvae make holes, isolating distal sections of the leaves in which latex flow is therefore negligible; this also works on figs (Moraceae) and Asclepiadaceae. Other 'defences' such as resins may be actively collected as in the case of those gathered by euglossine bees in their 'pollen baskets' from Anacardiaceae, Burseraceae and Leguminosae used in nest construction, sometimes with added bark fragments. Leaf-cutting bees in Malesia take resins from dipterocarps, *Chalicodoma pluto*, the world's largest bee, using resin to line its tunnels in the active nests of arboreal termites. The insects may be attracted by volatiles, in the way that heartwood borers are attracted from distances up to 2 km to felled *Shorea robusta* trees in India, as the resin is collected from wounds in the trunk. Such resins are known to inhibit the growth of certain fungi.

It is known that attacks on plants by herbivores reduce the quality of the tissues for subsequent feeding. In poplars, it has been found that there are increased concentrations and rates of synthesis of phenolics within 52 hours of attack. This is also true in sugar maples, where it has been claimed that undamaged plants near the damaged ones have increased levels of phenolics and hydrolyzable and condensed tannins, suggesting an airborne cue from the damaged to the undamaged, which then become less attractive to herbivores. Such increases in tannin in browsed *Acacia nigrescens* in southern Africa have been reported to take place in a few minutes and the production of ethylene during damage has been considered to be the promoter of tannin increase. With such rapid production, it would be advantageous for a particular plant to do this as browsers would move away from branches near the wounded ones. Certainly in the tiny weed beloved of geneticists, *Arabidopsis thaliana* (Cruciferae), wounding and other stimuli promote the expression of touch-induced (TCH) genes: within 10–30 minutes, messenger RNA levels associated with at least four genes increase 100-fold. Ethylene has been suggested as being involved in some but not all of this, the products including materials of the calmodulin type, materials involved in the metabolism of calcium ions and therefore many growth processes: in this case the change to 'bolting'.

Although twice as high a proportion of tropical plants contain potentially toxic alkaloids compared with extra-tropical ones and, although these are often concentrated in the young tissues of plants, insects specializing on such tissues may have rather unsophisticated problems compared with those feeding on mature leaves, where so much 'secondary chemistry' is manifest. It should be pointed out, though, that were it not for the attacks through evolutionary time of these herbivores, there would be none of the commercially significant heartwoods, latexes and resins so useful to humans. On Barro Colorado Island (6), herbivory rates and defence mechanisms of young and mature leaves of 46 canopy species were observed throughout the year: leaf properties such as toughness, phenolic content, pubescence, water, fibre and protein contents accounted for over 70% of the variation in herbivory, with leaf toughness most highly correlated. Mature leaves of pioneers were attacked six times as much as canopy species. In 70% of all species, young leaves suffered more than old ones, though they have two to three times the phenol concentrations. It has been suggested that pioneers are able to tolerate high rates of herbivory because of their 'cheaper' leaves and faster growth rates. For a given herbivore pressure, the advantage of defence should increase as the potential maximum growth rate declines. This is confirmed by nutrient-poor soils such as kerengas in Malesia having no poorly-defended, fast-growing trees. In Africa (19), comparison of forest on white sands in Cameroun with that on better soils in Kibale, Uganda, showed that the first had trees with higher concentrations of tannins and other phenolics, though in Kibale there were more species producing alkaloids.

In Bako National Park, Sarawak (20), it was found that in terms of percentage leaf area consumed, young leaves of *Eugenia ochrocarpa* (Myrtaceae) were eaten no more than mature ones, the damage being due to edge grazing, mining and holing (in that order in both), whereas the leaves of the dipterocarp *Shorea* species were not significantly eaten as adults. Young leaves of both had higher enzyme inhibition than older ones, though levels of condensed tannin were unchanged. With more nitrogen, less condensed tannin and less enzyme inhibition in the young leaves of *Shorea* compared with *Eugenia*, it is not surprising to discover that they suffered more damage.

When forests in the Malay Peninsula (Kuala Lompat) and northern Borneo (Sepilok) were phytochemically compared (21), and the levels of nitrogen, fibre, total phenolics and condensed tannin, and degree of digestibility by both cellulase and pepsin determined for both young and old leaves of the canopy species from both places, that at Kuala Lompat was found to be more digestible in having higher nitrogen but lower fibre and phenolic levels. Sepilok is typical of the dipterocarp forests of the region and the trees have high levels of quantitative defences, whereas Kuala Lompat, which is unusually rich in leguminous species by comparison with dipterocarps, has trees with less reliance on them. The study was carried out on trees of greater than 30 cm girth at breast height and it was suggested that the differences between the sites were due to the soils, the podzols of Sepilok being less fertile than those at Kuala Lompat.

6.1.2 Some mammals

The Colobinae are a group of monkeys with sacculated stomachs with associated bacteria and flagellates, and they show the best adaptations to leaf eating, allowing the use of coriaceous leaves, though all species eat at least some fruit and seeds. However, there are marked variations in biomass of *Colobus* spp. in Africa and *Presbytis* spp. in Asia in different forests (21) and these cannot be explained in terms of the overall degree of herbivory of an individual species or its body size, both of which are known to be correlated with biomass in primates as a whole. At Sepilok, the biomass is 64 kg per km^2 but 876 kg per km^2 at Kuala Lompat, 153 in Gabon and 1849 in Kibale. In studies at these sites, correlations were found between feeding and nitrogen levels in young and mature leaves, but a negative relation with fibre; the effect of condensed tannins was less obvious but colobines most frequently ate leaves low in this. There was no strong correlation between the levels of condensed tannins and palatibility, for, despite the debilitating effect of tannin on digestion *in vitro*, it is clear that the precise types of tannin, the substrate and probably other substances, besides the possibility of interspecific differences in the tolerances of them, all affect the edibility. Colobines select foliage maximizing nitrogen and mineral input and minimizing digestibility retardant fibre (and possibly some tannins): alkaloids seem not to be a deterrent as many can probably be detoxified by bacteria in the foregut. The biomass of colobines shows a close positive correlation with the ratio of protein to fibre at each site, while there is no evidence of predation or disease limiting the populations. It is likely that the protein/fibre ratios for mature leaves represent the degree to which colobines avoid them and rely on more limited and seasonal resources with a concomitant decrease in population density notably at times when the preferred foods are scarce, especially at Sepilok and in Gabon.

Presbytis melalophos (banded leaf-monkey) at Kuala Lompat and *P. rubicunda* (red leaf-monkey) at Sepilok spend about half their time eating leaves, favouring high-protein, low-fibre foliage found especially in young leaves. *P. melalophos* is able to eat foliage from many canopy species in its home range, while *P. rubicunda* relies on rare trees and lianes. As there are many dipterocarps at Sepilok, the rarity of food plants there is held to be the main reason for the greater home range and lower population density compared with *P. melalophos* at Kuala Lompat (22): the home range is some 70–84 ha compared with 31.5. It may be that the oleoresins of the dipterocarps have an effect on the bacteria of the fore-stomach of colobines and it is now necessary to study the interaction of these bacteria with fibre and the wide range of tannins, some of which shield protein from degradation in the fore-stomach, or denature them so as to enhance proteolysis. On white sands in Cameroun, where tannin levels are much higher (23), black colobus monkeys avoid all canopy trees and feed selectively on gap species and, unlike other colobines on other soils, have to feed on seeds, which make up half their diet.

Phytophagous bats take principally fruit or flower products, though some take leaves, extracting fluids and rejecting solids (24). The sloths, arboreal herbivores with stomachs and digestive systems showing convergence with ruminants (25), at least the two-toed (*Choloepus hoffmani*) and three-toed (*Bradypus infuscatus*) on Barro Colorado Island (25), feed on at least 31 species, not just one or two as was long-believed, and it is estimated that they consume about 14.7 g dry weight of leaves each day in the case of the three-toed sloth, a cropping-rate of 5.1 g of leaf per kg of sloth per day, whereas howler monkeys crop at seven times this rate. The low rate in sloths is probably associated with their lower metabolic rate (51% lower than other mammals of comparable size) and daily lowering of body temperature. The effect is to maintain more animals on the same food supply, but with less metabolically active individuals, so that retention or re-evolution of poikilothermy is part of a syndrome of cryptic, relatively sedentary individuals with relatively high opportunities for contact and mating: 8.5 three-toed sloths and 1.2 two-toed sloths per ha. Their biomass totals 22.9 kg per ha, 73% of arboreal mammal biomass on Barro Colorado Island and they crop 38 kg leaves per ha per annum (40–60% of their diet), whereas the howlers take 52.8 kg, reflecting their higher metabolic rates and less ruminant-like digestion: they are therefore less efficient, even though at a lower density and biomass and with a smaller percentage of leaves in their diet. By comparison with that of insects, sloth herbivory there is low: about 0.63% of leaf production, but some 'modal' trees, targeted by sloths can lose as much as 7.7% and one tree suffered 20% loss. Sloth faeces, estimated to contain half the leaf material, are returned to the ground around the trees: the three-toed sloth actually buries them!

The largest living lemur, *Indri indri*, weighs 9 kg and lives in the Madagascar rain forests, feeding mainly on leaves, 40–70% of sightings being of buds and young leaves being eaten, though fruits, seeds, dead wood and earth (possibly involved in improving the physical texture of the food) are also taken. Spider monkeys (*Ateles geoffreyi*) on Barro Colorado Island eat leaves, which account for about 20% of their fresh weight intake and provide most of their protein, notable being the protein-rich young shoots of *Poulsenia armata* (Moraceae) and *Ceiba pentandra* (Bombacaceae), which has 25.3% dry weight protein, and *Cecropia* leaves, especially their petioles, which have high mineral content. Leaves make up 15% wet weight of the diet of the omnivorous white-throated capuchin monkey (*Cebus capucinus*) on Barro Colorado Island. In Gabon, the prosimian *Euoticus elegantulus*, by contrast, has gums from trees and lianes making up 75% of food intake. These gums are highly polymerized pentoses that have to be broken down by bacteria, which release nutrients roughly equivalent to the yields from fruits.

Gorillas eat only, or at least concentrate on, 29 species including some with apparent deterrents, including the viciously stinging hairs on the shoots of *Laportea alatipes* (Urticaceae), but these 29 species lack condensed tannins. Important all-year-round foods of the lowland gorilla in Gabon (26) include the large ground-living herbaceous Marantaceae and Zingiberaceae, which are also consumed by chimpanzees, pygmy chimpanzees, mandrills and elephant as well as being browsed by duiker, bongo, buffalo and so on. They are important in the forest, especially in light gaps, representing about 230 kg of food per ha in the forest for such animals. In western Ghana, the elephant (the forest elephant, *Loxodonta africana cyclotis*, of central and west Africa, subspecifically distinct from the elephant of east and south Africa) browses some 138 different plant species (27), mostly lianes for leaves and stems, but also 35 different types of fruit. The number of tree species with high proportions of 'barked' trees is low, but bark of a few species is pulled off in strips up to 10 m long as seen in *Bombax brevicuspe* (Bombacaceae) and *Lannea welwitschii* (Anacardiaceae). Through their activity, browsing areas (gaps) are maintained as well as seeds dispersed. The Asiatic elephant favours rattan (28), as do squirrels, while young rattans are used in pig-nests.

The 'deterrent' compounds of many rain-forest plants are used (and have been fought over in major wars) by humans to flavour food, e.g. cinnamon (bark of *Cinnamomum verum*, Lauraceae), tea (alkaloids and tannin of *Camellia sinensis*, Theaceae) leaves, ginger (rhizome of *Zingiber officinale*, Zingiberaceae) and seeds such as nutmeg (*Myristica fragrans*, Myristicaceae), cardamom (*Elettaria cardamomum*, Zingiberaceae), and coffee (*Coffea arabica*, Rubiaceae), and cloves (flower buds of *Syzygium aromaticum*, Myrtaceae). Many others have to be removed (or have been bred out) during preparation of food, e.g. the raphides of many Araceae and the cyanide-containing compounds in cassava (*Manihot esculenta*, Euphorbiaceae). What such compounds 'taste' like to other animals is difficult to ascertain, but sometimes very different flavours are due to quite subtle chemical differences between compounds, the sweet taste of vanillin, for example, giving way to the hot and pungent zingerone when condensed with acetone and its double bond reduced, while pungency itself can be explained as to its relative strength by the replacement of particular groups in the molecule (29). Gymnemins from *Gymnema sylvestre* (Asclepiadaceae) remove the sense of sweetness in some mammalian tongues, but are not deterrent to caterpillars and have less effect on dogs or rabbits than they do on apes. They appear to have no effect at all on pigs, rats or guinea-pigs. Thaumatin, which seems sweet to primates is not perceived thus by dogs, hamsters, pigs or rabbits.

6.1.3 *Some invertebrates*

Other compounds have been used to fend off micro-organisms, e.g. 'fever barks' such as quinine from *Cinchona officinalis* (Rubiaceae) angostura bitters from certain South American Rutaceae, and timbers such as sandalwood (*Santalum album*, Santalaceae) and camphor (*Cinnamomum camphora*, Lauraceae). They are used to 'tan' leather and give rise to the 'termite-proof' and 'teredo-proof' timbers so prized in tropical construction. Humans also exploit the foliage of plants with natural insecticides such as those containing anti-feedant triterpenoids, notably the cucurbitacins in Cucurbitaceae, but especially azadirachtin and other limonoids from Meliaceae, that from *Azadirachta indica* being very important: even desert locusts will not attack it.

Some of the most spectacular herbivory, besides the outbreaks of caterpillars mentioned in Chapter 4, and other 'outbreak' insects of the last chapter, are due to the activities of leaf-cutting ants, the Attini, which occur between 25° north and south of the equator in the Neotropics. The species of the genus *Atta* and *Acromyrmex* cultivate (possibly different strains of) a basidiomycete fungus, *Attamyces bromatificus*, which allows them to overcome a wide range of plant defences and to be polyphagous in a diverse environment (30). They also take sap from cut leaves and this is the principal, if not sole, larval food, sometimes avoiding leaves that would be good substrates for cultivating the fungus, but which are variously repellent to ants. They attack high-protein leaves, more acidic ones having less available protein, the optimum pH for the fungus (pH 5) being reached when the material is mixed with the ants' acidic faeces: then the fungus can break down the hydrolysable tannin/protein complexes.

Atta cephalotes builds nests up to 150 m² in surface area, containing up to 5 million workers, ranging from minima 2 mm long to the largest soldier 20 mm long with medians about 10 mm long, which do most of the leaf cutting. The minima act as nurses and tend the fungus garden fed on leaves (31). The leaves are cleaned and the cuticle scratched, thus aiding fungal decay; the ants cut and chew the pieces into fragments, which are mixed with saliva as well as faeces before being put in the garden, the swollen tips (gongylidia) being collected by the ants for food. The gongylidia undergo enzymatic hydrolysis of their cellular components. This seems to be associated with the development of lipoproteic particles and the resultant material, unlike the ordinary hyphae, are edible (32). The fungus lacks some proteolytic enzymes, which are added in the ant faeces and are derived from the gongylidia, the ants merely acting as translocators, their digestive tracts not breaking them down. A small piece of fungus is carried by a founding queen in her buccal pouch before her mating flight.

Some plants are immune to attack because of their terpenoids, some of which are volatile and affect the fungus, i.e. plant protection from leaf cutting is achieved by the synthesis of an antifungal agent (29). The nests may persist for up to 20 years (33, 34) and such a nest in rain forest can completely defoliate a major tree in a single night. It takes some 5 or 6 years for a nest to produce sexual insects, so that it would appear that the complete defoliation of the immediate area or the removal of the more palatable species would lead to deteriorating conditions for the ants. In other words, in the long term, the maximum energy return for a given energy output in the short term would seem inappropriate. Evidence suggests that, on the contrary, grazing pressure is evenly spread throughout the area exploited, that area is related in size to the size of the nest and that it is defended against competing leaf-cutting ants and that a wide range of plant species is exploited. When the forest is felled, however, and replaced by monocultures of, for example, fruit trees, with a much reduced number of leaves per hectare, the system breaks down and there is often widespread crop damage. Among leaf-cutting ant populations under more natural conditions, inter- and intra-specific competition are some of the factors controlling overall numbers. Aggression leads to the death of founding queens and small colonies (or their emigration).

Ants, which may make up to half of the animal biomass in some habitats, are not always involved in such an apparently one-sided relationship with plants, as we have seen in the case of the myrmecophytic epiphytes. Ant plants are not necessarily found in nutrient-poor sites and many so-called ant-plant species sometimes have ants in them but sometimes do not, e.g. species of *Aphanamixis* and *Chisocheton* (Meliaceae) and many examples seem to show that the presence is due to 'casual' entrance through wounds suffered by the trees through damage by other animals or tree falls and so on. The most widespread plant feature that attracts ants, however, is extrafloral nectar (35), produced from special glands. Although attractive to several types of insects, such secretions are usually involved in ant attraction and the ants protect their 'host' from attack by other animals. The chemistry involved here is scarcely different from 'normal', when compared with the elaboration of 'secondary' compounds in defence. They are found at various points on plants, including fruits as on those of *Crescentia* spp. and in 15 other genera of Bignoniaceae (36), and the nectar may make up substantial parts of the ant diet as in the case of the extrafloral nectar of *Caularthron bilamellatum* (Orchidaceae) of Central America, which comprises up to 48% of the associated ants' diet at some times of the year.

The presence of ants allows *Bixa orellana* (Bixaceae), a dye plant with nectaries on the nodes and pedicels, to mature twice as many seeds as antless individuals, while *Aphelandra scabra* (*A. deppeana*, Acanthaceae) with nectaries on the bracts matures nine times as many fruits as plants deprived of their ants. Species of *Inga* (Leguminosae) with them on their rachis have herbivore damage reduced by a third or even a half. Moreover, parasitoids may also feed at the nectaries and then infect insect larvae by oviposition thereafter and, in *Inga*, this seems effective in control at altitudes where ants are uncommon.

Ants are also attracted by lipid- or protein-rich food bodies, e.g. lipid in *Macaranga* spp. (Euphorbiaceae), protein in *Acacia* spp. (Leguminosae) but carbohydrate in *Cecropia* spp. (*Cecropiaceae*). *Piper cenocladum* (Piper-aceae) in the understorey of Costa Rican forests has a petiolar chamber formed by the adpressed margins of a flattened petiole and this is usually occupied by *Pheidole bicornis* ants. Only when the ants are present, however, does the plant produce significant numbers of food bodies, which are lipid-storing single-celled ones developing from the adaxial surface of the petiole. In this and other ant-inhabited species of *Piper* in Costa Rica, the ants remove encroaching climbers and it is argued that the plants absorb nutrients from the decaying nest material of the ants (37, 38). However, where ant presence promotes food-body production, certain (*Phyllobaenus*) beetle larvae prey on the ants; such can be found inhabiting plants even in the absence of the ants and their presence causes the food bodies to grow, while herbivore damage seems to increase. The beetles are clearly parasitic on the system (39). The climber-pruning habit is known in several other genera and was first documented in Central American *Acacia* spp. (40). The origin of such pruning is possibly associated with the discouragement of other invading ants (41), for it is more frequently found in ant species with stinging rather than other chemical defences. In Peru, where uncuttable contacts were made to plants of *Triplaris americana* (Polygonaceae), *Crematogaster* ants invaded more frequently than otherwise, even carrying off larvae and usurping the resident *Pseudomyrex dendroicus*. On *Cordia nodosa* (Boraginaceae), however, invaders did not increase under these conditions, perhaps because the plant is so pubescent and therefore impassable to large ants in any case: only when lianes occupied by the smaller *Crematogaster* ants came near did the resident *Allomerus demerarae* attack. Liane deterrence may thus be a side benefit and it is possibly attained in other plants by different mechanisms as in the case of some tree ferns, where marcescent fronds (42) may deter both lianes and epiphytes.

In *Barteria nigritana* (Passifloraceae) in Africa, one coastal subspecies is inhabited by one species of ant; however, the more widespread other subspecies has another ant species, but again not all specimens are infected. In Nigeria, saplings occupied by *Pachysima aethiops* (43) have more leaves and branches and less attack than those without but, because the few larger workers have a ferocious sting and the habit of dropping from the crown, it is suggested that they may have been effective against large browsing mammals too; the ants also attack nearby plants and keep the host's leaves clear of debris and epiphyllae. In *Leonardoxa africana* (Leguminosae) in Cameroun, there are extra-floral nectaries and swollen internodes occupied by ants in some parts of its range with different ant species involved in different areas (44). The ant, *Petalomyrex phylax*, is apparently restricted to the plant and protects its stem apices from attack. Indeed, in some sites *Leonardoxa* trees are unable to reach maturity without the ants. Another ant species, *Catallaucus mackeyi*, does not protect the tree and also keeps *Petalomyrex* ants away: it is therefore a parasite on the system. The long-lived leaves are not patrolled but have an effective chemical defence.

Daemonorops verticillaris and *D. macrophylla* (Palmae) are rattans characteristically associated with resident ants in Malesia (45): their nests are made of plant hairs and sometimes spines as well and they can absorb 185% water by weight. ^{14}C in throughfall is absorbed by the rattan through the nest, and falling debris around its apex, particularly in *D. verticillaris*, yields nutrients in throughfall and these are also absorbed via the nest. In other plants, nutrients flowing down stems are absorbed through the root system, but here they apparently pass straight into the stem. If disturbed, the ants beat on the stem, alerting predators, possibly elephants, with an alarming noise. Other 'domatia' are more substantial structures and are known from over 400 species (half in America, fewest in Africa) in over 30 genera of trees and lianes (35, 46): in *Acacia* they are in stipular thorns but in the rest they are in the stems, where the stems are sometimes swollen as in some species of *Clerodendrum* (Verbenaceae). These are sometimes naturally hollow with ants biting entrance holes at thin-walled points while others have natural invaginations. In *Cecropia*, the pith stops growth before the stem reaches its full diameter and the *Azteca* queen enters through a thin unvascularized membrane at the top of an internode. As the brood grows, workers chew through septa in the shoots and kill or seal off other queens, after which the whole tree is taken over (46).

Even in the absence of ants, the tuber of the epiphytic *Myrmecodia* develops but the ants, usually species of *Iridomyrex*, keep broods in smooth chambers, placing debris in warted ones: the release of nutrients may be facilitated by the presence of fungi. *Myrmecodia* plants are patrolled by ants, especially along the stems, where they gather nectar and are protected by stipules or spines. Some ferns are similar, e.g. *Lecanopteris* spp. (Polypodiaceae) one of which is spiny, the spines being used in ant 'carton' runways. The neotropical fern *Solanopteris bifrons* has complex rhizomatous sacs in which ants leave debris and then roots enter while detritus traps of many ferns with polymorphic leaves are often inhabited by ants: the ants get nectar from nectaries on the fronds. Other such epiphytes include some bromeliads and several orchids, where ants live in hollow pseudobulbs or between the roots. In *Dischidia* (Asclepiadaceae), there is a range of form associated with increasing involvement between the different species and the ant, *Iridomyrex cordatus* (47). *Dischidia nummularia* has lens-shaped leaves with no ant-association, while *D. parvifolia* in the Malay Peninsula with similar leaves is associated with *Crematogaster* ants, which, in the mountains, make tunnels at the bases of *Leptospermum flavescens* trees and take *Dischidia* seeds inside such that plants germinate in the insect frass there, such that their distribution is thereby associated with that of the tree. In *D. cochleata* the leaves are arched with edges pressed against the trunk, adventitious roots and ant broods with debris underneath, while in *D. major* (*D. rafflesiana*), the leaf forms a sac into which ants and adventitious roots go, though the nests are elsewhere. Inside are stomata (the few outside are small and probably permanently closed), which probably gain CO_2 from the ants and decaying material, reduce transpiration loss and also improve the oxygen and moisture regimes for the ants. *D. pectenoides* (Philippines) has an inturned lip rolled back on itself such that inner chambers are produced in the sacs, though the adventitious roots enter only the outer ones.

The so-called myrmecophytes are found in all continents and are represented in about 20 familes of angiospers (35), notably Rubiaceae and Leguminosae. Almost all of them are found in regrowth, e.g. species of *Cecropia* and *Piper*, where the chewing off of competing plants is important. *Endospermum formicarum* (Euphorbiaceae) is the Arbor Regis of Rumphius, who named it thus because it is untouched by other trees. The ferocity of ants in attack is a consequence of their haplodiploid sex determination: sisters in a brood are genetically identical with half of one another and half identical with the rest (their fathers being haploid), so that, on average, they are $\frac{3}{4}$ identical to sisters and only half to offspring. Ant species that are not so aggressive avidly remove insect eggs from foliage and generally 'spring-clean' the host. The ants are broadly-based, e.g. almost all ant-fed epiphytes in Asia (47) are visited by *Iridomyrex cordatus*, while many different ant species service plants with extrafloral nectaries. Similarly, Peruvian myrmecophytes (48) appear to have a range of possible inhabiting ant species and these may compete for occupation. Plants with defence trichomes are suited to the smaller species as the larger ones have to cut a path through them and are therefore at a disadvantage. In the Malay Peninsula (49), there is just one ant species, *Crematogaster borneensis*, for all infected species of *Macaranga* (Euphorbiaceae). This ant does not sting and cannot live away from its hosts; although it does not bring in nutrients, it protects the host from herbivore and liane damage. The queen sheds her wings on entering the internode and seals the entrance, but coccids, attracted by food bodies on the leaves, are cultured within.

Some ant chambers contain fungi and these are kept in check by nematodes (cf. fig syconia discussed below) and many species tend sap-sucking Homoptera: aphids most commonly in temperate regions, coccids (scale insects) in the Old World tropics and membracids (tree hoppers) in the New, and these are a drain on the host-plant. Domatia on the abaxial surface of leaves (50, 51) are usually small invaginations or hair tufts in the axils of the veins. Of 32 species studied in cultivation (these representing 18 different families), all but one had their domatia ('acarodomatia') containing mites (Arachnida) and 75% of them held mite eggs. A high percentage had beneficial mites, a much lower percentage had harmful ones. It can be argued that these associations serve as shelters and nurseries for mites, which, in turn, eat herbivorous arthropods and pathogens, a general theory put forward by Lundström over a century ago. As domatia are found on crops like coffee, the phenomenon may be of agricultural significance. However, insects such as the harmful scale insects are sometimes found there too.

The weighing up of 'advantages' and 'disadvantages' here and in general is extremely difficult and will change with time and locality. Furthermore, defences—mechanical or chemical properties such as tannins, which make digestion hard (though the effect greatly varies in magnitude), or toxins such as alkaloids—are no deterrent to some animals. There may be variations in toxin levels with the season, and animals, in any case, eat different parts of plants with different levels of toxin and, at different stages of their own development, may be able to deal differently with such toxins (52). It has been suggested (53, 54) that alkaloid level fluctuations may affect the feeding patterns of orang-utans through the day.

The relationship between herbivores and plants has often been seen as some kind of arms race, but there have been opposing views in that, for example, spittle-bugs in the dry tropics of West Africa, rather than depleting valuable water resources may promote microbial activity beneath particular tree species, allowing them to take up nutrients in the dry season from the surface layers of the soil, provided that other roots can tap a deep water source. Such are the 'raining trees' of Africa (55, 56). This attractive hypothesis has been extended to other groups of plants and their 'consumers' suggesting that under certain circumstances plants may benefit from those animals that feed on them and, indeed, may positively 'encourage' such herbivory. The step from predation to this symbiosis is a short one.

One such hypothesis with respect to tropical trees argues that the solid bole of a tree locks up an enormous amount of nutrients and that, if this could be recycled by 'employing' micro-organisms to rot down the heartwood, thus providing a roost for animals whose nests and droppings would subsequently rot too, the roots of long-lived organisms like trees would be prevented from exhausting local resources and competing with other roots, and the chemicals produced to prevent heartrotting would be saved. The discovery of a species of *Guarea* (Meliaceae) with its own roots growing up into its hollow trunk, shortcircuiting the nutrient cycle, seemed excellent evidence in support of the hypothesis (57), though in Jamaica it was found that many such hollow trees were full of roots, but some of them were from trees that were not merely different individuals, but of different species!

6.2   Frugivory and dispersal

Compared with pollination, the process of seed dispersal is more closely associated with the depleting of plant materials. Indeed, the line between 'predation' and dispersal mechanism is hazy in that plants with a small numbers of seeds in any season may lose the whole crop to 'dispersal agents'. Rather than try to disentangle the scores of relationships examined between animals and plants in this area by putting them into categories of predation or dispersal mechanism, an attempt is made to look at the relationships of particular taxonomic groups of animals and their food plants.

The seeds of 50–90% of tropical trees and shrubs are dispersed by vertebrates (58) and much of the vertebrate fauna—up to 80% in Peru for example—is supported by fruit. The seeds of herbs may be dispersed by large mammals, which eat and pass plant material in bulk (59): many such seeds have up until recently been considered to exhibit merely 'gravity' dispersal, when looked at in isolation from the rest of the plant. The leaves of such plants may therefore have 'fruit features', even though the original megafauna has been removed from many tropical and temperate habitats. Such dispersal mechanisms are associated with small seeds, which suffer less damage in the mouth and enjoy a rapid passage through the gut, possibly aided by their laxative qualities as in *Plantago* spp. (Plantaginaceae, long-used medicinally by humans). Moreover, anti-insect toxins may become attractants for mammals as in the case of cabbage and onions in humans.

Fruits are generally protected from 'wrong' animals and microbes, and protect the seeds from climatic extremes and predation (60). During development, they expand from hard, small and inedible to large, soft and edible, often containing a wide range of compounds toxic to a wide range of organisms. As they ripen, the toxicity may fall as in *Passiflora* spp. where the cyanide content declines to 1% of its original strength and oxalates in other fruits may disappear altogether. Indeed, relative to other 'prey', fruits have evolved to be generally accessible, conspicuous and rather digestible when ripe. In frugivore diets in those Neotropical forests examined (61), species with black fruits outnumber those with red and, in Florida and Europe as well as Costa Rica and Peru, the proportion of species with fruits of one or both of these colours is some 62–66%. However, it must be remembered that red and orange are conspicuous to neither insects, nor, probably monkeys, and bats are colourblind (62). A complicating factor, so often seen in evolution, involves the 'cheats', for there are many species with non-nutritious seed appendages but coloured so as to mimic them, e.g. species of *Ormosia* and *Rhynchosia* (Leguminosae). Fruits contrasting in colour with unripe ones, or with bracts, peduncles or calyces are common, and seeds may contrast vividly with their dehisced fruits or be bicoloured: the combination of black with red occurs in 18% of all fruit 'displays' in both

Costa Rica and Peru, whereas red, orange, white and yellow tend to occur alone. Uniformly blue berries that are familiar in certain temperate Ericaceae seem to be particularly associated with the attraction of pheasants, at least in the Malay Peninsula (47), where they are found in *Pollia* spp. (Commelinaceae), *Peliosanthes teta* (Liliaceae) and *Labisia longistylis* (Myrsinaceae) in the undergrowth there.

At one extreme, some plants produce many small, nutritionally poor fruits that attract a wide range of 'poor quality' dispersers, while at the other, some produce a smaller number of large, nutritionally superior fruits, dispersed by a limited number of species. Rarely, though, has it been demonstrated that only one species is associated with dispersal, in contrast to several examples known of one-to-one pollinator-pollinated relationships. On the contrary, the majority of studies of birds, for example, show that a wide range of species visits particular fruiting trees. Even the 'specialized' high-reward fruits of some Lauraceae may have the seeds dispersed by more than 17 species of bird. Such birds may vary widely in morphology and taxonomic affinity. Most studies of this type have yet to evaluate the efficiency of such species as dispersers rather than visitors or feeders, but in cloud forest of Costa Rica, where three pioneer tree species were found to be visited by six bird species (63), only three left seeds in a viable condition, the 'seed shadows', found by radio-tracking, extending up to 500 m from the parent tree. Despite the apparent closeness of the relationship between the three plant species and the three bird ones, this is illusory in that all three plant species are geographically and altitudinally widespread in tropical America, whereas the birds are much more restricted; moreover, the food gained from these plants is a very small proportion of the birds' diet. Indeed, most frugivores are variously omnivorous (60), though oilbirds, manikins, bell-birds and cotingas in the Neotropics may be essentially obligate, while many 'carnivores' often take fruits. When non-exclusive fruit-eating passerines (in Israel) were kept on exclusively fruit diets, they lost weight as the protein digestibility increased: they ingested more energy and protein than was required but digested less, suggesting that the fruits may contain agents that reduce the efficiency of nitrogen metabolism, such that dispersal is improved because birds must move away from the fruit source to seek out insects or other animal proteins to balance their diet (64). Moreover, the effect of avian frugivores on insect frugivores may be to force the latter to eat unripe fruits or leave the host, lest they be eaten by omnivorous birds (65). Again, larvae in fruits may improve the bird diet!

In pollination, the target for pollen is a stigma, which pollinators are encouraged by a number of lures and mechanical devices to brush, whereas the target sites for seeds are particular points suited to germination. In few cases a dispersal agent does deliver a seed directly to such a target other than by chance. Exceptions would be certain Loranthaceous and Viscaceous seeds, which birds remove from their bills by brushing them through bark crevices, or the passage of seeds through an animal to germinate in its dung. Indeed, many seeds will not germinate unless they have undergone such passage or, at least, had the nutritious outer integument of the seed, an aril, removed. Again, the deposition of pollen is timed to suit its germination in that stigmas and anthers are synchronized, whereas the site for germination of a seed may occur randomly in time as well as space in the regenerating gaps of forest. A habitat specificity of a pollinator would improve pollination success; however it can be argued that in a dispersal agent this would lead to increased predation of seedlings, spread of disease and sibling competition, while it must be remembered that a site suited to germination of a particular tree species is no longer suited once an adult is in that place. Furthermore, pollinators are directed to suitable target sites for deposition of pollen by a series of rewards at those sites, whereas dispersers are not and indeed, jettisoning of the 'ballast' from their food as rapidly as possible is to their advantage: food may take 10 to 20 minutes to pass through a bat or small bird, though up to several months in animals like the rhinoceros (60).

A plant could in theory 'specialize' in a few generalist dispersers, just as a frugivore could specialize on a plant dispersed by many other species. In short, there is no theoretical reasoning for supposing that there should be mutual 'co-evolution' of dispersers and dispersed plants. On the contrary, a wide range of dispersal agents would allow a larger number of individuals to strip the tree of fruit at its prime and it would then be distributed to a wider range of habitats and be buffered against any disaster that might overcome the dispersal agent of a species relying on just one such agent. So far as the dispersers are concerned, being generalist would overcome the problems of poor years of any particular plant species, which, in any case, rarely fruit continuously throughout the year. Furthermore, the common dispersal agents of tropical seeds are much larger than the pollination agents: few of them complete a life-cycle that would be compatible with the short flower-to-fruiting cycles of plants, more suited in this respect to close associations with insects, as in pollination. Not surprisingly, therefore, a number of large, principally frugivorous birds undergo marked seasonal movements during the year, in response to local variations in fruit abundance.

6.2.1 *Birds*

With these general views in mind, the relationship between different groups
and their plants will be examined. Tropical bats and other mammals,
particularly primates, and also lizards, may take fleshy fruits from trees, and
other mammals and reptiles may disperse fallen fruits, but birds are most
important overall, and bird dispersal was probably critical in the advance of
the angiosperms. Mammal fruits are often rather dull-coloured but highly
scented, as opposed to the bright colours and relatively little scent of bird-
dispersed fruits. Some frugivorous birds that have been tested (four species
of tanager,which crush fruit) are known to be able to detect differences in
sugar contents of 8, 10 and 12% in their diets and prefer the sweetest, though
it must be pointed out that others like manikins, which swallow fruits
whole, are less sensitive to sugar content. Nevertheless, it is likely that birds
have selected for sweeter fruits through evolutionary time. Fleshy fruits of
the type birds eat are often thought of as primitive in many features, that is
to say that they share more characters with ancestral flowering plant fruits
than do fruits with wind-dispersed seeds. This hypothesis is supported by
the fact that most plant families that produce bird-dispersed seeds are of
very wide distribution and that several of them are held to be primitive, e.g.
Annonaceae, Lauraceae, while several genera conspicuous in this trait are
found in both tropical America and Australasia. Many of the families
involved are represented in temperate countries by species with wind-
dispersed fruits, e.g. the ashes, *Fraxinus*, in Oleaceae, many tropical species
of which have drupes.

   Although the families of plants involved are ancient, they are represented
in tropical rain forest to different extents in different continents. In
particular, the two most important families, Lauraceae and Palmae, are
richly represented in the tropical Americas and Indomalesia including New
Guinea, but poorly so in Africa. For example, the island of Singapore has
more native species of palm than has the whole of the African continent.
This probably explains why Africa is relatively rather poor in specialized
frugivorous birds. The tropical American cotingas (Cotingidae) and
manikins (Pipridae), and the birds of paradise of New Guinea, have no
African counterparts. Because fruits have a limited range of presentation
possibilities and in many respects mimic one another in attracting birds, the
frugivorous avifauna is less diversified than the insectivorous. For example,
in the New World the largely tropical groups, the cotingas and manikins,
are together represented by only 140 species, whereas there are 370 species
of flycatcher (Tyrannidae) and 220 species of antbird (Formicariidae).

Again, the insectivores in closely related groups can coexist, whereas this is rare in the frugivores.

Birds range from the most unspecialized, which regularly take other foods besides fruits and cannot subsist on fruit alone, to the most specialized, absolutely reliant on the fruit. Of these, the unspecialized are typically opportunist, taking fruits as and when they become available. The fruits associated with such dispersers are conspicuous, produced in abundance, and are rather watery, sugary and with little nutritive value, for they have little fat or protein. As with the plants visited by nectarivorous birds, these plants provide a source of energy but not a whole diet. The fruits and seeds tend to be small. These are features of the early successional plants of clearings and forest edge, and, in Europe, would be exemplified by elder, *Sambucus nigra* (Caprifoliaceae), with its small shiny-black sweet fruits produced in abundance. By contrast, the most specialized frugivores are attracted in general to the large-seeded fruits, a common feature of later-successional species. In such plants, the seed is attractive in being covered with a layer of nutritious material. As it must be swallowed, the layer is not too thick, which may account for the common occurrence of relatively thin layers of nutritious material over a single seed or a few seeds in a fruit. Such a layer may be derived from carpellary structures, it may be an aril (a third 'integument' of the seed), or a sarcotesta, which is a fleshy layer less readily removed, sometimes being an elaboration of one of the other integuments. Such fruits have high levels of protein and fat and are not very conspicuous. The specialized frugivore is typically a larger bird than the unspecialized one and may have a very wide gape, but also the capacity to strip the seed and regurgitate the ballast rather rapidly, while the fruit itself must be available all the year round. The specialized frugivores are, in general, foraging for only a short period of the day—in the cases of manikins and cotingas, for only some 10% of the daylight hours. (Such species also have social systems in which males spend almost all their time at display grounds while the females spend almost all of theirs performing nesting duties single-handed.)

In temperate plants, it has been shown that the proportion of two-coloured displays of fruits discovered by birds is greater than that of monocoloured, but that the consumption after discovery is greater when the distinguishing features are 'natural', e.g. representing different stages of ripeness, rather than differences due, for example, to artificial painting (69). The 'modal' fruit in both tropical and temperate regions is a 7–9 mm black or red berry (70), but as there is a greater gape range in tropical birds, it is perhaps not surprising to find a wider range of fruit and seed size there. In temperate regions, there is nothing like the exclusively frugivorous fruit pigeons of the tropics, but many temperate bird species eat little else in autumn and winter. There are generalists in both tropical and temperate regions and, in the diet of some temperate birds, a single plant species can be most important as in the case of the Sardinian warbler, *Sylvia melanocephala*, which is rarely found except near fruiting *Pistacia lentiscus* (Anacardiaceae), but this is highly unusual.

Within the range of a single tree species, there may be very distinctly different assemblages of birds taking the fruits (71–74). In *Casearia corymbosa* (Flacourtiaceae), for instance, there are ecologically distinct groups of birds on the different sides of a mountain range in Costa Rica. Overall, it is now generally held that the interactions between fruiting plants and birds in both tropical and temperate regions are loose, asymmetric, variable in time and space, non-obligate and apparently inefficient. In a moderately seasonal environment, in which several species produce fleshy fruits, there might be expected to be selection for a particular species to fruit out of step with the others, and thereby avoid competition with other species for dispersers. In *Miconia* (Melastomataceae) in an area of forest in Trinidad, a set of species with such staggered fruiting seasons, between them, produced available fruit all the year round but among the bird-dispersed (and insect-pollinated) Lauraceae in the lower montane forests of Costa Rica this was not found, although the trees shared many seed dispersers (as well as flower visitors). Over 7 years, the 22 sympatric species of Lauraceae showed great fluctuations in fruit production and this was apparently unconnected with weather or earlier reproductive behaviour. The fruits make up some 60–80% of all those eaten by some of the bird species, which, in years of scarcity, respond by migrating locally, turning to previously ignored foods or by delaying breeding.

In these forests, the number of bird species feeding on any particular tree was found to be correlated with its size and frequency but the Lauraceae with large fruits obviously have fewer visitors, though the birds with big gapes tend to be generalists and take fruits of all sizes. The chestnut-mandibled toucan, *Ramphastos swainsonii*, the largest toucan species in Central America (75), eats a very wide range of fruits. Small seeds pass through the digestive tract but the large ones are retained in the crop for 5–30 minutes and are then regurgitated. Unlike smaller birds which, though more numerous, go to cover to eat, this bird is 'fearless' and defends the tree against other frugivores and may thus, on two counts, be rather inefficient at promoting dispersal of seeds away from the mother tree. Those destructive birds that 'mash' seeds can take fruits larger than their gape, when compared with those species that swallow fruits, so are less restricted in their feeding than the more benign swallowers. Because of this behaviour and because large seeds may be more easily dropped than small ones, studies of faecal samples in work of this kind will be biased towards small seed sizes.

In Panama (76), 60% of the seeds from the understorey tree, *Guarea glabra* (Meliaceae), were removed by four species of North American migrants, though there was no evidence that any single bird species was dependent on the tree for nutrition, nor the tree reliant on any one species of bird for dispersal. In the north temperate zone, the latitudinal range of fruiting season seems to be associated with the arrival of the southward-shifting, migratory, frugivorous birds such as thrushes. In the Mediterranean, where many such frugivores overwinter, bird fruits tend not to ripen until winter, and show some of the attributes of tropical specialized fruits in that they are drier and more nutritious than northern fruits and can provide a complete diet for the birds. Notable then is the liane, ivy (*Hedera helix*, Araliaceae), the only member of its predominantly tropical family to reach northern Europe, for it produces unusually nutritious fruits that do not fully ripen until spring (it flowers in autumn) when these are fed to nestlings, whose diets are otherwise almost entirely of animal origin.

Another isolated tropical-type plant in northern Europe is the mistletoe, *Viscum album*, the only representative of Viscaceae, a pantropical family. The relationship between the ecologically similar Loranthaceae and their birds—mistletoe birds (Dicaeeidae) of Indomalesia and Australasia, certain small neotropical tanagers and certain small African barbets—seems to be close. The birds are small like the fruits, and have short, rather stout bills, quite unlike the specialized frugivores discussed above. The seeds are embedded in a sticky material undigested by the bird's gut and are either voided through the gut, or, as in mistletoe, the sticky seed is scraped off in crevices in bark by the bird, which has first removed the fruit structures. A similar parallel is seen in the epiphytic cactus, *Rhipsalis*, the only cactus genus to be found outside the Americas, while birds may also be involved in the dispersal of the fruits of other epiphytes such as aroids and bromeliads.

Many birds eat seeds and are often referred to as seed predators. Parrots and most pigeons, for example, are notorious seed consumers. Of the fruit-eating pigeons, obligate frugivores in tropical Australia (77), one group does not take grit and the seeds are passed through the gut while the pericarp is digested, but the brown and white-headed pigeons have a large muscular gizzard typical of columbids and take grit, which means that the seeds are ground up and later digested. An argument for the co-evolution of dispersers and their trees is that they, in contrast, avoid the seeds, though in the past such seeds may have been in some way indigestible, promoting the relationship of today. Indeed, the seeds of such specialized plants as the Lauraceae of Trinidad are not, in general, attacked by parrots. Palms seem to have a mechanical defence in the form of a woody endocarp, familiar as the 'stone' of a date. So far as is known there is no toxic substance in palm fruits, but by contrast, the seeds of Lauraceae contain alkaloids: the avocado pear (*Persea americana*), for example, has a poisonous seed. Seed-predators that hoard seeds may be effective as dispersal agents: in temperate regions, nutcrackers and jays (Corvidae) store large numbers of seeds and exploit them later, and death or 'forgetfulness' of the bird will result in successful dispersal.

Of course, other animals behave similarly. In tropical America, agoutis (*Dasyprocta punctata*), like squirrels, scatter-hoard seeds, including those of the Brazil nut, but there is no evidence that they remember specifically where they have buried them. However, the sites are usually near a buttress or a fallen bole (62). The agouti-dispersed palm, *Astrocaryum standleyanum*, produces seeds that have a much smaller chance of survival if the pericarp around them is not removed and they are not buried. Burying conceals the seeds from other predators, for which they are an important resource in Central America (78–80), while the agouti's peeling off of pericarp before burial may remove any invertebrate larvae therein, before they can get to the seed.

The guts of birds may contain seeds of more than one species of plant at any one time and, in eastern North America the greatest mixing of seeds and consequential passing of seeds in faeces is of seeds from fruits with low nutritional quality and, perhaps not surprisingly, of small rather than large seeds (79). Frugivorous birds, by eating fruits of preferred species in preferred proportions, co-ordinate the dispersal of particular species, so that in the long run, the birds will modify their habitats in terms of the nutritional configuration or the spatial arrangement of the local fruiting plants. This could lead to a mutual reinforcement connecting the dispersal mechanism to the composition of the vegetation (80).

6.2.2 Dispersal by several agents

Most studies have been carried out by watching just one plant species and drawing conclusions from that, though a study by Leighton and

Leighton (81) analysed a whole forest in eastern Borneo. Fruiting in general was found to be markedly seasonal, one dry period being followed by a massive flowering and fruiting. Predator satiation, particularly of primates and some squirrels, which seem to eat almost anything, was notably associated with some trees like many Euphorbiaceae. Of the predators, fruit pigeons were abundant on figs whose seeds are inviable thereafter, but they avoided, as in the Trinidad example, the large fruits, such as those of *Dysoxylum* (Meliaceae), which they would appear to be able to deal with (these fruits are dispersed by hornbills). Because of the seasonality of fruiting, the major question left unanswered is where the dispersal agents come from and where they go to. This has, of course, important implications for the appropriate sizes for nature reserves.

Of all large and medium-sized trees that fruit in the rainy season on Barro Colorado Island, 85% are animal-dispersed and 12% wind-dispersed (82). Of those that fruit in the dry season, 36% are animal-dispersed and 57% wind-dispersed. Of smaller trees, almost 100% of the wet-season fruiters are animal-dispersed as are only 35% of the dry season ones, of which 21% are wind-dispersed. In more seasonal regions, as in West Africa and South America compared with Indomalesia, for example, the percentage of wind-dispersed trees in the canopy is in general higher.

When the flora of fleshy-fruited species in a forest as a whole is examined, as at Cocha Cashu in Amazonian Peru (83), the fruits of the majority of the species (in this case, two-thirds of 258 species, representing a quarter of the total flora and half of the genera with fleshy fruits) seem to fit into one of two categories. Within many genera, there seemed to be little variation, but, within families, there was rarely uniformity, the two classes being: (i) large orange, yellow, brown or green fruits with a tough outer layer; and (ii) small black, white, blue, purple or mixed colours, corresponding to mammal and bird 'syndromes' respectively. Within large genera, however, a range of form is sometimes found. In *Aglaia* (Meliaceae) in the Malay Peninsula, this range seems to be associated with the different chemical constitution of the fruits offered: three species with dehiscent fruits with arillate seeds were taken by at least nine bird species and there was a high lipid content (27–60%); other species taken by primates were lower in lipid but higher in free carbohydrates (reducing sugars and sucrose) and sweet-tasting amino acids. Although only a handful of species in this large genus (the biggest in the family with perhaps 110 species) were examined, and only a single sample from each, while other nutrients such as minerals and vitamins were not assessed, the results are rather indicative of a pair of chemomorphological dispersal 'syndromes' within the genus.

Some plants, however, are known to be efficiently dispersed by two quite distinct types of agent, e.g. the fruits of *Faramea occidentalis* (Rubiaceae) on Barro Colorado Island (85) are eaten by a range of mammals and birds, but dispersal is said to be principally effected by howler and white-faced monkeys and by crested guans. At Los Tuxtlas in southern Mexico, the females of the pioneer, *Cecropia obtusifolia*, more or less continuously produce fruits and 48 species of animal use them: one insect, one reptile, 33 birds (both residents and migrants) and 13 mammals (flying, arboreal and terrestrial). Seeds from the faeces of three species of mammal were subjected to various light treatments and it was found that passage through two of them reduced the effectiveness with which exposure to far-red light overcame the effects of previous exposure to red light (86, 87). The pioneer *C. glazioui* of the coastal rain forest of south-east Brazil (88), is visited by three tanager species by day and three species of fruit bat by night, all in search of fruit. The minute hard fruits are ingested with the fleshy perianth and later dispersed through defaecation, though only 32% of those passed through birds would germinate compared with 78% from the bats. Fruit bats in Malesia press small-seeded fruits against their palates and drop the dried dross including the seeds. In this way figs, *Muntingia calabura* (introduced from the West Indies, but fruiting all year round) and species of *Melastoma* are dispersed. The exotic *Piper aduncum* from South America has probably recently been spread around Kuala Lumpur, Malaysia, in this way. Of species dispersed by birds and primates, the latter may sometimes surpass the birds in spreading seeds. For instance, in *Stemmadenia donnell-smithii* (Apocynaceae), spider-monkeys may excel over parrots and other birds in this respect (89) in Guatemala.

Fruiting *Dipteryx panamensis* (Leguminosae) trees on Barro Colorado Island are visited by 16 mammal species (90), taking pericarp or seeds. There is predation by peccaries and rodents, which eat seeds (most of the others eating the pericarp and leaving the seeds under the trees): occasionally primates, coatis, kinkajous and so on disperse seeds but only the bat, *Artibeus lituratus*, the agouti and the squirrel *Sciurus granatensis* disperse large numbers, the last two being seed predators most of the time, but sometimes dropping or scatter hoarding seeds. In a study of the Panamanian rain forest tree, *Tetragastris panamensis* (Burseraceae), visited by a wide range of animals, it was found that three mammal species accounted for over 97% of the seeds removed from the tree (91), the howler monkey (*Allouatta palliata*) alone being responsible for 74%. A number of other mammals and birds probably also disperse a very small proportion of the seeds, while two parrot species are the important seed predators among several mammals and birds that destroy seeds. From 19 study trees in one season, fewer than 4% of the total of more than 430 000 seeds produced had a chance of establishing, for 6% were killed by the parrots, 66% fell under the mother trees and 24% germinated in close competitive clumps in faeces. Even lower percentages are estimated for *Shorea ovalis* (Dipterocarpaceae) in Malesia, where there was 83% abortion, possibly due to the failure of pollination, while 90% of the survivors were killed by a single insect predator species before being dropped. Sixteen per cent of the survivors on the ground were then killed by three other insect species and although mortality declined with distance from the mother tree, it was not due to host-specific predation (92). In a study of a forest as a whole, the Budongo Forest in Uganda, it was found that 40% of all seeds were eaten by rodents before and immediately after germination and about a further 30% were killed in 2 years by browsing antelope, while there were further losses from seed rot, insect and fungal attack and drought. The result was some 2% survival of the seeds as saplings in 2 years.

Compared with most of the species mentioned so far, many figs seem to produce fruits almost all year round and are the most important foods of many specialized frugivores in Africa, Malesia and Australasia—the siamang (*Hylobates syndactylus*) in Malaya spends a quarter of its feeding time in fig trees. A *Ficus* liane in southern Cameroun was visited by one species of fruit pigeon, four of hornbill and three of greenbul, two of monkeys and two of squirrels over 9 days; the greenbuls alone ate 17 332 syconia (93). Their gizzards are gentle when compared with those of fruit pigeons, which generally destroy the seeds.

In the New World, bird species are less restricted to fig species (94), even if fruit is available continuously, as was found for individuals of two *Ficus* species in Panama, which, between them, produced fruit all year round (95). In both these species, though, the major fruiting peak occurred when most other species were not in fruit, and it is suggested that the two species fruit asynchronously at relatively short intervals. This would seem to be linked with the capacity to provide egg-laying sites all the year round for the obligate fig-wasp pollinators, as well as to the year-round provision of food for the specialized dispersal agents such as howler monkeys and bats, which may travel long distances, in the case of *Artibeus jamaicensis* in Mexico up to 10 km, to feed on figs (96). Janzen (97) calculated that in a fig 'seed' shadow in Costa Rica created by bats depositing 'splats' at the density of some 10 per m^2 over 2500 m^2, there were 367 500 fig seeds dispersed. By contrast, wind-dispersed species in a Mexican study normally fruited during the dry season, producing a large amount of seeds, which were simultaneously dispersed on days with low relative humidities. Such species thus give rise to wide and uniform seed shadows.

Megachiroptera are known to feed upon fruit of 145 plant genera, often secondary species and usually with sweet rather than oily fruits (98, 99). Large fruits like mango and papaya are consumed *in situ* and the smaller ones taken away, especially if the bat is disturbed. The large *Pteropus vampyrus* (weight 800 g) can carry up to 200 g, *Pteropus* travelling up to 50 km to forage. Nevertheless, the plants are dispersed by a range of other agents including primates and even water.

In forests with very seasonal production of fruits and seeds, as at Cocha Cashu, there are periods of scarcity when frugivores turn to a limited range of food sources, sometimes referred to as 'keystone' species. At Cocha Cashu, 12 species provided 80% of the animal biomass with food and it is argued that the carrying capacity of the forest is governed by the abundance of these species, the most important of them being figs.

6.2.3   Some marsupials and primates

In a secondary forest near Cayenne, French Guyana (100), five of the six nocturnal mammalian frugivores (effective dispersers rather than predators) were found to be marsupials: 13 of the 26 plant species with a major role as food sources for nocturnal mammals made up approximately half of the total basal area of the forest and 25% of the number of trees there. The trees were found to exhibit three types of phenology: (i) low fruit production over a long time (usually pioneers); (ii) synchronous but irregular fruiting (associated with predator satiation of rodents, the seeds being larger than those in (i); and (iii) synchronous regular cycles with successive and partly overlapping fruit production (providing a year-round supply for the frugivores). The lean period at the beginning of the dry season with few available fruit resources is tolerated as the marsupials store fat.

New World primates are red-green colourblind and spider- monkeys eat the seeds of *Virola* spp. (Myristicaceae) corresponding in their colours to a 'bird syndrome'; *V. surinamensis* is dispersed by both birds and water (62). Of capuchin monkeys in Amazonia, brown capuchins with heavy jaws can open palm fruits and can survive the dry season on them, though the white-fronted capuchins cannot, such that in the dry season, 90% of their food is figs. The figs are widely scattered in the forests, however, so a troop ranges over much more ground than does one of brown capuchins (101). The effect of passage through the primate gut has been examined in the tamarins in Amazonian Peru (102), where it was found that germination success rates were about 70%. The seeds tended to be large and heavy and passed through the gut in 1–3 hours, the passage time inversely correlated with specific gravity of the seeds and had an indirect effect on the distance that seeds were dispersed away from the mother tree. Of the 'preferred' fruit trees, the distributions of the adult trees of three of them closely resembled the seed shadows made by the tamarins in a way not unlike that of the birds mentioned at the end of section 6.2.1.

Howler monkeys can detoxify several chemicals that many other New World monkeys cannot: unlike others, they can also eat unripe fruits. Generally, unripe fruits are rendered unpalatable by the tannins that precipitate salivary mucoproteins and cause astringency. In ripening (commercial) fruits, the tannins form large complexes with pectins and the astringency is lost (103). In the Old World, chimpanzees can deal with unripe fruits, breaking open tannin-rich mature fruits or eating unripe ones. Howlers on Barro Colorado Island eat large quantities of fruit, notably figs, though different *Ficus* species have different amino-acid contents in the protein fraction of fruits and leaves (40% of their intake is leaves). At Los Tuxtlas, howlers spend half their time collecting fruits of 19 species with a preference for those of certain Moraceae and Lauraceae. Ninety per cent of the seeds in their faeces belonged to just eight species of tree and one liane, the rest to 15 other liane species. The germination success, compared with a control with 35% was about 60% for seeds from the faeces. Nevertheless, besides dispersing some species, howlers waste a large amount of fruit and seed material.

Figs are also important in Old World primates, troops of siamang taking up to 13 kg of small ones within 10 days in the Malay Peninsula. Although the orang-utan of Sumatra and Borneo (53, 54) is primarily a fruit-eater, it takes a wide variety of foods including leaves, young shoots, flowers, pith and bark and occasionally mineral-rich soil, insects and possibly small vertebrates and birds eggs. Plants in some 34 families were used in Borneo over a period of 16 months and all but 5 were sources of fruit. The apes fed on a wide range of foods at all times of the year, but when fruit was scarce, they spent more time travelling and turned to less nutritious plant materials as well. Of the 28 chief fruit species taken, 18 were rare (fewer than two trees per ha), six occasional (two to four per ha) and four had four to eight per ha. Individuals of a few species like durians provide food for several orang-utan meals, but most are cleared by one animal in a single meal. In rainy weather, they sometimes concentrate on the less nutritious foods rather than search for more fruit, but in fruiting trees seem not to pay much attention to other frugivores though they eat most, and that wastefully. Because of the scattered distribution of food, the orang-utan population is split up into small dispersed foraging units and the flexible nature of orang-utan society allows them to exploit irregular fruit distribution better than territorial monkeys or gibbons, for example. The monkeys live largely on leafy shoots and eat fruit when it is available, while the gibbons have become largely specialized for diets of figs. Furthermore, the orang-utan can tolerate a wide range of bitter and sour fruits and high alcohol levels and its strong hands allow it to open spiny or hard fruits like durians and legumes.

The Sumatran orang-utan seems to be an important dispersal agent for some *Aglaia* spp. (Meliaceae) but durian seeds are so damaged that ground-walking animals are thought to be the principal effective agents—possibly sun-bears, or even tigers. Man has also been suggested—it has also been pointed out that the fruits favoured by him and the ape are the same. Certainly the absence of commercially significant rain-forest fruits in the New World, compared with the Old, is because the primate fauna in the Old World has flavour preferences nearest to those of humans. Possibly, then, the orang-utan has been excluded from the ground through competition with man, and has become arboreal, perhaps also as a result of being hunted for 30 000 years by its competitor.

6.2.4   *Fish and ants, wind and water*

Dispersal agents not mentioned so far include fish. Some seeds of rheophytic plants are fish-dispersed and fish dispersal gets round the perennial problem of how plants can move 'upstream'. One such rheophyte is *Dysoxylum angustifolium* in the Pahang River basin in West Malaysia. In the same family (Meliaceae), in the allied genus *Guarea* in South America, *G. guidonia* is fish-dispersed. This mechanism seems to have arisen in parallel in these plants. Growing in the same river as *D. angustifolium*, but in other places as well, is another rheophyte in the same family, *Aglaia yzermannii*, which may also be fish-dispersed. In the case of the *Dysoxylum*, the fish flesh is rendered rather unpalatable by substances from the seeds. The indigenous people avoid these fish during the tree's fruiting season and, if other fish predators do the same, the tree could be said to be operating a Securicor system for its offspring. In the peat-swamp forest of Sarawak, the commercially important *Gonystylus bancanus* (Thymelaeaceae) is distributed by a small catfish, the strange flavour of whose flesh at different times of the year is attributed to the tree's arils.

There are probably well over 100 000 km² of flooded forest in Amazonia (105) with rather predictable seasonal floods, with periods of inundation in different places lasting up to 6 months. Some characin fish there are seed predators as they crush seeds, which are not destroyed in the gut itself. There are 30 predator species but many so-called predators have been found with whole seeds in their guts, e.g. *Colossoma brachypomum*, which can reach 20 kg in weight, with seeds of *Astrocaryum jauary* (Palmae). A given species or even genus tends to specialize in that at the same time in the same area, the same plant species is exploited, and this takes a great toll of water-dispersed seeds such as rubber. As dispersers in these forests, fish are probably more important than any other animal group, because of their larger biomass and feeding at night, as well as in the day. The important groups are the frugivorous characins and catfishes, though even piscivores have been found with viable seeds, perhaps eaten by the herbivore prey in the first instance. Fleshy fruits, including figs are taken and over 100 species are known to be dispersed in this way. Many of the fish species migrate from one river system to another: they can travel 20–30 km upstream in a single day and seeds can remain in their guts for 1–7 days. Characins migrate only upstream, but their young are displaced downstream once more; they are less good as dispersers than catfish, however, as the latter have weaker jaws. In these nutrient-poor waters, the fishes may be reliant on such vegetable input and, in turn, the homogeneity of the inundated forest of Amazonia may be related to this fishy connection.

Invertebrates are less conspicuously involved in seed dispersal, but there are many examples of ant-dispersed seeds in temperate countries, e.g. species of *Carex* (Cyperaceae), *Ulex* (Leguminosae), and *Viola*, whose seeds are attractive because of oil bodies, often rudimentary arils. Carunculate seeds are perceived by ants, which see in ultraviolet light, because the caruncle reflects ultra-violet light, the fluorescent compound probably being a phenolic (106). A recently reported tropical example (107) is of two arrowroot relations, *Calathea* spp. (Marantaceae) in the Mexican rain forests. Of the 21 species of ant attracted to the arillate seeds, there are predatory *carnivorous* ponerines, which bear off the seeds, like animal prey, to the nest, where the nutritious arils are removed. The aril-less seeds have higher rates of germination than the undivested ones and the distribution of seedlings matches the ants' behaviour patterns.

The hemi-epiphytic *Ficus microcarpa* of Malesia (108) has a caruncle around the fruit inside the syconium; when it is removed, the fruit is less likely to be ant-transported and those with it removed had a higher germination rate. Remarkably, the primary dispersal is by vertebrates, in whose guts the caruncles are not destroyed. After primary dispersal by a range of vertebrates in a dry-seasonal forest in Costa Rica, the fruits of *F.*

hondurensis are known to be re-arranged by several ant species. Ants may hinder dispersal, however, as in the example of the aggressive arboreal weaver ant, *Oecophylla longinoda*, which reduces the dispersal of *F. capensis* syconia in West Africa by deterring night visitors such as bats (109).

The plumed seeds of *Dischidia parvifolia* (Asclepiadaceae) are wind-dispersed but, if they land in the vicinity of ant nests, they are gathered and taken into tunnels, where they germinate and their roots gain access to nutrients in the insect frass (47). In Amazonia, taxonomically diverse groups of angiosperms grow on the carton nests of ants, forming arboreal 'ant-gardens', which may be up to 1 metre across and 2 m long. Several of the plant species have extra-floral nectaries but almost all the plants can be found growing away from the ant nests and the ants can form nests without the plants so that, despite the apparent advantages to both plants and animals, the relationship is not obligatory, though *Codonanthe* spp. are not found away from the nests. The seeds of the epiphytes have arils, while those of *Codonanthe* seeds, unlike all other New World Gesneriaceae, closely resemble ant pupae. The seeds are brought to the nest, though the gardens are sometimes colonized by other plants, e.g. *Anthurium* sp. (Araceae), which offers nothing to the ants (110). The seeds of most of the plants have a smell of vanilla; nine of the 10 epiphyte species examined in Amazonia had the same volatile oil and a second, unrelated, compound in six of them. Compounds similar to these are known to retard fungal and bacterial growth. They thus protect the ants against nest pathogens in a way similar to that of the fungistatic resins collected by bees for their nests (111).

Other dispersal mechanisms on the forest floor may include rat dispersal, which has been claimed for *Curculigo* spp. (Liliaceae) in the Malay Peninsula (47), while, on Fanning Island in the central Pacific, *Pandanus tectorius* is effectively dispersed by a land crab, *Cardisoma carnifex*, which moves fruits and discards the seeds (112). Ocean currents are important for other coastal plants such as some mangroves (though none of these has reached the isolated archipelago of Hawaii), while seeds of the forest lianes, *Entada* (Leguminosae), are washed even as far as Europe. In Australasia (113), seeds such as these germinated in North Island, New Zealand, after some time in the sea and 21 months in soil.

In the Amazon, there are many water-dispersed palms and legumes (62): propagules of *Montrichardia arborescens*, a characteristic aroid, and even rubber, stay afloat for 2 months. Again, in Amazonia, there are some water-ballistic understorey herbs, their seeds being bounced by raindrops, e.g. *Monolena* spp. (Melastomataceae), while in the Malay Peninsula (47), raindrops run down the longitudinally dehisced capsules of *Didymocarpus* spp. (Gesneriaceae) carrying away the minute seeds. Along the margins of the immense rivers of Amazonia, there takes place wind-dispersal with alate drupes of *Triplaris surinamensis* (Polygonaceae), winged seeds in *Couratari* spp. (Lecythidaceae), plumed ones in *Salix martiana* and *S. chilensis*, and seeds in carpel hairs as in *Ceiba pentandra*. Storms can disperse seeds and fruits via high altitudes as in dipterocarps in the Malay Peninsula (114), where one recorded storm deposited 3–4 per m^2 possibly over several km^2. Lianes tend to have a higher percentage of wind-dispersed species than do trees, though hemi-epiphytic lianes tend to be bird-dispersed (115). The rarity of lianes on truly oceanic islands, e.g. Mauritius and Galapagos Islands (one or two only), can be explained by wind-dispersal not being efficient enough to get them to islands. However, wind-dispersal is important in orchids, where air occupies much of the space within the testa (116), three orchids reaching Krakatau after its 1883 eruption, by 1896, 25 by 1933.

Air-dispersal probably accounts for the distributions of some liverworts, though many spores are not long-lived in a desiccating atmosphere, but the presence of certain hepatics on Ascension Island, St Helena and Tristan da Cunha (117) seems to confirm it. Within a climatological belt, the spread of bryophytes with spores smaller than 25 μm is probably by wind though birds and insects may be important over short distances (118). Vascular-arbuscular mycorrhizae are spread as spores by wind, water and animals both invertebrate and vertebrate, the longest distances perhaps being achieved on birds' feet, as well as by the more familiar root contact and infected plant material (119). However, for ectomycorrhizal fungi in southwest Australia at least (120), dispersal is by marsupials and germination of spores is unsuccessful unless they have passed through a digestive tract. In northern forests, the dispersers of e.g. Fagaceae also eat ectomycorrhizal fungi and seeds planted by them also have ectomycorrhizae. In mast fruiting, there are increases in dispersers in anticipation—pigs (flowers) or jays (green cones) but pigs also take fungi, e.g. truffles, which exude sex pheromones (119). In northern forests, the formation of most sporocarps is associated with fruiting generally and mast-fruiting in particular. Perhaps the mycorrhizae direct the pigs in their spectacular migrations in Borneo.

6.2.5 *'Anachronisms'*

Although parrots have been treated as seed predators, they are not always so, for they appear to be the principal dispersal agents of seeds of several species of *Parkia* (Leguminosae) in the Neotropics (121), as well as certain

Lecythidaceae there, even though they are partly destructive. It may well be that the gum that surrounds the seeds of these species of *Parkia* and not of the seeds of the Palaeotropical ones is important here. Some other *Parkia* species are dispersed by agoutis and it is possible that others may have been adapted to dispersal by large terrestrial herbivores that are now extinct. About 10 000 years ago (122), over 15 genera of Central American large herbivores became extinct so that the megafauna, previously comparable with that of Africa before the decimations caused by man, largely disappeared. Those that became extinct included gomphotheres (mastodon-like proboscidians), ground sloths and equids. The plants that these animals dispersed may thus be seen as vegetable anachronisms and the introduction of horses and cattle in historical times seems to have locally restored the ranges of such trees as jicaro (*Crescentia alata*, Bignoniaceae) and guanacaste (*Enterolobium cyclocarpum*, Leguminosae), which may fall in this category. The absence from tropical Australia of plants whose dispersal is, or was, associated with large mammals may explain why few, if any, indigenous fruit species there are worth developing for horticulture. However, in southern Australia, there are some southern rain-forest species still with large diaspores, and fossil evidence suggests that the extinct cassowary relations, the Dromornithidae, were there in the Pleistocene. These large birds could have been the original dispersal agents (123).

Of the extant 'megafauna', the smallest of the five living species of rhinoceros, the Sumatran rhinoceros, *Dicerorhinus sumatrensis*, eats wild mangoes and other fruits as well as browsing many gap species (124). In lowland Nepal (125), the greater one-horned Asian rhinoceros (*Rhinoceros unicornis*) takes the large hard and dull fruits of *Trewia nudiflora*, (Euphorbiaceae), which are unattractive to birds, bats or monkeys. The fruits are an important food source and the seeds remain in the gut for 3–7 days. Such treatment does not improve germination (they are protected by a hard endocarp during their animal passage), but seedlings grow up in dung piles in clearings: the species is shade-intolerant. The tree regenerates elsewhere, where there are bison and wild cattle or even domestic buffaloes and cattle.

Such anachronisms are likely to be common in those regions from which the large forest animals have been lately removed and may, in any case, be a common feature in evolution (126). Some of the woody lobelioids of Hawaii have thorns and toxins, apparently noxious to the type of herbivores not found in these islands, but possibly on the mainland where their ancestors came from (127). Surprisingly, Hawaii has plants with extrafloral nectaries, though no ants are native there (128). In a national park, 33 such species were found: one endemic (*Acacia koa*, Leguminosae) though six other indigenous species were pantropical coastal plants, but three others with such nectaries elsewhere in their range did not have them in the Hawaiian populations. In the archipelago as a whole 11 of the 1394 endemics have these nectaries as do a number of indigenous non-endemics. No other relationship with other insects has yet been demonstrated, but much of the fauna may be lost, in which case there is some short-term 'phylogenetic inertia'. If this is not the explanation, then the inertia is even greater, the nectaries being relict structures from the ancestral colonizing plants.

The Hawaiian endemic bird pollinators of *Freycinetia arborea* (Pandanaceae) have died out, but the plant's pollen is still to be found in the head feathers of museum specimens (129). Pollination is now effected by a Japanese white-eye, *Zosterops japonica*, introduced in 1929. Another Pacific anachronism associated with birds is the tree *Pisonia grandis* (Nyctaginaceae), found from Madagascar to Polynesia, but on Niue Island is in decline as it is guano dependent and sea birds are decreasing in numbers there (130).

Dispersal agents must have changed with time as in long-lived taxa such as *Ginkgo biloba* (Ginkgoaceae) and Podocarpaceae (60), which were probably dispersed by long-extinct groups of reptiles, while in the largest genus of plants with fleshy fruits in New Zealand, *Coprosma* (Rubiaceae), comprising 29% of all such species there and 50% of the species of fleshy-fruited shrubs, 52 species are dispersed by birds, bats, lizards and formerly the moas, now extinct (131). As with the Japanese white-eye pollination on Hawaii, seed dispersal on Three Kings Island, New Zealand, is now effected by introduced bird species.

Another example of a dispersal anachronism is the extraordinary palm, *Lodoicea maldivica*, now restricted to small areas of the Seychelles. This tree is familiar to botanists because it has the largest seed known, as well as one of the largest leaves. The leaves of palms in general are often used for thatching and are relatively durable compared with most tropical foliage. The leaf litter of this remarkable plant is therefore bulky and relatively persistent. A single fallen leaf can smother a wide area of seedlings and it is perhaps no surprise then to learn that it has such a large seed, which could not only penetrate this mass but have enough stores to supply a seedling as it grew up through it. Nevertheless, these large seeds are rather unlikely to be moved uphill by dispersal agents, particularly as there are no large animals on the island where the palm grows. It seems then (possibly retarded by a cotyledonary stalk up to 3 m long), to be slowly moving downhill as a species. The seeds are not resistant to salt water and immersion kills them. In short, this species is 'adapted' to going extinct. How did it reach the Seychelles? A clue may be found in the geology of those islands, for some of them are granitic and part of the Gondwanaland, which linked India and Africa when they were adjacent. Is it then a relic of the greatest antiquity (132)?

6.2.6 Specificity

Where there is great specificity in seed-dispersal, a state of affairs that seems in general very rare, the disappearance of the dispersal agent is the road to extinction for the plant. Elephants, which lack colour vision but have a strong sense of smell and favour large, smelly, dull fruits (133, 134), are the only animals able to swallow the pyrene of *Panda oleosa* (Pandaceae) and collected seeds do not germinate unless passed through an elephant, in whose dung they germinate. The seed content of droppings throughout 1 year in a rain forest in Ivory Coast comprised some 37 species of trees and five of herbs. Of the trees, only seven are known to be dispersed by monkeys or birds as well. Some 30% of the trees whose dispersal agent is known in these forests are dispersed by elephants. It may well be that other such cases as the *Panda* represent extreme forms of the 'anachronisms' and that, formerly reliant on a number of dispersal agents, only one of these is now left. Again, the flightless cassowaries to 1.5 m tall survive in Australia (one species), New Guinea and the neighbouring islands. In northern Queensland (123), their dung heaps (up to 1 kg each!) were examined and found to contain 78 species of seeds, of which 70 were viable. Although passage through the gut did not improve germination rates, it must be remembered that the birds are the only frugivores large enough to deal with some of the fruits, e.g. *Casuarius casuarius* in Queensland taking *Beilschmiedia* (Lauraceae) fruits and passing the seeds more or less intact even though they are 6 cm diameter and weigh 52g. A celebrated, final, example is that of *Sideroxylon* (*Calvaria*) *sessiliflorum* (Sapotaceae), only geriatric trees of which could be found in Mauritius, to which island it is restricted. An inspired guess was made that the failure of the seeds to germinate was because its seeds had not passed through the gut of the dodo (*Raphus cucullatus*), extinct since 1681. This led to forced feedings of turkeys, and, after being voided, resultant germination of seeds of the tree which was on the verge of extinction, though this may have been exacerbated by the depredations of introduced monkeys, which take unripe fruits (135).

By contrast, the flexibility of plants (as in the examples of the *Freycinetia* and the Three Kings Island flora discussed above) in terms of dispersal agents can be seen in two Neotropical studies. In a broad-scale study (136) of cloud forests, it was concluded that random colonization had played an important part in determining the species compositions of different ones and that the predominance of widespread species of successional habitats elsewhere reflected a high rate of generalist dispersal. In the second (137), three submontane tree species common to Trinidad and Tobago were examined, the assemblages of frugivores in the two islands being very different, with passerine dispersers predominating in Trinidad, parrots and other predators on Tobago. On Tobago, the fruits of the tree species had significantly more thickened layers and higher frequencies of multiple seeds and so on. Selection by the different avifaunas seems to have generated these infraspecific differences and the tree species survive in both islands. Although not immediately apparent in studies of 'variable' plant species, it must be expected, in view of the non-uniform disperser fauna in tropical regions, that wide-ranging species will probably be shown to have such patterns. Nevertheless, such patterns reflect not only the present fauna, but in the cases of long-lived trees, the fauna of the past. In short, in the context of the modern fauna, many trees may appear to represent the 'living past'.

6.3 Florivory and pollination

In the previous chapter, it was noted that some groups of trees employed in turn the same species of pollinating insects, suggesting that within certain constraints these were generalist in their flower-visiting. This contrasts with the often-repeated one-to-one association of certain orchids and their insects and of the hundreds of species of figs and theirs. How closely knit are pollinating animals and plants in rain forest? In the wet forests of Costa Rica (138, 139) it was found that some 90% of tree species were insect-pollinated, approximately the same percentage as in dry forest, so the insects will be considered first.

6.3.1 Insect pollination (entomophily)

For insects, plants represent humid but well-drained surfaces and increase the amount and diversity of living space while for some, they represent food to browse, or a repository to bore into in egg-laying. Early pollinators were attracted by pollen and probably had grinding mandibles: they also took ovules and developing seeds (140). They were probably derived from predominantly predacious groups adapted to seeking pollen, which is high in protein, and the foraging behaviour of modern predacious insects and pollinators is similar. Pollen and ovule eating may have begun in the Carboniferous, the first 'modern' group with such mandibles, the Coleoptera, appearing in the Permian, with Diptera and Hymenoptera in the Triassic, and Lepidoptera in the Lower Cretaceous. There appear to have been trends in the fossil gymnosperms leading to increased protection of the ovules. By the Cretaceous, insect damage is very pronounced as in the example of hundreds of cycad cones that have been examined: 22.2% of them had damage especially noticeable in those with mature pollen-bearing organs.

Many different groups are involved in the pollination of modern angiosperms and, in rain forest (138, 139), this seems to be rather bound up with the level in the forest at which the flowers are presented. In Costa Rica, the large and medium-sized trees are associated with generalist insects and not bats or birds, the almost exclusively tropical pollinators. Only 3% of species there were visited by bats: a similar percentage was recorded for wind-pollinated trees. Hummingbirds and hawkmoths are largely associated with trees producing flowers near the forest floor, where noctuid moths are found to be important pollinators in the smaller species of Meliaceae and butterflies in such families as Rubiaceae. The scent-sensitive Lepidoptera are thus associated with the strongly-scented flowers in the relative calm there. Almost all the insects examined are generalists, particularly small bees that visit a wide range of the trees with rather small 'unspecialized' flowers. The stratification could be summarized as follows: large bees in the emergents, smaller bees in the lower reaches with other generalist insects, and near the forest floor everything, including wind, as pollinator.

In West Malaysia (141), there seems to be a clear distinction between the pollinators above and those below the canopy. Below, the flowers have minimal visual lures and are produced over extended periods, or even continuously, while others are strongly scented and produced more periodically. In the canopy, the flowers are larger and more conspicuous, as in almost all of the emergents. The pollinators in the lower regions are non-specific and include meliponid bees, solitary wasps and butterflies and also unspecialized beetles, midges, flies and thrips, all rather short-range pollinators. Most of these are also found in the upper canopy, although that is dominated by the wide-ranging bee, *Apis dorsata*, with *Xylocopa* bees also conspicuous. Sunbirds and bats, which live outside the forest, also work the flowers at this level. *Xylocopa* and the sunbirds are mostly found in the forest fringes, foraging on gap-phase plants though they move to the mature forest trees during flowering periods. In the lower reaches of the forest, then, there is a constant but low level of presentation of a few flowers dealt with by low but more or less constant densities of animals, whereas in the upper layers, the flowers are more conspicuously advertised and draw such animals as birds, which are wider-ranging and with good sight, as well as the local, more stationary pollinators from lower in the forest.

The trees of the lower reaches are thus generalized in their attractions while trees like the dipterocarps get the local fauna and the bee *Apis dorsata*. Some trees such as Polygalaceae, two subfamilies of Leguminosae, *Papilionoideae* and *Caesalpinioideae*, have floral structures restricting access to the nectar or pollen to *Xylocopa* bees, a mechanism familiar enough in the temperate representatives of these families. Some of the dipterocarps, as was mentioned in the last chapter, are visited by thrips though these trees are strongly self-incompatible. Other ways in which outcrossing may result include the consequences of the aggressive behaviour of some of the pollinating bees, which may force other pollinators to withdraw from a particular tree and seek another, though short-distance pollinators are often territorial and may see off new arrivals, thus reducing crossing (142). In mass-flowering, the thrips with their rapid breeding cycles and buildup of numbers can form a very high percentage of the pollinators. Possibly such mechanisms can be used to explain where the pollinators come from in species-poor forests such as the *Shorea albida* forest in Sarawak when the dominants flower.

Despite this apparent stratification, certain groups of plants are very restricted in their range of pollinators. Palms, which may vary in size from emergents in the montane forest of the Andes to treelets 30cm tall in the Madagascan rain forest, are thought to be pollinated largely by beetles and sometimes bees (138, 139), and not, as was formerly believed, entirely by wind. Indeed, many trees that look as though they might be adapted to wind pollination are known to be pollinated by insects—like *Mallotus oppositifolius* (Euphorbiaceae) in West Africa, which has an attractive sweet scent—and it has suspected that even forest grasses are animal pollinated.

Examination of particular species seems to show elaborate systems for enhancing cross-pollination. Many tree species are dioecious, although sometimes the rudiments of the opposite sex are found in the flowers. Such is the case in the majority of the Meliaceae, for example. The dioecious *Xerospermum noronhianum* ('*X intermedium*', Sapindaceae) in West Malaysia (143) has a prolonged sequential development of flowers on each inflorescence as well as the sequential development of the inflorescences themselves, making a small number of flowers available at any one time but ensuring economy by not producing excess flowers within a short period (these would not be pollinated owing to the low density of pollinators). The major pollinators are short-distance generalists and the fidelity is maintained by the simultaneous flowering of trees of both sexes within close proximity. Furthermore, it was observed that female trees flowered less frequently: 22% of the known females in a 5-year study flowered more than twice, compared with 35% of the males. This could be argued as promoting the crossing between different parent trees in successive seasons. In Costa Rica, 333 of the approximately 400 tree species at La Selva were examined and 65.5% were found to be hermaphroditic, 11.4% monoecious and 23.1% dioecious; of 28 species examined in detail, 24 showed self-incompatibility, either in the stigma or style in distylous species, in the ovary in most of the monomorphic ones (144).

Insects collect pollen and, in the case of, for example, beetles, solid food bodies in or about the flowers. In a species of the palm genus, *Bactris*, the fleshy petals and sepals are eaten, while the fleshy bracts of a species of *Freycinetia* (Pandanaceae) are eaten by birds, which are thought to pollinate it. Insects also collect nectar as a 'liquid reward' (145, 146) a category that may include stigmatic secretions; these may be involved in insect nutrition as well as in pollen-tube germination. The extent of nectar production is apparently associated with pollinator size and there is variation in its sugar content and dilution. In dioecious species, the sugar proportions are different in male and female flowers.

Humming-bird flowers have high sucrose/hexose ratios, while passerine (sunbirds, white-eyes, honeyeaters and honeycreepers) flowers have a lower ratio—these birds also take fruit-juices, which are highly hexose-dominated. Hexose-dominance is found in bat flowers in the Americas, though is perhaps less obvious in the Old World: again there is a similarity with fruit sugars. Moth flowers are sucrose-rich and narrow-tubed butterfly flowers are similar but 'bee and butterfly' flowers with shallow tubes are hexose-rich. Bees may get sucrose- or hexose-dominated nectars, but this is complicated by the fact that some plant families have either sucrose- or hexose-dominated nectars. The attraction for the bees may be conditioned or due to 'taste' but it must be noted that nectars contain things besides sugars, particularly lipids, amino acids and proteins as well as alkaloids, non-protein amino acids and phenols. An example showing that such categorizations and generalizations are perhaps unwise is that of *Inga vera* var *spuria* (Leguminosae), which is first visited by moths and humming birds (high sucrose), but 4 hours later produces a sour odour attractive to bats, perhaps due to fermentation by micro-organisms, the resultant diminution of sucrose making the ratio suitable for the bats. Several Passifloraceae present pollen and nectar at different times of day and share the same pollinators (147), while certain nectar-less Bignoniaceous lianes mimic nectar producers, by flowering at the same time as, or just after, those that are pollinated. Mistletoe flowers appear to mimic nutritionally rich fruits, the fruits of the previous season being more or less contemporary with the new flowers. Again, *Epidendrum ibaguense* (Orchidaceae) produces nectar-less flowers all year round in Panama, apparently mimicking the flowers of *Lantana camara* (Verbenaceae) and *Asclepias curassavica* and attracting Monarch butterfly pollinators (148). On the other hand, the 'robbing' of nectar, familiar in temperate plants is also prevalent in the tropics and bees may attack bird-pollinated flowers.

Although it has been argued that nectar might contain repellents to ants that otherwise predominate on sugary substrates in the tropics, it is found that ants will take this nectar when it is presented to them, though the floral parts of such flowers may be less palatable (149). Indeed, pollen exposed to ants for a short time has reduced germinability and shorter pollen-tubes: such pollen leads to lower seed set. It has been suggested that in their nest-building and brood-rearing habits, ants secrete large amounts of antibiotics, which check pathogens and it is these that affect pollen, while bees and wasps do not, so it is not surprising that ant pollination systems are rare (150). Myrmicacin is found with plant hormones in a secretion produced by the thoracic gland which disrupts the flow of components to cell walls and cell division is disturbed. However, some of the many other glands and secretions on ants may also be involved.

Bees (Apidae) Probably the most significant of insect pollinators are bees. Within a single tree family (151), there can be a wide range of bee pollinators, as in the Lecythidaceae of Amazonia, one of the most important families there, with different morphological flower types and rewards associated with different bee sizes. The most primitive state is where the stamens are united basally and are numerous, curving inwards in a symmetrical flower as in *Gustavia* spp., some of which are self-sterile; small to medium-sized bees like species of the stingless *Trigona* leave pollen on the stigma and become dusted with new pollen from the stamens where they forage. In *Cariniana*, there are similar pollinators but the androecium is somewhat one-sided, while in *Couroupita*, it is very irregular in that, besides a ring of stamens with fertile pollen around the stigmatic surface, there is a one-sided ligule bearing stamens with sterile pollen curved back over the others. These flowers are visited by larger bees such as species of *Xylocopa*, which land on the hood-like ligule and forage underneath it for pollen. In so doing, bees rub their backs against the stigmatic surface and effect pollination and also collect fertile pollen. In the Brazil nut, *Bertholletia excelsa*, and *Eschweilera* spp., the hood is pressed down on the fertile stamens and bears only staminodes with nectar at their bases: only a strong bee can lift the hood—species of *Xylocopa* and female euglossine bees. Most specialized is the condition in *Couratari* spp., where the hood is coiled under itself, concealing the nectar, which can only be reached by the long-tongued female euglossines.

Generalist bees (152) are usually faithful over short periods, one species of trigonid bee near Manaus for example visiting 33 species in 21 plant families in 1 year, though at any one time, over 60% of the individuals of the species had only one pollen type and over 23% had only two. Trigonids are very common in south-east Asia and tropical America. They can make up to 141 foraging trips a day, comparing favourably with commercial honey bees, which can manage 150 (153). In Malaysia, they have pollen sources throughout the year, mostly in the understorey; they have a memory for food sites, can forage up to 1km from base and can recruit other foragers. *Trigona fulviventris* (154) usually forages alone but, if rewarded several times at a site, will bring recruits, their number dependent on the molar concentrations of sucrose—the first bee marking the surroundings with an oily pheromone from its mandibular gland. The recruits recruit others, colonies sometimes fighting over a resource, though individuals of this species usually give in to those of other more aggressive *Trigona* species.

In Amazonia, bees collect pollen from some species that have 'buzz' mechanisms as in *Mouriri* spp. (Melastomataceae). The bees shiver their indirect flight muscles, which ('buzz') vibration promotes the expulsion of a cloud of pollen grains from the tips of the anthers over the insects. Such systems are known in at least 6–8% of all angiosperms and are represented in 71 plant families with poricidal anthers, those notable in the tropics being Solanaceae.

On Barro Colorado Island, *Drymonia serrulata* (Gesneriaceae) produces more nectar per day than has been reported from any other bee-pollinated flower (155). The effective pollinators are a pair of *Epicharis* spp. but there are glandular trichomes inside the corolla, secreting oils that are deposited on the anthers, but this is collected only by *Trigona pallens*, an ineffective pollinator. However, the oil helps pollen adhesion on other insects and this may be important in 'buzz' flowers as well. Several plants, especially Malpighiaceae and Memecyloideae of Melastomataceae, offer oils from glands on the calyx (these are vestigial or absent in Old World plants). The oil is collected by *Centris* bees and is full of glycerides, which are later mixed with pollen in the hives. Such oil flowers and their bees (156) are most abundant in neotropical savannas and forests and are known in 79 genera of plants in 10 families. The oils are energetically twice as efficient as carbohydrates and, although the bees are highly specialized, flowers very like modern Malpighiaceae, complete with oil glands, are known from the mid-Eocene.

Most remarkable, perhaps, are the euglossine bees (157–160), of the neo-tropics, of species of which, mark-recapture studies show that the solitary males of *Eulaema* spp. can live for 2 months. Males and females can be kept in captivity for several months, the males surviving without the floral fragrance compounds to which they are attracted. They 'trapline', moving directly from one food site to the next, apparently remembering them from previous days, and are fast fliers, visiting plants producing (few) flowers over long periods. Some captured males with orchid pollinia are believed to have flown some 45–50 km in seeking the orchids. The males have dense tufts of hairs on the tarsi of the forelegs, picking up oils by capillarity; the hind tibia, corresponding to the pollen basket of the female, serves as a storage organ for scent. The odours are the sole attractant for the bees and this can be detected up to 1 km away across water, though perhaps less in forest. The scent is transferred to the tibiae in flight, the bees re-alighting and collecting for up to 1 hour.

Particularly important for these bees are certain groups of orchids, each species of which may offer 2–18 (mean 7–10) different compounds, some of which are unattractive on their own; however, eugenol, methyl salicylate and vanillin are highly attractive individually. The attractants appear to change geographically and seasonally, while it is perhaps the blending of compounds that restricts the number of bee species attracted, although there are overlaps between them and the bees' choices also vary geographically and seasonally. The geographic differences may be genetic or reflect the age of insects or the availability of resources. The seasonal and geographical differences in orchids on the one hand and bees on the other are incompletely congruent. On Barro Colorado Island, pollinator specificity occurs in fewer than half of the orchid species and host specificity of the bees is also rare. There are other fragrance sources and it is argued that the relationship is not mutually obligatory but as all the euglossine genera are involved, it is suggested that the orchids have exploited a pre-existent behaviour pattern in the bees. The scents are possibly modified to pheromones in mating, for bees carrying scent apparently attract other males, so that about five display together whereupon the female chooses her mate.

The flowers of some plants, e.g. some *Calathea* spp. (Marantaceae), do not open spontaneously and may be opened only by euglossines. Some *Eulaema* bees regularly pollinate, while others learn to trip the flowers without effecting pollination. Of orchids, euglossines visit some 625 species in 55 genera, especially in the subtribes Stanhopeinae and Catasetinae; they also visit some Gesneriaceae and a number of Araceae as well as rotting logs, to which they are possibly attracted by fungi. In many of the orchids, the bees slip and fall through the flower or into a trap, for example in *Stanhopea candida* in Amazonia, pollination is by *Eulaema moscaryi*, which 'falls through' the flower after collecting the scent; in so doing, it brushes against the pollinia and in the next flower these are lodged in the stigmatic cavity. *Eulaema ignita* also collects scent but, being the wrong size, does not brush against the pollinia and is therefore a robber.

Not all bee-pollinated flowers are obviously pollinated by bees. Bagging experiments (161) in Panama on *Luehea seemanii* (Tiliaceae), which has flowers opening at night, suggestive of a nocturnal pollination syndrome, showed that diurnal visits by bees were responsible for higher numbers of fruits and seeds set. Whether such, perhaps resulting from crosses with male plants not far away, would have survived competition with those possibly derived from crosses with more distant fathers, in a mixed infructescence is not clear but this example and several others show how unrestrictive certain 'syndromes' are.

Thrips, beetles, flies, butterflies and moths   Of dipterocarps, trees in many genera are first visited by *Apis* bees, later by meliponid bees, and then by various other insects. In *Shorea* sect. *Mutica*, a single tree can produce 4 million flowers over 2 weeks (162). The thrips eggs are laid between the petals and the larvae feed on these organs. By the time of anthesis, there is time for four generations and a single insect can by then have given rise to 4000 juveniles: they get trapped between the anthers and the petals as the flowers open and, as adults, feed on the pollen. They carry on average only 2.5 grains each and, as the corollas fall in the morning, they fly back up to the canopy, where they effect pollination, any slight wind drifting them to other trees.

Of beetle-pollinated plants in the tropics, the aroids visited by scarabs are thought to have near-specific volatile attractants like those used by the euglossines. In Amazonian Annonaceae and Nymphaceae, beetle pollination (151) tends to be fortuitous by contrast, e.g. scarabs on

Nymphaea spp. in lakes, though in *Victoria amazonica*, t
species of dynastid beetle attracted by scent and temperatur
ambient): these become trapped for 24 hours and feed on starcl
appendages before leaving with pollen. In *Annona sericea*, the f.
thick fleshy petals that never quite open. At about 7 p.m., the te
rises more than 6 °C above ambient and there is a strong smell like
chloroform attractive to chrysomelid beetles, which push their w , ιο the
stigmas, pollinating them. The stigmas fall, the stamens moving into their
place, dehiscing and powdering the beetles, which may be copulating; then
the stamens fall as do the petals, one at a time, releasing the insects.

In many palms, there is also a rise in temperature at the beginning of
anthesis, associated with beetle pollination: in *Astrocaryum mexicanum* at
Los Tuxtlas, there is a wide range of pollinating species, besides many other
herbivorous visitors (163) and this suite is shared with other palms in the
same and different genera represented there. The babassu palm (*Orbignya
phalerata*) in Brazil is pollinated by a beetle, *Mystrops mexicana*, perhaps
supplemented by wind. Several cultigens are similarly flexible, e.g. coconut,
in that pollination is thereby ensured in both open and closed habitats (164).

In Amazonia (151), flies pollinate flowers of species of *Aristolochia*
(Aristolochiaceae) and some orchids, and midges pollinate cocoa, in the
decaying fruits of which they breed. This also occurs in other Sterculiaceae,
notably *Sterculia chicha*. In northern Borneo, *Rafflesia pricei*, a species in a
genus including the largest-known flowers, *R. arnoldii* (*R. titan*,
Rafflesiaceae) up to 1 m across and all its species parasitic on lianoid species
of *Tetrastigma* (Vitaceae), is pollinated by carrion flies (bluebottles),
attracted by sight and scent. The flowers offer no reward and deceive the
insects by appearing to offer food and perhaps a brood place (165), so that
the plant is doubly parasitic—on the liane and on the bluebottle. The flies,
mainly female *Lucilia papuensis*, visit male flowers, enter the anther
grooves on the central column of the flower, the pollen being precisely
positioned on the thorax. In female flowers, the flies are wedged in tightly,
the thorax pollen being rubbed off on to the stigma.

Lepidoptera are frequent understorey pollinators as we have seen,
butterflies by day and moths at night, e.g. within the family Chrysobalanaceae (151) in Amazonia, butterflies visit *Hirtella* spp. (flowers
opening in the day, many with a showy appearance, usually pink or purple,
stamens few) whereas the related *Couepia* spp. are moth-pollinated (flowers
opening at night, few at a time, white, stamens numerous and more copious
nectar). Moths can be strong fliers, hawkmoths in Costa Rica, for example
(142), travelling up to 15 km with pollen and can live more than 2 months.

ᴊng-tongued species are particularly well-represented in Madagascar (166), and about 10% of all the orchids in that island, i.e. some 70 species, have nectar-spurs, up to 8 cm or longer. Five angraecoid species were found to share the same pollinator, *Panogena lingens*, but ethological and mechanical mechanisms apparently restrict interspecific pollination despite mixed pollen loads.

Fig-wasps The most celebrated 'one-to-one' interactions involve the genus *Ficus* (figs) and its pollinators, the wasps of the Agaonidae, related figs having related wasps. They cannot exist, or at least, procreate, without one another, and in some senses approach a dual-organism, like a lichen (167, 168). Very rarely is there more than one species of wasp involved with any particular fig species, an exception being *F. ottoniifolia*, apparently visited by different wasp species in forest and in open country in West Africa.

One or more females, possibly attracted by a pheromone, enter the young fig (an inflorescence known as a syconium), losing wings and antennae, and possibly fungal spores as well in forcing an entry through the tiny ostiole at the fig apex. The female removes pollen from pollen pockets and pollinates up to hundreds of stigmas; she attempts to oviposit down styles and then dies. In the oviposited flowers, the ovules are stimulated to produce only endosperm on which the larvae feed, a process not unlike more exposed gall formation in other plants. The CO_2 content is higher within the syconium than outside it and its level is believed to control the hatching of the larvae; it is affected by bacterial activity and this in turn is affected by nematodes in the syconium. The males, which are wingless and with reduced legs and eyes, emerge as the seeds mature, locate the females, copulate and help the impregnated females out of their florets. The females then fill their pollen pockets from newly-opened anthers and leave through holes in the syconium wall cut by the males, who then die. The co-operation is understandable in evolutionary terms as the males and females in a single-female syconium are brothers and sisters. If there is no pollination, the syconia (except in cultivars) are aborted. In isolated trees, however, intra-crown asynchrony could lead to (self-)pollination.

During the early male phase of *Ficus religiosa* (India to south-east Asia), the internal atmosphere has 10% CO_2, 10% oxygen and some ethylene (169) and the male insects are active under this régime, the females inactive. After puncturing the syconium wall, the atmosphere returns to normal, the males are inactivated and the females emerge. The change in atmosphere ensures development of both fig and pollinator: in other species, e.g. *F. carica*, the domestic fig, copulation takes place under normal atmospheric conditions, and in yet others, the females emerge while still enduring relatively high CO_2 levels. In short, there are many variations on the theme set out in the last paragraph.

The system may have developed from seed parasitism (170–175), but it itself is now parasitized by other wasps ovipositing through the syconium wall. Some other wasp species, apparently derived from pollinating agaonids, enter through the ostioles but do not carry pollen, though they have pockets; there are also weevils and moth larvae. In consequence, between 20 and 80% of ovules are killed by pollinators and other insects. In South Africa (171), ants visiting *Ficus sur* to tend homopterans deter parasitoids of the figs, thus enhancing pollination success, but some ants feed on arriving pollinators, while some birds eat them as they emerge. Moth and beetle larvae eat developing and mature larvae respectively in the fig. However, pollinators arrive and leave *en masse*, unlike the interlopers, who may thus suffer more from the predators than the pollinators do.

Pollinators trapped on *Ficus retusa* in Costa Rica, showed 99% fidelity (172, 173), giving strong support to the idea of there being a pheromonal cue. They arrived in a 1-day burst coinciding with the presence of receptive but unpollinated syconia. That dispersal agents can travel further than the pollinators can be attracted is the explanation for the *Ficus pubinervis* population of Krakatau (174), where it is a canopy dominant, never setting ripe fruits. In New Zealand (175), the introduced Australian *F. rubiginosa* produced no ripe fruits for many years, but its pollinator has now been introduced, perhaps by wind, and fruits are set. Whatever the volatile attractant to the syconia may be, it is likely that the parasitoids find it attractive too.

6.3.2   Bird pollination (ornithophily)

Bird pollinators (67) include unspecialized nectar eaters, which opportunistically visit many different kinds of flower, generally small ones, and specialized nectar eaters, which concentrate on a few kinds of flower, mainly larger ones. Nectar never provides a complete diet for any of these birds, which also eat insects or other animals but, as nectar is replenished as it is removed, birds may defend clumps of nectariferous flowers. In different continents, different groups of birds are nectarivorous—largely hummingbirds in the New World, sunbirds and sugarbirds in Africa and honey eaters in Australasia. There is little overlap in the families of plants the different groups visit and, in those families, the flowers may look very different. For example, in Australasia the flowers are of an open type compared with the tubular form, brush-type or explosive mechanisms of bird-pollinated flowers elsewhere. In short, it would seem that the relationship must have evolved over and over again. This is borne out by work on more temperate plants, for many of the hummingbird-pollinated plants of western North America belong to genera that are predominantly (and it is believed more anciently) insect-pollinated.

That bird flowers are similar in colour and nectar to fruits might suggest otherwise, however. It is often repeated that hummingbird flowers are red, part of the ornithophilous 'syndrome', and indeed, red does seem to be preferred by these birds. Nevertheless, it may well be that such birds are conditioned to this colour and it has been shown that nectar quality can overcome colour prejudice. Again, it has been argued that colour contrast and other factors may be important because red contrasts well with green and many hummingbird plants with bluish leaves have yellow flowers, the complementary colour.

In Puerto Rico (176), long-tubed flowers with high nectar levels are associated with long-billed hummingbirds, while short-tubed ones produce enough nectar for small birds, but not big ones. Any amino acids in the nectar are probably of little importance and nectars high in them are avoided by the birds. The nectar is merely a high-energy source.

Hummingbirds are large pollinators and potentially, therefore, very destructive: mechanisms reducing damage to the flower include the development in many groups of the inferior ovary, concealing the delicate ovules well away from probing beaks, as in *Fuchsia* (Onagraceae), and the red fruit-like spurs in *Norantea guianensis* (Marcgraviaceae) in Brazil, the rest of the flower being inconspicuous (151). The unspecialized hummingbirds range over herbs to trees, which produce many flowers with little nectar to attract pollinators. In a mass, however, the birds sometimes defend the flowers and thereby reduce cross-pollination.

Few canopy trees in Costa Rica are bird-pollinated and those that are, are visited by passerines such as orioles, which travel in groups: humming-birds would turn trees into individual feeding territories and thus reduce pollination (177). The specialized hummingbirds do not normally maintain feeding territories or defend patches of flowers but 'trapline' to flowers on other plants of the same species. These plants are mainly large herbaceous ones or lianes or epiphytes, mostly different in form, then, from those attracting unspecialized hummingbirds.

In the more seasonal forests, hummingbirds may be migratory, but in all forests they tend to breed at the time when the flowers on which they depend are most abundant, though as with the thrips and the *Shorea* species, there is some staggering of flowering times and avoidance of competition for pollinators between plants. Bird nectar thieves like the flower piercers are found in many species, but the bromeliads with thick protecting bracts are almost immune to their attacks. It has been argued (178, 179) that the dilute nature of the nectar in such flowers deters some bee robbers that may be unable to concentrate it: the birds would prefer concentrated sugars. It has more frequently been suggested that dilute sources will promote visits to many flowers and thus out-crossing, though it is difficult to see how such a mechanism could arise, and perhaps it is more satisfactory to surmise that visits by a pollen-dusted vector to several flowers on one plant are promoted, leading to cross-fertilization of more. Nectar flow in the red-flowered *Brownea rosa-de-monte* (Leguminosae) of the Panama forests (180) is known to be influenced by interference from stingless bees and lepidopteran larvae, though, in general, flow in this species decreases through the day during which any flower is presented. Although the inflorescences last just 1 day, their production tends to be synchronized within localized groups of trees. Other nectar-thieves include tamarin monkeys in north-eastern Peru: these feed on nectar of the bird-pollinated *Symphonia globulifera* (Guttiferae, 181).

In *Erythrina* spp. (Leguminosae), the differences in the morphology of the flowers and inflorescences are associated with the different, passerine or hummingbird, pollinators (182). Flowers visited by passerines are twisted such that they face back to the peduncle, on which the perching bird stands: hummingbird flowers differ in being tubular and sticking outwards making them readily accessible to hovering birds. The passerine species have nectar with low (about 7–10%) sugar content, while hummingbird flowers have up to 38%, though the passerine ones have higher amino-acid concentrations. A very specialized example is found in passerines in India, where *Dendrophthoe (Loranthus) longiflora* (Loranthaceae) is pollinated by sunbirds (183). The flowers remain closed until the bird squeezes the bud, which explodes open, offering nectar (but possibly no insects unlike most sunbird flowers) and is pollinated; if not visited, the flowers fall off.

6.3.3 Bat pollination (chiropterophily)

Bats visit a range of plants and these are visited by a range of bats; plants transferred from the New to the Old Worlds and vice versa are visited by the local bats. Of those that pollinate a small percentage of tropical rain forest trees, a minority feed exclusively on floral resources. Of these, in West Malaysia (184), two species roost singly in trees and fly short distances to feed upon trees which produce a few flowers all year round, and the third, *Eonycteris spelaea*, roosts, in large colonies in caves and can fly up to at least 35 km from them to feed and take food from a wide range of flowers scattered in space and time.

Flagelliflory, the presentation of flowers on long peduncles, occurs in several unrelated groups and is generally associated with bat pollination, but a whole family of plants in a wide range of habitats in tropical America has all its species bat-pollinated with the flowers projecting above the canopy; namely the Caryocaraceae (151). Within a genus exhibiting flagelliflory, *Parkia* (Leguminosae) has the greatest variation in its pantropical range in tropical America. The unit of pollination is a capitulum of flowers; of the 11 species studied (185), nine were bat-pollinated, the capitula with flowers opening for 1 night, secreting copious nectar and being bright red or yellow or both, with characteristic scents: they are visited by eight species of bat, some of them nectar thieves. One section of the genus *Parkia* has no nectar-producing flowers and is probably wholly entomophilous: the capitula are visited by insects, especially trigonid bees. The internal classification of the genus is founded on the arrangements of the flowers: heads of only fertile flowers for one entomophilous section; similar but with a distal zone of nectariferous ones in another; the distal ones are fertile and separated from the proximal staminodal ones by a zone of nectariferous ones in a third. These latter two are bat-pollinated and the capitulum can be pendulous or erect; the staminodal flowers sometimes form a fringe acting as a landing platform for the bats. They land head downwards in America, upwards in the Old World (it would be interesting to know the behaviour of bats in America when confronted with *P. globosa*, long ago introduced from Africa). Red capitula held against the sky appear black to the colour-blind bats, while yellow ones (seen as white) are those displayed against the foliage. Of sympatric species, those flowering together deposit pollen on different parts of the bat body, thus avoiding interspecific pollination, and some have staggered flowering times. Concurrently, though, the destructive bruchid beetle populations which damage seeds moved from one fruiting species to another. It would therefore appear that the ensuring of successful pollination must override the importance of losses due to seed predation. Indeed, such compromises may be widespread.

Bat-pollinated plants, e.g. species of *Oroxylum* (Bignoniaceae) and *Duabanga* (Sonneratiaceae) in Borneo, are possibly most conspicuous along river banks where, incidentally, wind-pollinated trees such as species of *Octomeles* (Datiscaceae) and *Casuarina* (Casuarinaceae) are also found. Wind pollination in Peninsular Malaysia (186) is rare and the introduced wind-pollinated *Pinus caribaea* does not set seed there, except on mountain ridges or on the coast. This is reflected in the pattern of distribution of the indigenous Coniferae, *viz.* species of *Agathis, Dacrydium* and *Podocarpus* (*sensu lato*), which are concentrated in such places with one species of *Podocarpus* on the tops of steep limestone hills. These are the only places with regular air movements, though this raises serious questions about how the trees of the gymnosperm forests of the past were pollinated.

6.3.4 *Other mammals*

Non-flying mammals have for some time been suspected to be pollinators in a number of vegetation types outside the tropics and, in recent years, this has been proved to be so in both South Africa and Australia. It is now known that three species of forest plant in the Amazonian forest of south-east Peru (184) are visited by up to seven species of diurnal and six species of nocturnal non-flying mammals in the dry season. Heavy deposits of pollen were found on the facial fur of these animals. They might be considered likely to feed on flowers at a time, such as the dry season, when there is a scarcity of fruit. Of the species observed, the canopy liane, *Combretum fruticosum* (Combretaceae), was seen to be visited by primates including the pygmy marmoset, while the tree, *Quararibea cordata* (Bombacaceae), was visited by large primates, opossums and kinkajous. It was notable that in the Peruvian forest some four genera of marsupials were involved, members of the group that comprises the non-flying mammalian pollinators of Australia.

Three species of rodent are reported to pollinate *Blakea* species (Melastomataceae) in the cloud forests of Costa Rica (188): the nocturnal anthesis and nectar production suggests bats but the lack of scent and cryptic flowers hidden in the foliage do not and, indeed, the flowers are not visited by bats in the harsh, windy environment.

6.4 The milieu for 'mutualism'

The intergrading interactions between plants and animals, herbivory, dispersal and pollination, present several conflicts summed up as defence versus attraction. For example, in *Parkia* (discussed above), the sequential use of pollinators versus the sequential susceptibility to bruchid attack on the seeds. In continuously flowering species such as *Muntingia calabura*, trees are attractive to beneficial agents, pollinators and dispersers, at the same time. The potential conflict in this species is removed because the flowers are pollinated above the branches by bees, the pedicels lengthening and becoming erect, to be lowered in fruit, when bats disperse the fruits from below the branches (189). This tree is one of the most commonly planted shade trees in India and elsewhere, yet only very lately has its phenology been fully investigated.

The various animal–plant interactions cannot be isolated from one another, therefore. In reviewing some of them in this book, it has been found that several complicating features arise, beside the straightforward failings of human beings wishing to pigeon-hole phenomena or experiences into neat categories: 'strategies' or 'syndromes' with classical-sounding names. Firstly, then, as we have seen, tropical rain-forest ecology is fundamentally dynamic; it is not 'stable'. Even in the short term, dependent on local circumstances, the numbers and behaviour of dispersal and pollination agents can change such that short-term observations are sometimes very misleading. An example of dispersal agents, which is the male toucan in Costa Rica (75), which defends sections of fruiting trees against all other frugivores besides his mate, displacing them by threat; however, when food is abundant, this does not occur. Again, the behaviour of predators may affect the pollinating activities of animals such as bats (190), encouraging them to make short visits repeatedly, rather than long ones. The ramifications in terms of outcrossing are clear.

Pollinator assemblages for any particular plant vary in space and time (191), both seasonally and year by year, which is perhaps not surprising in view of animal cycles and the fact that the environment 'perceived' by a plant species is irregularly and somewhat erratically structured, i.e. has a strong stochastic component. The inevitable consequences in terms of the frequency of pollen deposition on stigmas, the numbers of grains and the 'quality' of those grains from the 'point of view' of the maternal genotype may lead to selective abortion of embryos. In a Mediterranean example (*Lavatera latifolia*, Malvaceae), Hymenoptera delivered more pollen-grain more often than Lepidoptera or Diptera; Lepidoptera tended to fly longer distances between consecutive flowers than Hymenoptera or Diptera. Bees pollinated frequently and with large loads but generally promoted geitonogamy, whereas butterflies pollinated less often and with smaller loads but usually cross pollinated; flies pollinated infrequently, with small loads and generally promoted geitonogamy. Within each group, however, there were exceptions. In years with more bees, there is clearly a different quality of pollination and this varies with the time in the season considered, but poorer quality pollination at other times is better than none, so that specialization with one pollinator per plant species is unlikely to evolve. Nevertheless, specialization in some groups of plants does occur, as it does in some of animals, such that the broadly different 'syndromes' of pollination and dispersal have evolved repeatedly in different families and genera of plants and are apparently incipient even within some species. However, when observed over long periods, many species are found to be generalist in that individuals of them are capable of taking advantage of 'specialist' interactions at any one time: the 'anachronisms' show this, as do the generalized pollinators on islands where plants have 'left their pollinators behind', e.g. a cetoniid beetle, *Mauseolopsis aldabrensis*, visits 58% of all flowering species, native or introduced, of all colours and morphology, on the coral atoll of Aldabra off the African coast, but it shows a remarkably high degree of constancy in its foraging flights (192).

Secondly, many of the 'interactions' are actually mediated by other organisms, usually micro-organisms, whether they be involved in allelopathic phenomena in succession, or are gut microbes in primate stomachs, or mycorrhizae, or fixing nitrogen, or living in ant nests or fig syconia or the guts of termites, or perhaps involved in promoting pig migrations. Clearly we need to know more, for example (193), it is reported that the red leaf-monkey (*Presbytis rubicunda*) in northern Borneo, like other primates, eats soil, usually from termite mounds, but the soil has been analysed only chemically with no clear-cut conclusions: what about the micro-organisms and their products?

Thirdly, is the inevitable consequence of good arrangements—the presence of cheats, including mimics, thieves, parasites, parasitoids, nectar-robbers, toxin-users, and, of course, the whole pyramid of carnivores based on the herbivores, some of which (as in the case of the hapless fig-wasps) have been mentioned above, notably birds and ants. The relationship between ants and other animals, particularly their farmed Homoptera feeding off the host plants is complicated in itself but, in the Old World Tropics, mosquitoes of the genus *Malaya* are associated in their habits with certain ants, which run up and down tree trunks carrying the honey-dew they have taken from their aphids. The mosquito (194) hovers a couple of centimetres from the trunk and suddenly alights in front of the ant. It does not touch it but vibrates its wings: in response the ant opens its jaws, between which the mosquito pushes its proboscis and steals the honey dew. And what eats the mosquito? So far as ants are concerned, a single pangolin can eat up to 73 million of them, or termites, in 1 year (195), while ant eaters on Barro Colorado Island attack termite and *Azteca* ant nests; of arboreal nests, 91% of the termite ones and 37% of *Azteca* nests are attacked each year. In addition some 16% of the termite nests are abandoned annually, usually following tree falls.

A much-investigated system is that of the fig syconium, which is a microcosm until its wall is punctured. Few systems are closed for so long: bromeliads, epiphytes of the Neotropics, with 'tanks' of up to 2 litres (196) can contain up to seven species of algae, as well as bacteria, protozoa, rotifers, flatworms, oligochaetes and other invertebrates such as crabs, insect larvae etc. as well as tadpoles. *Nepenthes* pitcher-plants bear pitchers with fungi and slime-moulds, protozoans, desmids, diatoms, rotifers, oligochaetes, crustaceans, insect larvae and even tadpoles, with spiders positioned above the liquid awaiting a catch and the small primate, *Tarsius spectrum*, in Sabah, feeding off trapped insects. Some animals live only in the pitchers, others (a minority) are also found living elsewhere. *Misumenops nepenthicola*, a spider, lives there and captures flies; if these are distasteful they are (sometimes) thrown back into the pitcher; if disturbed, the spider goes down into the liquid on a thread, its armour and a bubble of air making it immune to the digestive juices there. But most of these ramifications are beyond the scope of this volume, though they provide the 'milieu for mutualism'. To take a final example, leaf-cutting ants have been extensively investigated but *Atta* nests have conspicuous mounds of refuse nearby (197) and this is derived from the fungus gardens; when the nests are near river banks, streams are used as disposal sites. In slow-flowing streams in the seasonal areas of Costa Rica, the debris accumulates and affects the frequency of the characid fish, *Astynax fasciatus*, a generalist feeder. Pools with the detritus would be expected to harbour more fish, but then predation from the belted kingfisher, *Ceryle alcyon*, increases. Conversely, refuse on the surface occludes predator vision and predation can sometimes be low.

The ripples from each ecological 'splash' overlap and intermingle, the ripples varying in speed and height with time. Nevertheless ecology has started with that which is obvious to the eye and is increasingly concerned with what is not. Until recently, the ecological significance of pheromones and olfactory clues in insects, as in the pollination systems of figs and orchids discussed above, was little considered. However it is not merely the invisible that points to our ignorance. Our knowledge of the basic natural history of some groups is still lamentable. For example, some of the most memorable features of fieldwork in the Old World Tropics from Madagascar eastwards, leeches, are so little known that their behaviour towards animals other than humans is unrecorded (198). There is much to do.

CHAPTER SEVEN

SPECIES RICHNESS

Not only trees but almost all groups of animals and plants exhibit their greatest variety in the tropics (1) in terms of the number of species recorded there. How is it that the tropics generate, or at least harbour, so many species and how is it that these can coexist? The answer to the first question involves an analysis of speciation, which in turn leads to a consideration of the ecological conditions under which such speciation might have taken place, and the answer to the second also involves an analysis of the physical and biotic influences on the organisms concerned.

7.1 Speciation

A prerequisite of speciation is variation, and if the variation patterns within species are examined, some ideas of the possible modes of incipient speciation may be gained. As was explained in Chapter 5, some tropical tree species seem to be very homogeneous in their morphology over wide ranges, even including populations widely separated from others with little possibility for gene exchange. Examples of this are species now in populations in southern India and far away in Malesia, like the yam ally *Trichopus zeylanicus* (Dioscoreaceae). Other species seem to be separable into geographical races based on their morphological features. These are usually given subspecific rank. The segregation may be more precisely ecological than geographical in (say) forest and savanna races of trees in Africa. Yet other species exhibit variation patterns that defy analysis of the sophistication of present-day biology. Such are the ochlospecies (p. 115) of White, a term coined to convey the annoyance such species cause taxonomists, and including some of the ebonies (*Diospyros* spp., Ebenaceae) of Africa.

Although there is much to be said in favour of the notion of evolution by discrete jumps in plants, particularly through allopolyploidy involving small numbers of individuals, current orthodoxy holds that the greater part of speciation events occurs through the isolation of fragments of an initial population, the change of these fragments in response to the conditions in their isolation and, with the breakdown of isolation, the co-existence of the newly speciated populations in one locality. Clearly, many partially speciated populations will come together and then merge once more. The ochlospecies is not readily explained, however, while apomixis is another phenomenon to be taken into account. It occurs in several dipterocarps (see later) and also some *Garcinia* spp. (Guttiferae) in the understorey of Malesian forests: these set seed without pseudogamy (2).

Already we have seen microspeciation occurring in trees in Trinidad and Tobago (section 6.2.6), but it has also been suggested that in crossing major faunal boundaries, plant species may undergo so-called 'cryptic' speciation (3) with major differences in fruit sizes and so on, associated with different dispersers; i.e. obscured rather than obscure in that herbarium material is rarely preserved with fully ripe fruits and complete field notes. Of course, such examples may represent changes undergone since the arrival of different faunas rather than the arrival of plants in different faunal regions. In view of the conclusions of the last chapter, it is likely that such patterns will be archaic and also confused by local fluctuations in dispersal agents and this may be exacerbated in regions of geological instability, e.g. New Guinea with confusing variation patterns as in the tree *Chisocheton lasiocarpus* (Meliaceae (4)) and the herbaceous *Impatiens hawkeri* (Balsaminaceae (5)). Certainly within tree populations, there are family groups in mosaics of genetically and phenotypically distinct patches, making up 'populations'. These distinctions are marked by morphological features such as leaf shape, thickness and pubescence, bark properties and biochemical features such as relative amounts of sugars, amino acids, terpenes and tannins (6). In turn, this makes them differentially susceptible to pathogens and herbivores. Such patterns may be the result of Sisyphean fitness and not represent irreversible tendencies to speciation, i.e. the formation of 'populations' of plant genotypes, representing the best available to arrive at a particular point under the prevailing biotic and physical features at any one particular time. Crossing with other such genotypes inevitably leads to the breakdown of this optimum (hence the name) and formation of new genotypes to be filtered by the new conditions.

Nevertheless, speciation appears to be greater in regions of geological instability or great physiographic diversity with rapid climatic gradients as in Madagascar. New Guinea has both such features, whereas Africa has been very stable in this regard, while contact with Laurasian floras has been restricted by desert barriers and this contrasts with the intermingling in other tropical regions. Other places, particularly the tropical Far East have been very unstable over the last 15 million years. The richness of the Malesian flora, for example, may be partly accounted for by the Laurasian elements in the lowland rain forest and the Gondwana elements in the heath forest and in montane communities. In the Neotropics (7), richness is due to the combination of Amazon-based groups and those based in the northern Andes, these latter being poorly represented in Amazonia. Almost all lianes and canopy trees in the lowland neotropics are in Amazon-centred taxa; epiphytes, understorey shrubs and palmettos mostly belong to Gondwana groups with Andean-centred patterns. Of trees with Laurasian affinities, almost half of the families represented have wind pollination, while none of the Gondwana ones do; many of the northern-Andean ones are hummingbird-pollinated, while those with pollination by large- or medium-sized bees tend to belong to Amazon-centred groups. The woody, wind-pollinated Laurasian groups have not speciated greatly in South America, but almost half of the neotropical flora belongs to the Andean-centred groups of epiphytes, understorey shrubs and palmettos, all of which are poorly represented in Africa and, to a lesser extent, Asia.

Highly frugivorous birds like toucans and cotingas have radiated with the Amazon-based canopy-trees and lianes (8), while the Andean fleshy-fruited understorey flora has been associated with the radiation of manikins, tanagers and bats, making these faunas greater than those in Africa or Asia. The low patchiness of fruit associated with the richer 'fruit-floras' in America has resulted in the evolution of more diverse communities of frugivores with reduced dietary overlap, relatively small body size and sedentary ranging patterns, whereas in Asia, where fruit seems to be rarer in space and time, with fewer species, there are fewer frugivores with broadly overlapping diets (e.g. terrestrial and brachiating primates), larger body size and more mobile ranging patterns. Africa is somewhat intermediate in this respect.

7.1.1   Refugia

It has been widely argued (but see later) that within the lowland Neotropics, the fragmentation of populations took place as a result of climatic and associated vegetational changes in the last 2 million years. The changes seem to have been alternating periods of humid conditions and aridity. It is

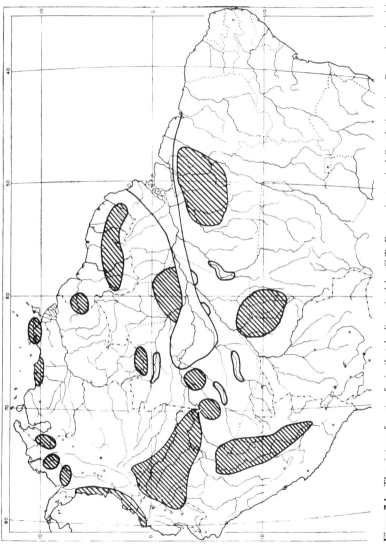

Figure 7.1 The sixteen forest refugia (hatched) posited by G.T. Prance in tropical South America. Reproduced with permission from *Acta amazonica* **3**, 3 (1973) 23, Fig. 24.

argued that during the dry phases, the humid forests would have been fragmented into island-like refugia (Figure 7.1) of isolated blocks of forest. The organisms would then come back into contact, merge again or coexist as new species (9).

There is some (disputed) geomorphological evidence for such contractions, supported by palaeobotanical work. Much evidence comes from the mapping of the distributions of rain-forest organisms, most notable being those of various tree families by Prance and his co-workers. Although the detail of the distributions varies, the broad pattern is the same. Similar schemes have been drawn up for several mammalian groups, certain butterfly genera, various reptiles and amphibians while a good deal of work has been carried out on the distributions of the avifauna. All these give some idea of possible refugia, sometimes referred to as centres of dispersal. That the modern distributions of organisms could be explained in terms of ecological rather than historical factors was rejected, as no clear-cut correlations between the distributions and any ecological factors such as precipitation or soil factors could be discerned.

In Mexico and central America, eight refugia have now been recognized; in the *terra firme* rain forests of lowland south America, 26 centres of endemism have been identified (10, 11), 15 in the great block of Amazonian-Orinoco-Guayanan rain forest. Today, the forest is at a maximum, so that it is the drier forests that are today's 'refugia'. In these, some species appear to be uniform, e.g. *Tabebuia impetiginosa* (Bignoniaceae), while others such as *T. ochracea*, are divisible into several subspecies indicating active differentiation. It has been argued that differentiation in refugia is less marked in trees than it is in insects or birds, because of their different life spans, but such theory provides an explanation for the now sympatric occurrence of 12 species of *Eschweilera* (Lecythidaceae) near Manaus. Similar explanations have been put forward for the distributions of herbaceous plants, as in *Ischnosiphon* spp. (Marantaceae), where cycles of dryness have been invoked to explain species relationships and distributions (12).

The distributions of forest passerines in Africa show centres of endemism (13), centres of richness and disjunctions, which have been fitted to refugia theory, but close examination of the geographical ranges of subspecies of the Heliconiini and Ithominae groups of butterflies gives 44 centres of endemism in tropical America, though the two groups have very different ecological characteristics. The richest centres have the most complicated mosaics of different habitat types reflected in topography, soils and vegetation. However, the patterns of species diversity do not coincide with those of endemism and one of the present 'refugia' seems to have been underwater 5000 BP and its endemism must have arisen subsequently.

7.2   Species diversity

Several hypotheses to explain the coexistence of so many species have been proposed. Some place emphasis on biotic, others on environmental factors. Of biotic arguments for explaining greater tropical diversity, one of the most familiar is that there are more 'niches' than in temperate latitudes, arising from more habitats or more resources, but these hark back to environmental arguments ultimately. Usually, however, it is argued that there may be a finer division of these habitats or resources or, in short, that tropical species are more specialized. Added to this is the historical argument that in the tropics more of the available 'niches' are filled and that, because of ice ages or other environmental factors, temperate habitats are not yet saturated with species.

Much of this 'niche theory' is rooted in zoological studies, so some animal examples will be considered first. Species co-existence levels in rain-forest mammals (and reptiles) are lower than in birds, whose locomotor abilities allow them access, at a low-energy cost, to food unavailable to species confined to terrestrial or arboreal substrates (14). Moreover, there can be many bats and small- to medium-sized mammals (arboreal) from the smallest marsupials to the smaller monkeys, in contrast to the larger mammals confined to the forest floor where food is always scarce, and therefore there are fewer species of them. Between closely related taxa, resource partitioning is found as in the case of squirrels in Gabon, where two of the nine coexisting species are found in special habitats, the other seven being taxonomically rather diverse: four of these are essentially arboreal, the other three ground foragers, and in each group there are differences in size such that food is partitioned by hardness and size. The weights of 66 species of mammalian primary consumers in Gabon are nearly uniformly spread over a range of five orders of magnitude, the 11 shrew species alone covering three. The ways in which resources are partitioned are many, as, for example, on Barro Colorado Island where there are as many as eight variables of importance in just the food of co-existing 'surface-gleaning' bats.

Among primates, orang-utans, gibbons and siamangs in west Malesia have differences in body size, permitting coexistence (15). The gibbons are chased off by siamangs as they dominate through size, though they require a lot to eat over a small area; the gibbons travel longer distances for small resources. Of the 46 species (16 genera) of American primates, all but 10 (2 genera) occur in Amazonia and up to 14 (12 genera) can occur sympatrically in Amazonia: the importance of size in resource partitioning is evident (16). In the most diversified locality examined, in south-east Peru, size has effects on not only diet and prey search but also capture methods, vertical position in the forest, habitat use and locomotory patterns, the larger primates apparently differing in diet as to the percentage of fruit *versus* leaves for example.

A 'community' of bats in Queensland comprised two subunits (17): four species in the closed canopy and five in gaps, while three other species used both. The gap specialists are fast fliers whereas those in the canopy are slower and more manoeuvrable. The bat community on Barro Colorado Island is perhaps the most complex assemblage of sympatric mammals anywhere in the world, comprising 35 species, which can be arranged in nine feeding 'guilds', based on diet and mode of food gathering. Within the guilds, food resources are partitioned spatially and temporally as well as by size and quality (18).

At the Brown River (19) in southern New Guinea, 83 coexisting bird species were examined: resource partitioning allows a high number of frugivore species (few mammals there) and there is a size range among them. There is a similar range of related carnivores but these have no great bill-size differences. The frugivores seem to have wider 'niches' in rain forests than elsewhere, but the converse theory that the habitat lends itself to finer division into niches seems to hold for the carnivores. Also in New Guinea are sympatric birds of paradise apparently dealing with fruit in a similar way, staying in the trees for only a few minutes, perhaps through fear of predators. On examination, it was found that two species were largely insectivorous, two largely fig-eaters and five were generalist frugivores (20).

Fundamental to the idea of finely divided 'niches', however, is the notion that there are finite or limiting resources, whereas in trees in the tropics with abundant light, water, CO_2 and a conservative nutrient-cycling system, this seems as inappropriate as it does in the case of tropical corals making up species-rich reefs, where food is not readily seen as limiting when compared with space to occupy. Another and very fashionable biotic argument is that of 'predation'; that there are more predators and greater predation in the tropics than in temperate regions, and that these are matched by more defensive or avoidance mechanisms. The prevalence of predation appears to be correlated with the relative lack of seasonality, for, in temperate countries, winters prevent a good deal of predation. Nevertheless, it would seem that this 'explanation' is merely a restatement of the facts of tropical high diversity in that if there are more species, there will be more species of predators.

Of environmental arguments, attention is focused on those features of the environment that are peculiar to the tropics. Firstly, the tropical belt is a large region and for marine organisms, for example, the sheer size of the available area for diversification is greater, though, as we have seen, tropical rain forest occupies only a small part of the tropical land-surface, and thus a very small fraction of the tropical global surface. Nevertheless, the range of environmental heterogeneity in the tropics from the foot of the Andes to their peaks is greater spatially than in temperate regions, even if the environment in temperate regions varies greatly temporally. The absence of the rigours of an annual drought in tropical forest is correlated with high species-richness of both forests and within individual genera of plants (21). We are left, furthermore, with the heterogeneity in the tropics engendered by disturbance through storms or other violent environmental factors such as fire and the seasonal changes of weather, some of which, like short droughts, are unpredictable. Cyclones are seen as a force that prevents the ecological succession of coral reefs from terminating in less species-rich assemblages (in those reefs, the earlier stages are more species-rich). Furthermore, the importance of predation can be seen as yet another element in the disturbance picture.

In summary, time, space, resources, competition and predation, may play a part in explaining the biological diversity of the tropics.

7.2.1   Environmental heterogeneity

In Amazonia, the density and species-richness of non-flying mammals is positively correlated with soil fertility and undergrowth density (22). What is the evidence for some microenvironmental differentiation of rain forest supporting a diversity of species, each in the case of trees with very specific requirements in terms of nutrients, soil water content and so on? Firstly, individuals do not all occur randomly in a forest and often there is much clumping as we have already discussed. Furthermore, a monograph of any large tropical group will show varying degrees of recognition of the ecological requirements of different species. Of the 51 species of *Chisocheton* (Meliaceae), in Malesia for example, one species is restricted to peat-swamp forest and another to limestone, while, in dipterocarps, *Shorea curtisii* is always found on ridges and slopes apparently associated with particular water relations in the forest. Similarly, in the *igapó* forests, that is those inundated by black or clear water, in Brazil, the species distributions are thought to reflect differential tolerance of flooding (23).

In Sri Lanka, the abundance of *Dipterocarpus zeylanicus* is associated with particular levels of soil potassium, that of *D. hispidus* with nitrogen and phosphorus, while *Artocarpus nobilis* (Moraceae) and *Campnosperma zeylanicum* (Anacardiaceae) are found only where *D. hispidus*, the local dominant, is not (24). In the dipterocarp forests of Borneo (26), the variation in species patterns is correlated with HCl-extractable phosphorus and magnesium where these levels are lower than about 220 ppm of phosphorus and 1200 ppm magnesium. Above these levels, such correlations are less evident and over-ridden by a correlation with topography. Below these levels, species richness is linearly positively correlated with soil magnesium, but above it negatively so. It is suggested that the critical levels coincide with those below which the rate of litter accumulation exceeds its breakdown and that the ecological optimum may represent that of the dipterocarps' mycorrhizae. At higher nutrient levels with higher water-retaining capacity, trees with a higher heat load in the canopy, i.e. with densely arranged large leaves, are at an advantage and shade out others, depressing species-richness. This example shows that environmental factors cannot be cleanly isolated from biological ones. When 46 forest sites in Costa Rica were examined (27) it was also found that there was a negative correlation between soil nutrient availability and the richness of tree species. In other words, the highest species richness was found under 'poor' growth conditions. Under very low nutrient conditions in general, however, species numbers are low as they are, contrariwise, under greenhouse competition experiments with high nutrient levels.

An example that pinpoints the problem was exposed when a map of all trees over 20 cm diameter at breast height in $\frac{1}{2}$ sq km of the semi-deciduous forest of Barro Colorado Island was made (28). It showed that most species are patchily dispersed, while many appear to be randomly distributed and a very few are uniformly distributed. Most of those that are clumped, however, are in groups not associated with the topography, while some species are restricted to particular kinds of site: *Ocotea skutchii*, for example, is always on slopes. In short, some heterogeneity and aggregations may have a physical explanation, others not.

In the neotropics, the question of the maintenance of diversity of the rich epiphyte flora has been considered to have a partly physical explanation, though the germane physical features are generated by living hosts, i.e. the nature of bark surfaces (29).

7.2.2 Biotic factors

Among seven species of fruit pigeons in tropical Queensland (30), the diets of the birds change with the seasonal fruits, each species with a different suite of plant species. Coexistence is reinforced by nomadism and migration, but the diets overlap at times of both high and low fruit availability. Indeed, of the biotic explanations of how existing plant species avoid competition, some are based on time-sharing of animal agents. For example, over a season, six species of *Shorea* are visited in turn by species of thrips, which are believed to be their major pollinators. Thus with non-specific pollinators, competition is avoided and there is no problem of the build-up of pollinators at any one time of year to satisfy any one plant species. The problem comes with the first-flowering species. However, apomixis through pseudogamous agamospermy has been confirmed in a species of *Hopea* (Dipterocarpaceae) and inferred in two of *Shorea* on the presence of polyembryony or triploidy, e.g. *S. macroptera*, which is the first of the section to flower, when the pollinator numbers have not yet built up (see section 6.3) and would experience the greatest vicissitudes if dependent on insects (31, 32). Similar time sharing is known in dispersal (33). In Trinidad, 19 species of the pioneering *Miconia* (Melastomataceae) fruit successively, providing fruit for dispersing birds throughout the year and staggered batches of seedlings that avoid interspecific competition. Although this example has been questioned, there could be 'shiftwork' in other examples—in that there is a segregation of animals between night and day.

Biotic explanations for both clumping and diversity have been adduced. The habits of some animals lead to the deposition of many seeds in one place. This is especially noticeable with frugivorous bats, which, in a dry forest in Costa Rica, produce seed shadows of a mixture of species around fruiting trees in which they roost (34). Again, agoutis in south-east Peru (35) store fruits of *Astrocaryum*, a palm, in hoards, which may lead to clumped distribution as well as to the 'clumped' behaviour of white-lipped peccaries, which in turn disinter this food. By contrast, it has been argued by Janzen that species-specific 'predation' of juveniles growing around their mother will lead to a progressive decline in density of successful saplings towards the mother. This might then account for the apparently non-environ-mentally controlled distribution pattern in which other species could insinuate themselves in the holes of the forest 'lattice' devoid of the first species. This has been extended in the case of specific fruit and seed predators to account for the frequency of dioecy in some forests. In dioecious species, regulated in numbers by such predators, a high level of gene exchange could be maintained by the presence of a number of male trees that would not be attractive to these predators.

'Predators' can include fungi and vertebrates as well as invertebrates (36–42). Of insect attack, that by bruchids of *Scheelea zonensis* (Palmae) in central America is greater, the nearer the mother plant and early survival of *Virola surinamensis* there is enhanced by being away from the parental crown, as seedlings closer to the bole suffer attack from weevils. *Faramea occidentalis* (Rubiaceae) seeds dispersed as little as 5 m from the mother there have a distinct advantage over those beneath her. Seedlings of *Platypodium elegans* (Leguminosae) also in central America die of damping off as a function of distance from the mother in their first year, when the mortality rate can be up to 81%. Experimental work on Barro Colorado Island showed that decreasing density of seedlings and increasing distance from the mother reduced the incidence of damping off. Seedlings of another legume in Panama, *Dipteryx panamensis*, survive in light gaps best, under adults worst. Damage to the apical meristem shows similar patterns with respect to distance from adults and density of seedlings as the pathogen damage in *Platypodium*, though in *Dipteryx* it is due to insect herbivory and tree or litter fall. In other parts of the tropics (43–45), similar patterns have been found in *Aglaia* seedlings (Meliaceae) in Malaysia and *Pandanus tectorius* on Fanning Island in the central Pacific. Such mortality at La Selva leads to uniform distributions of *Iriartea gigantea* and *Socratea durissima* (Palmae) and *Pentaclethra macroloba* (Leguminosae) from initially clumped ones.

At the 'community' level, most of the 300 or so species populations studied in the large plot on Barro Colorado Island discussed above show positively correlated survival with distance from reproductive adults up to 15 m. In plots in the Queensland rain forest (46), seedling recruitment of subcanopy trees and understorey species was found to be lower for common than for rare species: for canopy species there was no such correlation. It was found that growth and mortality were no greater in dense stands of conspecifics than in others. At very small spatial scales, there seemed to be trends in recruitment, growth and mortality that fit the general hypothesis in promoting species richness, but at larger scales, besides the smaller species mentioned above, there were no such discernible trends.

There are certain general observations that might favour the predator argument. In chalk grassland in Britain, for example, the removal of grazers, such as rabbits or sheep, has a devastating effect on the fine-leaved bouncy turf rich in species. In a few years, the sward may become dominated by the coarse tussock-forming grass, *Brachypodium pinnatum*, and eventually the vegetation turns into some form of scrub and possibly woodland. On the other hand, an increase in grazing pressure leads to the elimination of all but a few very tough species. At the maximum plant-species diversity, the grazing lowers the competitive ability of certain species, allowing, under the same conditions of soil, aspect and so on, the coexistence of the smaller species that otherwise could not exist there. What is perhaps most remarkable about this grassland is that it is only maintained under active management—even rabbits were introduced by the Normans—and, although it resembles some periglacial assemblages of plants, it must have been reconstituted after the forest maximum unless there were extensive grazed glades in that period. In short, it is possibly not very old.

Palatable seaweeds in North Carolina have improved fitness if growing with unpalatable competitors less susceptible to herbivory, i.e. one competitor seeming helpful to another, the abundance of the less palatable leading to increases in abundances and numbers of other species (47). The edible species grow at times of high predation as epiphytes on the less palatable. In the absence of herbivores (fish and urchins among them), the palatable species grow less when epiphytic, though the less palatable are little retarded.

A series of observations first put together by Gillett (48) is perhaps nearer our subject. It is well known that it is difficult to grow plantation rubber (*Hevea brasiliensis*, originally from Brazil) successfully in the forest lands where that species grows wild, and it grows best in Malaysia; cocoa, which is also native to the New World, grows best in West Africa, cloves from Indonesia, in Zanzibar, and so on. The argument runs that in the absence of their predators, performance is better and that when such predators, be they animal, fungus or other pathogen, catch up with the crops or evolve to attack them, disaster follows, as occurred in the coffee harvests of Malaya at the beginning of the century and in the banana-growing countries of Central and northern South America before that. The corollary is seen where alien plants, contrary to expectation, appear to oust native plants in species-rich environments, where every 'niche' would appear to have been filled by species evolving more or less *in situ* to fit those environmental factors. Examples are *Hakea* (Proteaceae, Australia) and pines that have been able to take hold in the Cape. Exotics may appear to compete better with indigenous vegetation in Britain: possibly the success of sycamore can be seen in this way. The principle is, of course, the essence of biological control: the introduction of the predators or other pests of plants getting out of control may lead to their collapse in numbers. The most successful example is probably the reduction of American prickly pears (*Opuntia*, Cactaceae) in the rangelands of Australia by the introduction of the moth whose larvae feed on such cacti in the Americas.

7.2.3   Combinations of factors

A biotic effect may be attained via an environmental one as in the case of allelopathy favouring particular associations of plants, mentioned in

Chapter 3. Furthermore, of those explanations that incorporate biotic and microenvironmental as well as disturbance effects, the most balanced account is that of Grubb (49), who used the term 'regeneration niche' to cover all the factors that affect a plant during its career from fertilized egg to death. According to this view, differences at any point in the natural history of the plants will allow their coexistence and an examination of related species' response to potential germination sites, shade tolerance, pollination or dispersal mechanisms and so on will disclose these differences. These then are features of species or subdivisions of them, by geography or sex for example, leading to a consideration of the form of organisms, so that differences in morphological features may reflect different lifestyles in closely related species, which can coexist.

In the neotropical Lecythidaceae for example (50) all species have large circumscissile capsules, but such uniformity disguises a wide range of variation in the texture of the mesocarp, degree of dehiscence, and development of the aril into a fleshy or winged structure, with the result that different species are dispersed by birds, bats, monkeys, rodents, wind, water and possibly fish. Another example, the Brazil nut (*Bertholletia excelsa*) has an operculum or lid that opens inwards so that animals have to get the seeds out, while in *Gustavia* the mesocarp is brightly coloured so that, once damaged, it is attractive to animals. In others the seeds hang out on funicles and are dispersed by bats; others have persistent calyces that allow the fruit to float. Germination features may differ, as in species of *Shorea* (Dipterocarpaceae) where, in Malaya, *S. curtisii* requires moister conditions than do *S. leprosula* or *S. parviflora*. It is also more tolerant of low light intensities following establishment and is thus able to regenerate under the shade cast by the palm, *Eugeissona tristis*. Clumps of particular species may reflect past tree falls in this way. In Costa Rica, the size of clumps of *Cryosophila guagara* palms corresponded to the size of light gaps created by fallen trees (51).

In the lower montane forests of Costa Rica, however, 23 tree species within the same family, Lauraceae, have the same insect pollinators and bird dispersers. There is little evidence for phenological character displacement and thus avoidance of competition, though fruit removal rates decline when many species are in fruit, i.e. there is competition for dispersal agents (52). In a similar case of a montane forest (53), between 56 tree species in several families, this time in Jamaica, there was, again, only slight interspecific temporal separation in the use of animal pollinators and dispersal agents, thus 'niche' separation in this sense is slight and the coexistence of the 56 species cannot be explained in this way.

Nevertheless, in ordinations of vegetation, the regeneration characteristics of assemblages of species may give meaning to the observed distribution of vegetation types, rather than the terrestrial features of the environment. This was revealed in Whitmore's classic study of the vegetation of Kolombangara in the relatively species-poor rain forest of the Solomon Islands (54). The aim was to make a classification of the forest types of the island using 12 selected large tree species when, as a blessing in disguise, the forests were hit by a violent cyclone. The regeneration strategy of each of the 12 species was worked out and the percentage occurrence of certain of these differentiating the forest types on the island reflected different meteorological regimes.

The element of environmental compared with predation disturbance is undoubtedly one that maintains diversity and if this is removed, the species total may fall. Since 1923, on Barro Colorado Island the numbers of species of reptiles, amphibians and birds have fallen since the effect of the secondary habitats associated with agriculture, prevalent there beforehand, has diminished. A similar effect is seen in British woodlands where the maintenance of traditionally disturbing management (like coppicing) keeps up species numbers of animals and plants, whereas abandoning of such management leads to a decline.

Geomorphological data support the idea that the western parts of Amazonia are under severe and constant erosion with surface losses of 2–3 cm per century, or 3–10 cm in the life of a tree, leading to great instability in the forest (55, 56). Lake sediment records in Ecuador suggest that there was regional flooding of western Amazonia 1300–800 BP, i.e. major hydrological disturbance, which raised the water level as far east as Manaus, while savanna regions seem to have persisted on the northern edge of the basin throughout the late Quaternary. The areas to the south-west and east, today enduring fluctuating intensities or durations of dry seasons, have a history of such. In northwest Amazonia (Venezuela), there is evidence for fire disturbance over the last 6000 years, with charcoal deposits under *terra firme* forests as well as caatingas, and some of the fires may not have been anthropogenic. The effects of drought, floods, erosion and fire on the forest of different types—savanna-like in the north, seasonally-flooded in the centre, with a seasonal climate in the south and south-west, a dry one in the north-east and the perhumid climate with species-rich forest in the west—has not yet been fully assessed. There is no reason to think that disturbance was any greater in glacial times, though there was a cooling and 10–20% less precipitation in the monsoon rains that could have allowed savannas to take over in what are today ecotonal areas. Unlike Africa or Asia, however, there have been throughout the great diversity inducing system of the great rivers driven by Andean precipitation.

Although outside the Amazon basin, there is evidence for glacial or late-glacial aridity, in the Galapagos, the Guyana Highlands and the Caribbean coast, but the sites are far away. Refuge theory would predict that where forest now stands today, there would have been savanna and that the higher wetter country would have been the 'refugia' in glacial times. The only site so far fully examined, at Mera (Ecuador), seems to show that there was a minimum temperature depression of 4°C between 35 000 and 26 000 years BP when the forest was replaced by a moist forest with Andean elements, implying forest descent of at least 700 m, the pollen spectrum there being unlike any in modern Amazonia; there are, for example, high percentages of alder (*Alnus*, Betulaceae) pollen. Colinvaux therefore suggests that the so-called refugia were thus denied to some rain-forest species presently there, i.e. the supposed migrations to them of lowland species in inducing today's diversity is questionable. The refuge theory has also been questioned in terms of the botanical information originally used to formulate it, in that it has been suggested that the 'refugia' may represent regions where collecting has been at its most intense. It has also been argued that the highest diversity is found in elevated sites most protected from fluvial erosion.

7.2.4 An element of chance?

In our attempt to understand the diversity of tropical forests, using largely the example of the trees, we have had to take into account all the features of the plant that would appear to reflect the forces of natural selection in terms of physical factors: meteorological, topographical, pedological, and biotic ones. Added to this must be an element of chance in that not everything can be evenly distributed nor is disturbance predictable, as reflected in the 'first come, first served' stands of pioneers. This is best exemplified by islands, however, and by Krakatau, with its suspended succession, showing that with a chance introduction of some late successional species, diversity would probably increase. Again, an examination of the isolated neotropical cloud forests by Sugden (57), shows that under certain conditions, the role of chance in terms of long-distance dispersal may be as important as the environmental factors in promoting diversity. The woody floras of those forests that are surrounded on all sides by arid woodlands consist of species with wide neotropical distributions and many of them are found elsewhere in a wide range of habitats while few are restricted to montane forest. Many of them are pioneers, characteristic of early successional sequences elsewhere, and are well adapted for long-distance dispersal, so that under isolated circumstances such pioneers can form 'mature' communities, even though the level of disturbance is low.

Birds in central Panama seem to track microclimatic optima, such behaviour producing dynamic (non-equilibrium) assemblages or 'communities' in space and time, but this is not stochastic as each species seeks optima in the current context of environmental conditions (58, 59). Even in chalk grassland in Great Britain, non-equilibrium models may best explain the coexistence of plant species and, in lowland rain forest, there is still disagreement about the importance of chance establishment of trees. Moreover, there is some confusion about the level of selection in that a genetically linked population is composed of particular trees on each of which selection acts, and differing in space and time in terms of biotic neighbourhoods as well as genotypes. In terms of genotypes, competition between individuals of a very polymorphic species may be less than between related 'jordanons' (selfing lines) or apomictic lines, even though these be isolated genetically from one another. Outcrossing promotes diversity suited to high biotic uncertainty and does not invariably lead to specialization of species but, on the contrary, may lead to greater generalization of a species and therefore similarity between species of a tree 'guild'. Species may therefore come and go from guilds without causing havoc there. Hubbell and Foster argue that such thinking could lead to a return to the classical views of rain forest and speciation espoused by Corner, Fedorov and van Steenis (60).

In conclusion, it may be remarked that, besides chance effects, the greatest diversity is found where there is an unseasonal climate, topographic and edaphic heterogeneity, and where the nutrient levels are low, but not too low, and where predation and other disturbance is high but not too high. As for an explanation of local heterogeneity, however, T.C. Whitmore at the end of his study of the Kolombangara forests, may be quoted. 'It can therefore be seen that variation from place to place in climax tropical lowland rain forest, which has for so long intrigued and challenged ecologists, is not open to any single or simple explanation. Rather, it is due to a complex interplay of extrinsic and intrinsic factors, which are not unique to the tropics and which have to be resolved individually for any particular forest.'

7.3 Practical problems

That there should be about 400 woody species in a hectare of South American or Malaysian forest has led to problems not only in accounting

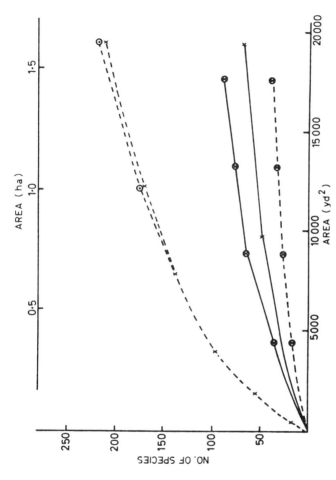

Figure 7.2 Species/area curves in tropical rain forest. Top to bottom: trees over 10 cm diameter at Bukit Lagong, Malaysia; the same at Sungei Menyala, Malaysia; the same in Guyana; trees over 28 cm diameter at Sungei Menyala; trees over 30 cm diameter in Guyana. Reproduced with permission from M.E.D. Poore in *Journal of Ecology* **52**, suppl. (1964) 221, Fig. 3.

for such diversity, the main subject of this chapter, but also in actually measuring the richness of such forests. Sampling is likely never to be comprehensive in terms of the total number of species recorded from an area (Figure 7.2) unless almost the whole area is used as the sample.

The time involved is also a major factor: in the 10 m × 10 m piece of forest in Costa Rica, discussed in Chapter 1, the analysis took 192 man-hours, in other words, a full hectare would take one man a decade (61)! In a recent study, 10% of Barro Colorado Island was used in sampling and only 80% of the known flora was found (28). The problem then arises that a sample of such size, or indeed any large plot, will mask any microtopographical diversity in the forest and thus any local species assemblages under particular microenvironmental conditions.

In an attempt to reach a practical solution, Brünig and Klinge (62) worked in Amazonian forest that was estimated to have some 700 plant species, excluding cryptogams, representing 70 families. In view of the problem of the masking of environmental heterogeneity, they found that small plots of 40 × 50 m were the most profitable in terms of time and information gained (even though they recorded only 500 common tree species in 35 families), and that for phytomass studies small plots were adequate. Nevertheless, they stressed that any sampling design should be rigorously stratified according to site conditions and the stage of regeneration of the forest, and that measurements of parameters such as stand height, basal area and tree density in the areas around plots should also be taken. In this way, they argued, a reasonable picture of forest diversity and structure could be built up.

CHAPTER EIGHT
TRADITIONAL RAIN-FOREST USE

The forest is a peculiar organism of unlimited kindness and benevolence that makes no demand for its sustenance and extends generously the products of its life activity; it affords protection to all beings, offering shade even to the axeman who destroys it. (Gautama Buddha)

Primate societies are remarkable among relatively long-lived and large-brained animal groups in that they have the capacity to store and retrieve much independently acquired information about their environment (1, 2). In the case of hominids, this has traditionally been attributed to selection for hunting efficiency in the Pliocene. It has been plausibly argued that this, perhaps too male-centred explanation, begs the question of what led to the increasing mental complexity necessary for hominid evolution in the first place, particularly when the older ancestors seem to have been tropical forest primary consumers. Milton has suggested then that the extreme diversity of plant foods in tropical forests and the way in which they are distributed there in space and time has been a major selective force in the development of advanced cerebral complexity in certain higher primates.

The capacity to use young leaves with a high protein content may have permitted the emergence of large primates in the first place, but primates capable of remembering the sites and fruiting and leafing patterns of a wide variety of plants, could move more directly to such food, when and where available, more efficiently than those searching randomly. Cohesive social units foraging a particular area over several generations would enhance efficiency by transmitting such information among closely related animals. Similar selective pressures associated with foods from the second trophic level may also have been of importance in the later development of the hominids themselves (2).

8.1 The fossil record

Most fossil evidence of early man comes from East Africa and, until recently, Africa has been almost unchallenged as the place of divergence of man's immediate ancestors from other hominids. New evidence from the Middle East (3) suggests that it may not be so simply explained, while a recently published account of a biologist's view of the myths embedded in Genesis lends some colourful support to a broadening of outlook (4). The thesis of that view is that man originated in rain forest and not in the more open habitats associated with the African finds. Possible early human ancestors existed in Pakistan two million years ago, while the much earlier *Sivapithecus* from India and Pakistan is held by some to be ancestral to all the great apes including man, or perhaps just orang-utans and man (5). An even older *Sivapithecus* (17 million years old) has been claimed for Kenya as being ancestral to the hominid–chimpanzee–gorilla lineage—a relationship which is confirmed by molecular studies—and the Asiatic orang-utan lineage; whether this animal is to be considered congeneric with the later species of *Sivapithecus* is disputed, however. The fruit-eating primates of Africa seem to have been the origin of the human stock, one way or another, whether the immediate ancestors were derived from an Asiatic '*Homo erectus*' or not.

Olive baboons in Gabon defaecate in their home areas, thereby concentrating edible-fruited species, and the Liebermans (6) speculate that the early hominids did the same because there is evidence that there were proto-gardens around occupation sites and that animals were unaware how edible plants came to be there; a phenomenon known as the 'Eden Syndrome' (7). Today (8), rambutans (*Nephelium lappaceum*, Sapindaceae) are distributed this way, growing up around deserted encampments of forest people in the Malay Peninsula, and several other fruits gathered there can only be swallowed as the flesh is either difficult to remove or the seeds are too fiddly to extract (as in blackberries, blackcurrants and tomatoes in modern society—sewage farms raise good tomato plants): for example the mata-kuching (*Dimocarpus longan*) and the rambai (*Baccaurea motleyana*, Euphorbiaceae). The distribution of such trees in the forest represent, at least in part, old cultivation sites, camps or once-only latrines. This is, of course, scarcely different from the effects of frugivorous birds as set out in Chapter 6.

The early Malays may have learnt the 'Malay knot' made by birds, perhaps by copying the animals (9), while their medicinal knowledge may have been common with that of other species: monkeys (10) in the Malay

Peninsula eat jelly surrounding the germinating seeds of *Scaphium* spp. (Sterculiaceae) and seeds are soaked overnight today by humans and used as a febrifugal mucilage. That from related plants is an important ingredient of modern western bowel-regulating preparations.

The forest does not offer the grazing typical of animal husbandry, however, and neither the plants nor the animals of domestication are forest organisms (11). Pre-agricultural man as a forest dweller would have left no trace, although, like D'Arcy Thompson, who argued that sacred groves, as revered in many tropical countries today, were the inspiration for columned buildings (hence the Greek temple and all modern architecture), it has been entertainingly argued by Corner that traces of forest life are embedded in modern society. He considers that the range of form in Pacific war clubs shows a gradation from the wrenched-up sapling, with its roots as spikes, to the mace of regalia. The digging stick of the forest becomes a spade, still with a wooden handle; lianes lead to ropes and string, bark and palm fibre to textiles, palm leaves to thatching and bamboos to piping. But the trick of firing the forest and thus a mechanism for massive clearing, is associated with seasonal forest, where man must have first made his mark.

Modern human fossils 100 000 years old are known and humans were widespread in southern Africa by 50 000 years ago at a time when Europe was still populated by the related Neanderthals (12, 13). *Homo sapiens sapiens* seems to have replaced the Neanderthals in a few thousand years about 30 000 years ago. Until lately, it was held that America was not reached until about 12 000 years ago and the southern tip of that continent in less than another 1000 years, but evidence recently gathered suggests that humans were in the Americas 32 000 years ago. There have been people in Australia for the last 40 000 years and claims for their presence there in the last interglacial have been made. Emigrations to both America and Australia were overland but after the ice-caps melted some 10 000 years ago, those populations became isolated from the others.

All these groups seem to have been nomadic or partly so. In Malesia, they seem to have made use of the limestone caves for shelter and may well have forced the orang-utans of the region to live further up in the canopy, for it is certainly true that in the Pleistocene, the orang-utans were much more terrestrial in habit (14, 15). It is difficult to assess the effects of the first true humans in rain forests but it has been argued that the Australian aborigines may have been responsible for the removal of some of the *Araucaria* forest of tropical Queensland, leading to the advance of *Eucalyptus* there, as well as locally exterminating much of the megafauna.

It may have been that the Palaeoindians moving through America from

Eurasia removed much of the Neotropical megafauna of the Pleistocene, but if humans were there in significant numbers as long ago as recent evidence suggests, then it is difficult to explain away the collapse of that fauna. Nevertheless, less than half of the mammal fauna of Java, for example, survived the mid-Pleistocene and the arrival of humans (16, 17).

8.2 Early agriculture

Truly forest people still survive in the dwindling rain forests of the three tropical masses: the Amazon Indians, the pygmies of the western part of Central Africa, and groups in the Andaman Islands, parts of West Malaysia and Borneo as well as the Philippines. Forest people are generally small and hence have a low maximal work capacity. They suffer from many parasitic diseases, and have high levels of infant and child mortality. An example of forest knowledge is that of the 'hunter gatherers' on the Philippine island of Mindanao, who recognize 1600 categories of plants. Although the Andaman Islanders still do not know how to start fire from scratch, wholesale clearing of other parts of the tropics has long been practised (18–21).

Hominids have been using fire in Africa for some 1–1.5 million years, but whether the early firing was the work of species of *Australopithecus* or *Homo* is not clear. There is evidence for forest clearing 11 000 years BP in Taiwan and of agriculture between 14 000 and 8 000 BP in Thailand, and New Guinea by 9000 BP, though tools much older than this have been claimed for the island; rice was being grown in Sulawesi by 6000 BP. The 'Hoabinhian' culture of southern China and Indochina to Peninsular Malaysia and Sumatra seems to have brought into cultivation candle nut (*Aleurites moluccana*, Euphorbiaceae), betel nut (*Areca catechu*, Palmae), species of *Canarium* (Burseraceae) and *Terminalia* (Combretaceae) for their fruits, cucumber (*Cucumis sativus*, Cucurbitaceae), the bottle gourd (*Lagenaria siceraria*, Cucurbitaceae) and the betel pepper (*Piper betle*, Piperaceae) between 8000 and 3000 BP. In Sumatra, the northern highlands seem to have been begun to be cleared about 17 500 BP but the firmest evidence is from 6200 BP, while the central regions may have begun being cleared some 7500 BP but certainly by 4000 BP; both areas had major clearances by 2500 BP, perhaps associated with irrigated rice culture. The clearances in neighbouring Java started at least as long ago as 4800 BP. In Borneo, Neolithic man (4000 BP) grew crops such as sago and kept domestic pigs (*Sus scrofa*) and dogs. Pollen evidence shows that even in the Amazon basin, maize was being grown some 6000 years BP and that there was forest clearance a thousand or possibly three thousand years ago at Lake Victoria in Central Africa. It must be remembered, though, that the majority of modern African forest dwellers are secondary invaders with knowledge of agriculture and iron tools, and may have been there only some 2000 years (22).

The form of farming known as shifting cultivation has gone hand-in-hand with the clearances in New Guinea over, at least, the last 5000 years. As with the clearance of the forest of the British Isles since the last glaciation, those areas with the 'least resistance', i.e. with shallow dry soils, have been cleared first. In such places regeneration is slowest, as can be seen from those areas of dry monsoon forest, long cleared away in East Java or around Angkor in Kampuchea. In time, plants were domesticated and selected so that they differ greatly from their wild ancestors, some so much so that their ancestors cannot now be recognized. In Asia, such 'cultigens' include the coconut, which may have had its origin on the Great Barrier Reef, and the mango, the talipot palm, *Corypha umbraculifera*, its leaves used as writing paper, and the betel nut with narcotic fruits. Plants now known only in cultivation but whose ancestry is fairly clear include sugar cane and rice and bananas, which are sterile triploids, many of those involving hybridity between a rain-forest species, *Musa acuminata* and *M. balbisiana* of drier forests.

Man may have reached coastal New Guinea as long ago as 50 000 BP and penetrated to the Highlands there by 26 000 BP (23, 24). The 9000-year-old agriculture there is now based largely on cultigens, including taro and bananas of south-east Asia, arriving before 4500 BP, but earlier phases could have been sustained using native species, including indigenous bananas, yams, pandans and sugar cane. These are largely propagated vegetatively and such 'vegeculture', based on cassava, may have antedated seed culture in northern South America. Vegeculture in the Neotropics as well as in south-east Asia, was replaced by seed culture, which is more suited to territorial expansion but originates in the more seasonal subtropics. Moreover, for foraging humans, tropical rain forests are too food-poor, although such foods are available to other primates, because they are too far up in the canopy, deep in the ground or the human gut is not capable of digesting them (25–29). The earliest colonists of the Amazon seem to have practised farming as well as hunting, gathering and fishing. The pygmies of Central Africa cannot find enough food in the Zaire forests and, indeed, there seems to be no unambiguous ethnographic account of any people living solely off forest products and the only tropical areas where it is believed that such people might have lived are in central Malesia. Anthropologists increasingly agree that until agriculture had evolved there could have been no modern 'rain-forest man'. For a very long time, western humans have been blinkered by a Rousseauian notion of national purity with no pollution from outside.

8.3   Hunter-gatherers?

The belief in 'isolated', 'primitive' hunter-gatherers, rather like the notion of 'pristine' forest favoured by biologists, is now being replaced by the theory that there seems to have been hundreds and thousands of years of interdependent contact with farmers and even 'states'. That the 25 groups making up the 15 000 hunter-gatherers of the Philippine Negritos practise minor desultory cultivation and intensive trading with non-Negrito agriculturists is not recent as has often been argued, for eighteenth-century reports indicate trade of forest products for rice, tobacco, metal tools, beads and pots, as well as shifting cultivation by the 1740s at the latest. Trade with China from the Malay Peninsula and Borneo by the fifth century is well-documented; the Philippines were involved shortly afterwards, while rice had been cultivated on Luzon, not far from the Negrito sites, since at least 1400 BC; linguistic evidence also suggests long interdependence.

The Late Pleistocene humans in the Philippines seem to have lived in the savannas, so that the pre-historic Negritos probably did not move into the forest until they had seasonal access to cultivated starchy food unavailable in the forest itself and they evolved a lifestyle to collect forest produce for trade. The pygmies seem to be similar, though some groups may have been forced by warfare into occupying the forests of central Africa. The Punan groups of northwestern Borneo may have taken up their lifestyle initially from demand for various forest products prized by the Chinese over 1000 years ago. Indeed, combined with the elevated notions discussed above, has been an intellectual imperialism for the 'primitiveness' perceived in such peoples becoming an excuse for colonialism and extermination. The controversy over the Tasaday, a group of 26 allegedly 'stone-age' cave-dwelling people 'discovered' in 1971 and pronounced a hoax in 1986 is based on the all-or-nothing 'cultural purity' argument of anthropologists. Headland has persuasively argued that, just as with the pygmies above, there was just not enough food for such groups in the forest itself; in this case, the people seemed to have been trading wild meat for goods including carbohydrate and, indeed, may have grown it themselves in the past.

Although pre-hominids left the forest, the first wave to return could only live there in symbiosis with peoples in monsoonal areas. Nevertheless, they modified the forest by nomadic behaviour and shifting cultivation, if they became truly independent. As they could no longer live in trees and needed the plants with which they had co-evolved in the more seasonal habitats in the beginnings of agriculture, at least vegeculture, those who did not cultivate had to trade. The second wave to return was European and could not manage without seeds and brought commensals and trading aspirations as well as religion. This second wave was most successful in the 'Neo-Europes' of temperate America, Australia and New Zealand but was perhaps much less so in the tropics because of micro-organisms, which the societies based on shifting cultivation there, had evolved to avoid.

Purseglove has argued (30) that Indian crops were known in China 4000 BP and African ones in India and *vice versa* a thousand years later. The Malesian region saw a rise in navigation and seamanship unrivalled elsewhere until recent times, so that by 1500 BC, or earlier, the south-west Pacific had been reached. Later these people reached all the inhabited areas of the Pacific and possibly America, it having been argued that they may have had a considerable influence on the rain-forest peoples of the New World, evidence coming from great similarities between certain groups in Borneo and central America, similarities greater than between these peoples and their geographic neighbours. They possibly introduced certain banana species and there are similarities in terms of longhouses, head hunting and blow pipes. Such contacts may also explain similarities between aspects of 18C BC China and the Olmec, the first great American civilization, notably in the use of jade (31).

In the New World, there was movement of crops between Mexico and coastal Peru by 1000 BC and, a few hundred years later, the people of southern Arabia were trading down the coast of East Africa. Important commodities were spices, the control of which was largely responsible for European involvement and exploration leading to the Spanish and Portuguese colonizations and ultimately to the Dutch monopoly of nutmegs and cloves in Indonesia by the end of the eighteenth century. Contact with mariners from China from before AD 300 led to trade in a number of commodities in the Asiatic tropics, notably resins, birds' nests, poisons, medicines, latex, pelts, ivory and rhinoceros horn as well as rattans. All are still traded today, the rattans being the stems of climbing palms largely used in 'cane' furniture-making and sometimes referred to erroneously as bamboo. By AD 500, there were probably bananas in Madagascar, imported from Asia, though some authorities believe Madagascar was not in effect colonized by man until 500 years after that. By AD 700, there were coastal settlements in East Africa, made by the Arabs on slaving enterprises into the interior, and, along their routes, the mango, initially from India, became naturalized.

The next phase of exploitation of the forest, the export of materials, is bound up, then, with the collecting activities of the early forest people. Contrary to what was to happen in the modern period (and still continues), the first objects of trade were not logs but what are sometimes rather dismissively referred to as 'minor forest products'. Intensively studied in this respect (32, 33) has been the Malay Peninsula, where there was trade with China, particularly in ivory and incense wood (*Aquilaria malaccensis*, Thymelaeaceae), the decaying heartwood of which is saturated with resin (the aloe or ahaloth of the Bible), in the period AD 420–589. By 960–1126 there was trade in these between China and south-east Asia generally and it was extended to caradamons, which had been exported to Europe from India since Roman times, rattans, camphor, dammars, bananas and *Dalbergia parviflora* (Leguminosae) a liane with heartwood used in joss sticks; the seeds of *Adenanthera pavonina* (Leguminosae) came into use as weights for goldsmiths and a system (Ganda) based on seeds half as heavy (another legume, *Abrus precatorius*) in India is derived from it. With the arrival of the first Europeans, the Portuguese, the all-important spices were traded at source: pepper from the Malabar Coast of India, cinnamon from Sri Lanka and cloves, nutmeg and mace from the Moluccas. With trading settlements and ports of call at Goa, the Cape and Lisbon, the Portuguese undermined the economy of the Mameluke Empire to such an extent that it was taken over by the Ottomans in 1517. The Portuguese also took Indian Ocean coconuts from their settlements in Mozambique to the Atlantic at the end of the fifteenth century, then from the Cape Verde Islands to the Americas: there is a great worry now that such a genetically narrow-based population could collapse in an epidemic (34, 35).

The original forest peoples of the tropics have been greatly reduced or, at least, the territory through which they could formerly move has. It is alleged that in the Philippines, for example, it is being lost at a rate of 3 ha per minute (36). At the time of the Spanish contact, there was about 90% forest cover and the population was about half a million; by 1945, the forest cover was some 75% and the population 19 million; by 1976, the cover was down to 38% and by 1980 27% and that partly disturbed. On the island of Luzon, the Agta people in the 1980s were down from about a thousand in 1936 to 609. Their high death rate is due to malnutrition, through ecological change, as well as introduced diseases. With an increasing non-Agta farming population, the Agta now view themselves as landless squatters in their own ancestral lands. Their most important protein sources are from the sea as land mammals are now in decline. New forces they have had to contend with include, as Headland so graphically puts it, 'obvious imported elements as the "green revolution", new diseases, multinational corporations, government roads, World War II, missionaries, government agencies sent to "civilize" the Agta, and international market price fluctuations of copra, as well as less obvious items such as modern medicines, radios and televisions, dynamite, insecticides, shotguns, alcoholic beverages, baby bottles, assassinations of national politicians and the influences of changing Euroamerican furniture fashions.'

8.4 Some rain-forest societies

Amazonian 'pre-contact' settlements (37, 38) were, in many cases, large and sedentary, particularly along rivers. The introduction of diseases by Europeans much reduced the population, perhaps by as much as 95%, the epidemics spreading ahead of the 'contact' via a network of trading routes, which connected the peoples to the Andean civilizations. The sites of many former settlements, on 'black earths', are now covered with forest but the savannas of Roraima are partly due to man-set fires and the open campina scrub on sandy soils was once cleared. Today the large populations are in *varzéa* rich in fish and turtle protein, much smaller ones on the *terra firme*, but they have now been in decline for some 500 years since the European invasion. A number of the more resilient groups have been intensively studied. It is thought that there are some generalized cultural patterns that can be recognized (39, 40). They have extended families of a man, his wife or wives, pre-adolescent children and married sons (patrilocal residence) or daughters (matrilocal), and their children, in a communal house. The house acts as a cemetery as well as a communal living and working space; it is built near a river for bathing, travel and food, and often in a cassava garden with *Guilelma gasipaes* palms and other fruit trees. Old house sites are regularly revisited as important food sources. Young of many animals are kept as pets, notably macaws and other birds including chickens.

A group of such households makes up a village and between such groups there is usually hostility from simple avoidance of one another to open warfare. At puberty, males become warriors and killing an enemy is often a prerequisite of attaining full adult status. Women and children may be captured from other groups and incorporated.

It is instructive to compare the ways the forest is exploited by two groupings of Amerindians far apart on the *terra firme* of Amazonia, the Jívaro, in the west under the slopes of the Andes, and the Kayapó in the east. The Jívaro occupy some 65 000 km² in the eastern lowlands of Ecuador and comprise some 20 000 people, women outnumbering men by about two to one. In spite of common language and culture, there is no permanent social or political cohesion. In fact, blood revenge and warfare are more intense between the constituent subgroupings than they are between them and other groups. Each village consists of a single house of a patrilineal extended family with 15–46 members. The house is abandoned when hunting is unproductive, the local fields are exhausted or the head of the village dies, so that in general one site is used for about six years. The Jívaro depend largely on agriculture and do not eat many wild plants, though they hunt a lot. Inter-village hostility may include the abduction of women or the murder of a member passing through alien territory, while a series of deaths in a village may be interpreted as the result of sorcery on the part of other villages. Such events require that the adult males take blood revenge and a single death restores the balance. Total warfare, though, aims at the annihilation of the enemy village, the death of all its inhabitants, the burning of the house and its contents and even the uprooting of the crops. It is plain, then, why men are in the minority. Sorcery and warfare lead to a thinly distributed, but quite large, mobile population. If the population declines, then so does the level of revenge killings and warfare. The population is also regulated by periodic suspension of sexual relationships for 3–6 months after the taking of a head and also from the time of birth of a child until he is weaned, i.e. 2–3 years. Each man has two or more wives and because adultery on the part of females carries the death sentence, the birth rate is rigorously controlled. As the whole system is headed by a man, it is likely that the village has to move more frequently than if it were headed by a woman. What seems barbarous is admirably suited to the rain-forest environment, for the scattering of the groups, their rigorous population control and their constant movement mean that the forest is nowhere completely depleted and recovers through cycles of regeneration before being farmed once more. The system represents a remarkable evolutionary adaptation of human behaviour patterns to the conditions of the rain forest.

The Kayapó of Pará, Brazil, unlike the Jívaro, live in a seasonal climate with a marked dry spell each year. The general pattern is the same in that the village is the largest political unit and that the cultural and linguistic inter-village similarities do not prevent hostility. In this group, the extended family may include some hundreds of people, while the time of greatest status for a male is that between puberty and parenthood. This has led to the adoption of a wide range of contraceptive techniques including the use of oral contraceptives (the biological origins of which are now eagerly sought by pharmaceutical companies). They also practise mechanical forms of abortion. Monogamy is universal but adultery frequent. The group subsists on gardening and hunting. The full status for the male is achieved only after killing an enemy, who, by definition, is anyone not a member of the home village. In this group, though, not every man need kill, for a blow dealt to the body during a war is sufficient. In major assaults, the village is burnt and the women and children captured and incorporated in the community. The Kayapó population density is 20 times that of the Jívaro and the villages are maintained indefinitely in one place only, at the cost of a periodic temporary increase in community mobility. In the dry season, groups of families travel out into the forest, moving camp every few days and living off wild foods, thus more uniformly exploiting the environment at a time of scarcity and the risk of irreversible depletion of the local forest.

Parallels between these two patterns of human behaviour and those of pioneer trees on the one hand or leaf-cutting ants on the other are clear. Moreover, there is a coincidence between centres of dispersal and enclaves of diversity among Amazonian languages and the plant and animal 'refugia' (41), but humans have been there for perhaps only 6000 years. Archaeological evidence shows that occupation of the lowlands in the last 3000 years was achieved by long-range dispersals from the north-west of groups with three ceramic traditions, though these declined in complexity as they could not be sustained by the *terra firme* environment.

8.5   Humans as ecosystem modifiers

The forest peoples of Borneo include groups who plant rattans as part of the shifting cutivation (swidden) cycle, the palms being planted as the gardens are left fallow after 1 or 2 years of crop production (42, 43). After 7–15 years the cultivators return to a mature rattan garden: the rattans are harvested and foodcrops planted. In the 1970s, the rattan industry based on *Calamus manan* and *C. caesius* was worth US$2 billion a year in the furniture industry. The financial return has disrupted the system in that the cycles have become shorter and shorter and palms were increasingly being taken from uncultivated forest. Indonesia has banned the export of raw rattan and this has encouraged other countries, like Sri Lanka, to emulate the 'rattan swidden' method.

Shifting cultivation in the Malay Peninsula is known from at least 6000 BP (44, 45). The 6000 surviving *Orang Asli* are divisible into three groups: the 2000 nomadic foraging and trading Semang (who may have been the first to arrive), with temporary camps in the forest; the Senoi with shifting cultivation; the Proto-Malays who replace the forest with their own gardens and fruit trees. The Temiar Senoi, for example, will stay up to 15 years in one settlement until travelling to more and more distant areas of cultivation becomes too troublesome, when they will move. Although there is a Temiar area over which this is carried out, there is no personal land tenure, though particular fruit trees may have owners. Such is the case with durians, which are deliberately planted and may be inherited. When a new site is chosen for cultivation, the undergrowth is removed, and later the trees, with an adze. The stumps and roots are left as are many big trees for superstitious or practical reasons (the two may be the same in many cases). Such a tree is the mighty *tualang* (see Figure 5.7), *Koompassia excelsa*, which has hard useless timber but harbours bees' nests.

In Sarawak (46), figs are considered to be homes of spirits and the Iban people will not fell them, even when clearing forest for rice culture; they are also preserved in Africa for similar reasons. Also left are useful fruit trees, which are thus under selective pressures. The dead material is left to dry, is then burnt and left to cool; planting in the fertilizing ash takes place after the first rain. The Temiar transplant young petai (*Parkia speciosa*, Leguminosae— grown for their edible seeds) seedlings into their gardens after the rice harvest and are reliant on bamboo for many purposes: house building, blowpipes and cooking pots for example, as well as knives to cut the umbilical cord in childbirth. Some bamboos are cultivated, some cared for, others gathered from the wild, e.g. *Bambusa wrayi*, a montane species, which is much favoured: clumps are owned by particular Temiar groups. Trading in dammars, incensewood and rattans sometimes yields financial returns so high that the people abandon cultivation in its favour. Over-exploitation of rattan does not indicate a high 'conservation of resources' sense, nor does their attitude to animals, of which they keep a wide range of pets from rats to tigers. The protein part of the diet is supplemented by hunting using bamboo blowpipes with the nerve poison strychnine, an alkaloid from species of *Strychnos* or from *Antiaris toxicaria*, the upas tree (Moraceae), or venom from toadskin or snakes. Trapping and fishing are also important.

Temiar gardens consist of a wide range of plants, apparently grown haphazardly and far from the regimented uniformity of plantation agriculture. This is typical of shifting-cultivation systems throughout the world. A similar system obtained in North America, where the Indians lived in villages, making clearings and foraging for firewood and other produce, setting fire to much of the forest to drive and enclose game. They planted food and medicinal plants, migrated seasonally and moved their villages (48). The villages were moved because of the depletion of firewood stocks (the Indians believed the English had arrived because they had run out of firewood at home). American chestnuts (*Castanea dentata*, Fagaceae) were planted in the villages; *Prunus nigra* (Rosaceae) and possibly *Gymnocladus dioica* (Leguminosae) were also planted. Most of what the first Europeans saw as original vegetation in north-eastern North America was manipulated regrowth. Moreover, shifting cultivation was still being practised in Czechoslovakia, for example, until the late 1970s at least (49).

Shifting cultivation is very variable in terms of the numbers of species grown. In central New Guinea, for example, there are monocultures of taro (*Colocasia esculenta*, Araceae) in small gardens; there is no general burning but the plantings are of several cultivars and the cropping period is brief, the fallows long. In contrast, certain tropical peoples grow an enormous range of plant species. An outstanding example (51) is provided by the Lua of northern Thailand who grow at least 120 different crops: 75 for food, 21 for medicine, 20 for ceremony and decoration and 7 for weaving and dyeing, a diversity that mimics the diversity of natural vegetation. The breadth of the food base is also seen in settled communities in the Malay Peninsula, where villagers may collect some wild fruit species but grow many in and around the village. Whitmore (15) recorded 29 types of tree around such a village in Trengganu, noting that 12 of them were identical with wild forest trees, such as the rambutan (*Nephelium lappaceum*, Sapindaceae) and the sentul (*Sandoricum koetjape*, Meliaceae), a further six appeared to be selected and improved forms of wild trees, five were cultigens like the betel nut, and three were exotics, like the papaya (*Carica papaya*, a New World cultigen). Even a formerly nomadic group, the Paumari in Amazonia grow 14 cultivars of the easily transported cassava (52) and some other groups there grow up to 400 different types of plants (cultivars as well as cultigens); some clear fell the forest, others merely 'weed' it.

Returning to the Kayapó, we find a well-developed ecological sense, as these people understand animal–plant interactions and the similarities between natural and man-made gaps (53–56). They use over 250 species of fruits and many hundred others provide edible parts; they cultivate some 17 cultivars of cassava, 16 of sweet potato, 13 of bananas, 8 of maize, 6 of annatto (*Bixa orellana*, Bixaceae, a dye-plant), 6 of arrowroot and 6 of rice.

The ant parts used to paint women's faces in the maize festival are from the ants that are attracted to the extrafloral nectaries of cassava and keep insects off the beans that are trained up the more robust maize. The Kayapó also create 'forest fields' of semi-domesticated plants, at least 54 species of them, which are collected in the forest and replanted near established camp sites. Some are planted to attract birds, some alongside paths in the forest and others are planted while the Kayapó squat to defaecate. Such plantings are transitional to full-blown gardens. Old gardens are not abandoned but are visited and cropped for up to 30 years; these also attract game and facilitate hunting. Such amounts to a kind of agroforestry type of manipulation of the environment and is known to be practised by other groups, e.g. the Bora of the Peruvian Amazon. With the higher populations of Amerindians in the past, such groups must have had a major effect on the structure of the forest.

The woody monocotyledon, *Cordyline fruticosa* (Agavaceae), is involved in ritual and magic from eastern Asia to eastern Polynesia and is planted around gardens, being readily propagated from stem fragments. It has accompanied humans in the western Pacific from the earliest times. It is particularly commonly seen in highland New Guinea, an island where spineless forms of the sago palm, *Metroxylon sagu*, have been selected; several fruit-trees are encouraged if not actively cultivated there. These include *Finschia chloroxantha* (Proteaceae), *Terminalia kaernbachii* (Combretaceae, possibly naturalized in Polynesia) and *Inocarpus fagifer* (Leguminosae), while species of *Artocarpus* and *Pandanus* are also planted in bogs and gullies. Widespread in the Pacific is *Broussonetia papyrifera* (Moraceae), originally from Asia and now grown from China and Japan through Malesia to America, where it is naturalized: it is the basis of Polynesian tapa cloth, which is made from its bark (57). Ornamental plants have also been selected and sometimes become feral, notably the forms of 'croton', *Codiaeum variegatum* (Euphorbiaceae) and the red-leaved forms of *Acalypha tricolor* (*A. wilkesiana*, Euphorbiaceae) and other plants.

The relationship with humans may be such that plants are no longer clearly 'wild' or 'cultivated', the forests including several 'cultigens' like the betel nut or *Uncaria gambir* (Rubiaceae), an important tan source in Asia; similar could be said of the date in north Africa and Arabia. Other plants, though not cultivated, may be encouraged as in the case of *Madhuca longifolia* (Sapotaceae (58)) with edible flowers and valuable oilseeds, owing its persistence and wide distribution in India to human activity.

The history of the domestication of many of the fruit trees of the tropics may therefore never be disentangled, though it is known that some of them at least, like the duku and lanseh, forms of *Lansium domesticum* (Meliaceae), are apomicts. It is interesting to note that among the fruits taken by birds in southern Europe (59), it has been shown in the case of *Smilax aspera* (Liliaceae), there is selection in favour of greater pulp:seed ratio (with a loss in seed numbers) in those conditions of high competition for dispersers, which, faced with an apparent glut, select the most nutritionally valuable. In the tropical cultigens, the same effect is achieved in a different way, but such selected lines of fruit trees become symbiotically associated with human beings.

The effects of humans can go far beyond this, in moving plants away from their natural range so that they appear 'native' in their new homes, often with the advantage of no pest pressure. Besides the mango mentioned above, there is the tamarind, *Tamarindus indica* (Leguminosae), which has been spread through the Old World tropics, such that its origin is now obscured. In the highlands of New Guinea, the nitrogen-fixing *Casuarina oligodon* (Casuarinaceae) is much planted as an essential part of the sweet-potato fallow but is probably not native there (60). Again, species of *Gigantochloa* (Gramineae) in western Malesia were probably taken by humans from Burma, where the genus is certainly native, and other bamboos may also have travelled along the ancient maritime spice routes between China, India, Sri Lanka and Indonesia. *Bambusa vulgaris* (61) has been carried throughout the tropics so that its home, too, is no longer recognizable. Indeed, coastal trading may be the only explanation for the peculiar distributions of some species, e.g. *Exotheca abyssinica* (Gramineae) in tropical East Africa and Vietnam and may account for that of *Stylosanthes humilis* ('*S. sundaica*', Leguminosae), which was perhaps taken from Brazil to Malesia by the Portuguese. Spreading to New Zealand by Maoris of the karaka nut, *Corynocarpus laevigata* (Corynocarpaceae) from Vanuatu and New Caledonia is the only reasonable explanation for its disjunct distribution (62).

It is not only the botanist who is troubled by the culture of tropical civilizations before the intrusion of western man. Remarkable disjunctions in the distributions of neotropical birds have been explained by their deliberate introduction to new areas by pre-Columbian bird-fanciers. Thus Haemig (63, 64) has argued that between 1486 and 1502, the great-tailed grackle (*Quiscalus mexicanus*) was introduced from its original home in Veracruz to the Valley of Mexico by the Aztec Emperor, Auitzotl. Similarly, the painted jay, *Cyanocorax dickeyi*, in western Mexico is found some 4000 km from the possibly conspecific white-tailed jay, *C. mystacalis* of Ecuador and Peru, and may thus be another pre-Columbian exotic introduction. Again, tiger teeth found in Bornean caves are believed to have been placed there by man and not to indicate the former presence of the animal on that island (15).

CHAPTER NINE
THE CHANGING FOREST TODAY

We have already observed, that the most luxurious vegetation of spontaneous growth affords us certain proof that the soil which has produced it will prove equally favourable for the production of the usual objects of culture (George Finlayson, *Mission to Siam and Hué... in the years 1821–2* (1826) p. 57)

By the time of his explorations in South America (1799–1804), von Humboldt's suggestion that the Amazon should be deforested on a large scale to improve agriculture was part of the European view of how tropical territories might be developed. With the establishment of colonies as opposed to trading posts, came the need for production and exports over and above what had sustained the earlier economic relationships between the West and the tropics. In the Amazon, the numbers of Indians declined, ceasing to be a threat to western expansion, so that by the second half of the last century, they were becoming more of a curiosity, regarded as anachronistic impediments to progress, which had to be educated on Christian lines on the one hand, the true lords of the territory to be given their freedom on the other. In Brazil, there came a romanticism of Indian life, though by the time of the establishment of the Republic in 1889, they had few rights and the rubber boom brought entrepreneurs to the uppermost headwaters of the Amazon followed by railways and telegraph lines. Developers hired murderers to exterminate Indians while in the coastal cities a pro-Indian movement gathered strength and by 1910 an 'Indian Protection Service' was formed. By then the population of Indians, about a million, was just over a quarter what it had been at the time of the European colonizations of the sixteenth century. Many groups have entirely disappeared except as blood mixed with immigrant blacks and whites, but some tribes resisted, to the despair of well-wishers imbued with the inevitability of Euro-Christian civilization. They rejected the new ideas offered to or forced upon them, convinced that their ancient ways were more appropriate to the forest in which they lived (1-6).

In general, however, the needs of tropical territories meshed into the modern economic order, have meant that the first wave of human invaders in the rain forest have had to take second place to the exploitation of the second wave. The effect on the indigenous peoples has been debilitating in that their loss of faith in, for example, indigenous medicine has been combined with a realization of the ineffectiveness of western medicine in treating psychosomatic illness. Besides the introduced diseases, they have suffered demoralization as qualities such as prowess in hunting, traditionally so significant, are no longer valued. The dominant cultures have very often not understood their ways, have gaped at their nakedness and laughed at the rituals they do not understand. But most significantly they have lost the use of their traditional lands. This has been due to agriculture and forestry activities, which must now be considered.

9.1   Forest conversion

9.1.1   *Farming and gardening*

The slash-and-burn type of shifting cultivation has been used continuously for some 2000 years in Indomalesia, each patch being cropped for only 2 or 3 years, the people at a density of some five or fewer cultivators per km². It is a practical and successful way of using land used throughout the tropics under different names, where poor soils, steep slopes and heavy rainfall obtain. It is successful where temperate zone type methods are not, and is generally held to be the most efficient in terms of soil recovery. Formerly considered the bane of tropical forests, perhaps in part as a way of keeping the 'peasant' economy distinct from the (colonial) plantation one, despite the fact that such was freely integrated in the system of exporting natural products, e.g. most of Indonesia's pepper, coffee, coconuts, tobacco and rubber is grown thus, it is now praised, possibly too much so, as a result of the increasing romanticization of forest living, for shifting cultivation is changing. With the arrival of western man, who took spices and introduced new crops, has come plantation agriculture and a reduction in the mortality rate. Smallholder agriculture has led to the concentration of people being three times that that would support shifting cultivation in some areas. There is insufficient time for the land to recover between cultivations and, largely through the agency of western man, exotic weeds become established and

Figure 9.1   Land being cleared for agriculture in the rain forests of northern New Guinea (October 1989).

encroach on the smallholdings. Inevitably, more forest land is brought under cultivation to counter this (Figure 9.1).

Before man started burning the forest, fires were the result of lightning, which is common in the tropics, and possibly of falling rocks and, in wooded savanna, fermentation under compaction. In New Guinea, in the seasonal forests of the Gogol Valley, is evidence of major fires, for *Intsia* spp. (Leguminosae), which readily regenerate after fire, make up some 25% of the log volume there. In Africa, it is argued that man was using fire at least 50–55 000 years ago, taking coals from camp to camp and using the fire to smoke out bees from their nests in honey-hunting, or driving game. This probably also occurred in the drier types of forest and savanna. Nevertheless, tropical rain forest may be flammable at the edges, so that savanna may spread. It has been calculated that because of burning, only some 60% of Africa potentially covered with forest actually bears it today.

Fires set to control weeds on degraded pastures (7, 8) also spread, particularly to selectively logged areas as in Pará, Brazil, as the gaps left have tracks connecting them and plenty of fuel: such have had devastating effects in Borneo during seasonal droughts (see section 2.2.2). The effects of fire are to increase light intensity at the ground, affecting germination but also the fauna and flora of the soil. There is an increase in evaporation and rain has more impact. Nitrogen is rapidly lost but immediately available potassium, magnesium and calcium increase as input from the ash. Run-off is increased, especially on slopes, and there may be marked compaction or other deterioration of the soil structure. Burnt sites improve grazing and tend to concentrate herbivores such as elephant and buffalo or antelope, some of which, in turn, attract concentrations of carnivores. Increased clearing in the Ituri forest, Zaire, seems not to have much affected the abundance of 16 out of 19 mammal species examined, though okapi, yellow-back duikers and leopards are less abundant.

After burning in the Indomalesian region, grasses appear and, if these are left without grazing or burning, will slowly be replaced by the forest. If there is grazing or cutting, the grass composition may change and, if burnt, it may become dominated by *lalang* (*Imperata cylindrica*) a grass that is very difficult to eradicate. Continuous use of the land for cropping leads to declining yields, such that rice production, for example, may be halved in 3 years, partly due to nutrient removal, erosion, the physical deterioration of the soil and partly to the multiplication of pests and diseases and an increase in competition from weeds. Relic terraces and raised fields in the Mayan lowlands of Central America have been interpreted as indicative of a settled, sophisticated, intensive, prehistoric agriculture, rather than shifting cultivation, which collapsed for just these reasons.

Of the many explanations (9, 10) for the collapse in the ninth century after such intensive cultivation without metals for 6–16 centuries, the most plausible is that it resulted from sustained failures of maize due to a leaf-hopper-borne virus, maize mosaic virus, which may have originated in northern South America at roughly the same time as maize was brought to the Caribbean by the Arawak about the time of Christ. Indeed, archaeologists now believe it was due to agricultural mismanagement combined with a booming population, for investigation of a major palace in the site at Copan, Honduras, has revealed that there was only a small contingent of specialist workers employed in constructing the elaborate structure, in other words that this was not enough to overburden the economy as is sometimes alleged to be the cause of the collapse of that civilization.

Certain soils, like the Kalahari and Benin sands of Africa carry good forest, but, if cleared, only two worthwhile crops can be grown. The luxuriance of tropical forest deceived many colonists from Europe, their efforts being commemorated by scrub and bare rock, nowhere more markedly than the 'Joden Savanna' of Surinam, the result of a short-lived farming attempt by Jewish refugees 150 years ago.

The traditional forest cultivators are now being joined in many parts of the world by subsistence peasants, new arrivals, pushing further and further into the forest, often along paths or tracks cut by lumbermen. In Peru, for example, people from the Andes are moving down and across as a front into the Amazon plain. It has been estimated that in the early 1980s there were 240–300 million people involved in shifting cultivation who occupied perhaps half of the total tropical land surface. They were converting the forest at a rate of some $100\,000$ km^2 per annum. The greatest effect is in Indomalesia, where some $85\,000$ km^2 are thought to be lost each year. In short, farming is reducing the rain forest by 1.5% of its area annually. This rate is likely to increase as many of the countries with moist forests—Brazil, Colombia, Indonesia, Kenya, Madagascar, peninsular Malaysia, Peru, Philippines, Thailand, Uganda, Vietnam and all of West Africa and Central America—have high population growth rates. Furthermore, pressed by urban problems and sometimes apparently with other motives, several countries, such as Brazil, Colombia, Indonesia (where, since 1905, at least 2.5 million people have been moved from crowded Java, Bali and Lombok, to the 'outer' islands—an official figure where two or three times this number may have emigrated unassisted (11)) and Peru are promoting transmigration schemes so that it is projected that the number of forest farmers could double or increase even more than that.

It has been argued that continuous agriculture could be maintained if a closed nutrient cycle could be achieved, the canopy not perforated so that leaching would be prevented and the forest floor would not deteriorate, and if nutrients were added to equal those exported as crops and the diversity of species maintained. The problems to be added to this list of desiderata, making this theoretical model almost impossible to attain, include the problem of all year-round warmth, so that there is no respite from pests and diseases, a rapid breakdown of litter and subsequently enhanced leaching after the harvest of the crop. Furthermore, the social aspirations of the cultivators themselves in a climate where actual and potential production per man per year is less than in temperate countries (12), and such simple facts as the rapid rotting of produce, have to be taken into account. The mixture theory is also matched by the fact that, in practice, many crops have had their natural resistances, whether they be chemical or mechanical, or indeed temporal like staggered germination, bred out. The question then comes back to tackling the problem of how yield could be improved using traditional methods. Such ideas as the use of more legumes or other nitrogen-fixers and the ingenious use of nitrogen-fixing bacteria in the floating fern, *Azolla*, in paddy fields in southern China offer a ray of hope. Remarkable systems used by Chinese farmers in western Borneo merit further attention (4). Here grains, pepper and rubber and many vegetable crops are interspersed with fishponds and livestock grazing, enabling the people to make permanent use of impoverished soils.

Farming, then, is the biggest devourer of rain forest. Perhaps the region that has received most attention in the press in this context is Amazonia. Formerly the Amazon region was more thinly populated than the Sahara, containing perhaps some 50 000 people (13), and importing food. The government's plan was to cut a 70 m wide swathe through the forest for a major highway and that 100 km both sides of this was to be developed. Every poor family that moved there from the crowded east of Brazil was to have about 100 ha and a small house. There was to have been an 'agropolis' every 50 km and a 'ruropolis' every 150 km. All the colonists were supposed to feed themselves, though each of the plots was supposed to be left half-covered with forest. Such an idea led to isolated fragmentation of the forest, but in the 10 years up to 1977 over 11 000 km of roads were built and settlers moved in. Land was given away and there were tax incentives to exploit the area, though increasingly it was large-ranching projects and multinational interests that took more and more land. By 1977, then, there were 300 cattle ranches and 6 million head of cattle, while a fifth of the forest had disappeared. In the 7 years up to 1977, 7389 families were established as opposed to the goal of 100 000 by 1974, largely because of poor agricultural yields, in part due to the use of the wrong rice cultivar for Amazonia, problems of storage of produce and of predators and pests, in an imported technology that was too temperate-oriented (14). The timber was burnt rather than used or sold. Such large companies as Volkswagen, which burnt down 120 000 ha, became involved. It was found that certain valuable cash crops like coffee, cocoa, oil palm and pepper grew well along rivers and that under the forest there was probably enough iron ore to meet world demand for 400 years, large deposits of bauxite, gold, nickel, copper and tin.

Thought of as a model for Amazonia (15), the great Jari scheme, where 4 million ha of land were bought along the Jari, a tributary of the Amazon, was planned to combine the mining and smelting of bauxite with logging, rice-growing and the establishment of plantations for pulp, as a world shortage of paper was envisaged. Over 4500 km of roads, four towns, an airport, railway, a port, schools and hospitals were built, the pulp mill came from Japan and was floated thence up the Amazon. Timber had been removed without nutrient replacement and there were major weed problems in the rice. Bulldozers were used to clear 1.6 million ha for *Gmelina arborea* (Verbenaceae), which can yield up to 30 m^3 per ha per annum, but the best stands yielded only 14 t, the scheme overall only 3–6 t. Pines and

Eucalyptus were tried but the pines produced only 6 t. Planting began in 1969, but in 1981, 15 years after the scheme had begun, the whole was sold off for a fraction of almost the billion dollars that had been invested in it. The stress on mineral extraction has left the settlers (now almost 10 million as was estimated early in 1982) in effect abandoned, and their fate is reported in terms of disputes over land, the sowing of their land with weed seeds from the air and even the poisoning of their drinking water.

A scheme, more successful from the settlers' point of view, was that in part of Ecuadorian Amazonia (16), where a 'settlements first; roads second' philosophy prevailed. It had more egalitarian holdings, less interest by speculators (because of its remoteness) and, as the government authority tended to be weak, the settlers were more self-sufficient in setting up schools and cutting tracks and so forth; moreover, it was less costly for central government. However, to lay claim to the land, settlers were compelled in such a milieu to be seen to clear it of trees, so the result was ecologically undesirable.

A further factor in the clearing of forest leading directly to farming is warfare. From 1962 to 1971 during the Vietnam War, the Americans (17) spread some 72 million litres of herbicide from the air, mostly Agent Orange and similar preparations. About 36% of the mangrove forests were destroyed; 10.3% of inland forests besides cultivated and other lands were sprayed, some of them repeatedly. In many areas, the recovery has been so slow that cultivators have been able to move in, while fires have been an added retardant to recovery, sometimes resulting in the familiar *Imperata* savanna.

9.1.2 *Logging and silviculture*

Wood is the most important commodity in international trade after gas and petrol (18), though most of it is used in the country of origin and much of it very close to source, being the only source of fuel for warmth and cooking for a third of the developing world. In the 1980s, the world wood requirement was some 3000 million m^3, of which, following earlier figures (4), some 47% was used as fuel (80% of it in the developing world), 43% for building and other 'solid wood' purposes (two-thirds of this in the developed world) and 10% for paper (some seven-eighths of this used in the developed world). It is estimated that the demand by the turn of the century will be 6000 million m^3. So far, tropical moist forests provide little more than 10% of the total wood used as solid wood and pulp, but as temperate forests become depleted or increasingly under pressure of the 'environmental' lobby to be managed for functions other than production, attention will inevitably be focused on the tropical resources.

Indeed, temperate forests (19) are worth up to 50 times more in return per unit area and some tropical countries are net importers of timber. Nevertheless, Japan is two-thirds wooded but imports half her pulp, and the United States grows half as much again timber as it needs and could easily become self-sufficient in hardwoods. In 1950, 4.2 million m³ were exported to the developed world, in 1980 some 66 million m³, and this could be 95 million m³ by the end of the century. Even now exports surpass local consumption, and Japan takes over half of them. Until lately, her major source has been Indomalesia, which is the origin of some 75% of the trade in tropical hardwoods. It is alleged that by 1989 30% of the world's tropical sawlogs in trade were from the Malaysian state of Sarawak, where forest was being lost at a rate of 3 ha per minute and the resistance of the local Punan to this had been contained. Certainly, world production and export of tropical logs and wood-products is dominated by Malaysia, Indonesia and Philippines and, in 1984, this trade was worth $4 billion dollars a year. The bulk of it was unprocessed but there have been moves towards processing it away from Japan and the industrialized countries of the region (20).

Recently it has become possible to convert simultaneously wood chips of a hundred or more hardwood species into paper pulp, which is important, as it is estimated that paper consumption will increase more than that of timber. The average citizen in the developed world uses over 155 kg of paper per annum (325 kg in the United States), whereas in the developing world, the average citizen uses less than 5 kg. It is estimated that demand in the developing world will increase by some 2.75 times per citizen and that in the developed world some three times, so that the proportion consumed in the tropical world is going to increase. Some parts of the developed world, much of western Europe for example, are self-sufficient in pulp, but Japan is heavily dependent on foreign sources.

At first, very few species were extracted for export. In South America, the mahoganies, *Swietenia* spp., were much sought after, so that today, they have suffered through genetic erosion to such an extent that often they are not worth exploiting, only those of poor form having been left in the forests to set seed. Some species have been almost exterminated, e.g. *Persea theobromifolia* (Lauraceae), formerly *the* 'mahogany' of Ecuador is now known from only 12 trees (21). In other forests only the greenheart, *Ocotea rodiaei* (Lauraceae), was exploited, the rest of the forest unused. In the Brazilian Amazon, exploitation was first in the *várzea* and *igapó*, but, since the 1970s, in the *terra firme* as well. Before the building of roads, it was peripheral and resembled that practised for the previous 350 years. In 1951, six species accounted for 89% of the harvest, 24% by volume being *O. cymbarum* and another 40% the meliaceous timbers—*Cedrela odorata*, *Swietenia macrophylla* and *Carapa guianensis*—cut close to the high water mark and floated downstream on the rising waters. Although there were losses to insects during the wait for the rise, some 60% of the logs travelled 1500 km. By 1973, there was still only 20% of the total from *terra firme*, but towards the end of the decade the proportion was over 40%, mainly *C. odorata* and *S. macrophylla*; only some 20–40 species were acceptable to sawmills. Only some 50 of the thousands of species in the Amazon as a whole are commercially exploited, though up to 400 may have some value. Africa exports only 35, of which 10 make up some 70%. So poorly known were the timbers in West Africa, that at the beginning of the century the houses and offices for British colonial civil servants were built from imported coniferous timber (22). In Indomalesia, there is concentrated interest in fewer than 100 species with exports of 12 or so. It is remarkable that despite these low numbers of species, little is known of their biology and one, the sebastião-de-arruda, a well-known timber export from Brazil since the last century, has only lately (1978) been botanically identified and given a Latin name: *Dalbergia decipularis* (Leguminosae).

In Trengganu, in the Malay Peninsula, the percentage of trees used in the 1970s (23) was 4.5–26.3%, the number of actual trees being 2.5–12.5 per ha. Because lianes often tie together several trees, felling of one tree may lead to considerable damage to others, while logging tracks (Figure 9.2) may take up to 10% or even 30% of the forest area (24). It is estimated that in southeast Asia, some third or two-thirds of the residual trees are damaged irreparably while up to a third of the area is left as bare ground, often compacted by the forestry machinery. It is argued that greater care in extraction would lead to a higher price of the timber but, in hill-forest in Sarawak (25) planned systems using directional felling are said to reduce logging damage by half without incurring additional costs. Nevertheless, logged forests are fireprone and, in Borneo, this has led to the devastation discussed in Chapter 2. In those fires in Sabah, 38–94% of the trees in logged forests were killed, 19–71% in unlogged nearby, with more than 80% mortality of saplings in both.

The effect on arboreal animals is largely to force them to leave logged forest and, as forests diminish, to promote their overcrowding. This is indicated by more frequent calling and, in the orang-utan, to fewer young, as the secondary forest is less acceptable to the apes, which prefer the

Figure 9.2   Cutting and extraction track for selective logging, made by bulldozers in New Caledonia (1984). Note the depth of the soil.

strangling figs of the 'primary' forest (26). Where selective logging is at a low density (10 stems per ha) in Pahang, West Malaysia, infant mortality in all primates is greatly increased, perhaps because of injuries due to the female adults having to make greater leaps from crown to crown across gaps or abandonment in the face of human presence or possibly reduction in available food. Nevertheless, the indications are that the populations recover if not disturbed again (27). Of ground-living mammals in Malaya over half of the species are forest animals and the elimination of the forest leads to the elimination of almost all the mammalian fauna except for a few rats. Added to this is the pressure from wild-life traders for skins, eggs, live birds, and butterflies, much trafficking in which is now illegal, but nevertheless rampant.

Perhaps even more destructive than these has been the effect of orchid collectors, particularly in South America in the last century. Large horticultural firms sent out or subscribed to the activities of professional collectors, who often employed local people to bring orchids to them and there was little hesitation in felling a tree to get a good specimen. It is alleged that one British collector had 4000 trees felled in tracking down one particularly rare species in Colombia (28).

Despite resistance of markets to novelty and indeed diversity and the fact that few mills can handle a diversity of timbers, the advent of techniques for pulping mixtures of species means that the diversity of the rain forest becomes less of a problem. Small areas can be exploited intensively and more and more may be capable of being processed; for example, one plant in Colombia produces pulp, adding tropical hardwoods to long-fibre coniferous chips (30%). Although timbers with high levels of latex, silica or high density are avoided there; another plant in New Guinea can now utilize wood from up to 200 species, including dead and defective trees, so that yields have already increased by some 300%. On the other hand, desolation in these areas is total. This plant exports the chips to Japan for processing but the scheme to replace the original forest with plantation to sustain the plant indefinitely has foundered because it has not been demonstrated that timber-growing is a worthwhile economic activity that does not threaten ownership of the land (29, 30). Moreover, with more recycling in Japan, long-fibred coniferous pulp from the *Pinus radiata* plantations of New Zealand and Chile is necessary for this, and increased imports from USA in the form of chips to help redress the balance of trade, there is little incentive for the plant owners.

It has been suggested that rather than rely on plantation, natural regeneration should be encouraged by clearing 200 m swathes through the forest. Narrower strips (4), perhaps 50 m wide have been suggested as appropriate in Amazonia, such strips being gardened for some years and replanted with trees as well; seed and mycorrhizae sources are close by.

Until fairly recently, tropical silviculture has largely relied on natural regeneration. 'Refinement' or the reduction of competition from non-usable species, often leads to a considerable increase in the area occupied by the crop, so that once this is cut, there are lots of gaps for pioneers. Sometimes these are valuable, as in the case of *Funtumia elastica* (Apocynaceae) in Uganda, a source of rubber. Systems involving the repeated return to the forest and removal of the best trees are referred to as 'polycyclic'. These cause repeated damage to the forest, including saplings of desirable species, and lead to the problem of genetic erosion of the crop, and are also difficult to administer.

Compared with the continuous damage, though small clearings, of polycyclic systems, monocyclic ones involve clearing large areas at one coup but there is then less disturbance in the time thereafter. In Malaysia, for example, there was a change from polycyclic buffalo extraction to monocyclic mechanical extraction: now, dependent on the site, the range of possibilities with these as extremes is considerable. When natural regeneration is difficult or inadequate, it may be supplemented artificially by planting out nursery-raised trees—'enrichment'—in gaps, lines or groups. In most extreme cases, close planting may require the clear-felling of the forest.

In 1980 (18), only 4.4% of tropical forest was under active management, and half of that was in India. Of the 828 million ha of productive tropical forest left in 1985, only some 850000 or so were demonstrably under sustained-yield management (31, 32) in the producer countries of the International Timber Trade Organization. Indeed, there has been widespread belief that management is not really feasible despite the fact that, in terms of producing a naturally-regenerating stand of timber trees without any sign of soil deterioration, there has been success for over a century, beginning with teak in southern Asia. Dawkins has suggested that, for success, the desired species should be commonly occurring pioneers, that there should be relatively easily controllable climbers, while ground-cover must be maintained. However, 'mega- mechanization', short-term financial calculations discounting the sustention of biological and soil factors (and thus favouring plantations), inadequate provision for silvicultural programmes or even the removal of the successful forests themselves have made the systems seem not to be viable. To establish a tropics-wide base for sustainable timber production, permanent 'forest estates' have to be established in the producer countries, with high public profiles and proper controls and more co-operation between the trade and environmental movements to create the best conditions for such management in appropriate forest lands.

Ecologically-based silviculture classifies the regeneration characteristics or syndromes into three major but overlapping categories, viz. (i) the pioneers (partially fire-tolerant) and the gapfillers (fire-sensitive), both incapable of establishment or growth in the shade, e.g. *Albizia, Tectona* (teak), *Ochroma* (balsa), *Macaranga;* (ii) trees of the consolidation phase, which are only capable of rapid growth when exposed, i.e. are light-demanding later in life, e.g. *Swietenia, Shorea, Triplochiton* (obeche), *Khaya, Carapa* ; and (iii) the trees that are shade-bearers throughout life and are incapable of rapid growth in the sun. e.g. *Ocotea* . The pioneering syndrome was discussed in Chapter 5, but in terms of silviculture, its important features are the relative longevity of seeds, the large annual growth increments—2–10 m in height (10 m in *Paraserianthes (Albizia) falcatoria*), 10–20 cm in bole diameter—the short lived nature of the tree and the light, perishable wood. The second group has shorter seed longevity and rarer seeding, germination in the open in some or in the shade in others, trees living for 100–300 years and producing light to dark woods of medium weight. The last group is the least known: the trees grow very slowly, live at least 300 years and produce dark, heavy, close-grained wood, much of which is unusable.

The manipulation of forest successional cycles can be seen as the ecological basis of tropical silviculture. The first-stage trees are largely used for pulp, while the second-stage ones are the bulk of the tropical timber industry. Many of these timbers are referred to as 'mahogany' of which some 200 sorts, many not true mahoganies (Meliaceae) but with timber features like them, have been listed (33, 34). The third-stage timbers have specialized, if any, uses. Notable is greenheart, used for locks and harbour fittings and ports, as in the Panama Canal, for it withstands sea water well.

Restrictions on silviculture in the tropics are most striking in the area of soil properties; for example, the pisolites of parts of West Africa are some 10 m deep with no obvious horizons and, although trees grow well at first, they later collapse and the removal of the forest leads to desolation resembling a gravel pit. Building on the experiences of mixed tree and herbaceous cropping, or agrisilviculture, elaborate ecologically-based schemes have been proposed, using a succession from grass and herb crops to shrubby and mixed perennial crops with an understorey of shade demanders. Such a system in Central America (35) begins with maize and beans followed by cassava and banana and palms, then cocoa with rubber and, finally, timber trees of the Meliaceae.

Pressures for increased production are forcing out the naturally regenerating forests as commercial propositions, though very specialized woods are difficult to grow otherwise. Plantations favour pioneers such as the pines, *Pinus kesiya* and *P. merkusii*, in the Indomalesian region. Plantations outyield by almost 10 times managed 'natural' forest and these pines have a long fibre suitable for pulp, but hardwoods, notably *Eucalyptus* spp. yield higher total amounts of dry matter, and the most productive may yield up to twice that of the best pines. Much research is now directed to the vegetative propagation of high-yielding clones, despite the potential danger of growing genetically identical monocultures over wide areas. Though increasing rapidly, plantation (4) probably still occupies less than 18 million ha in the tropics (about 15% of the world's plantations), an area of the same order of magnitude as that lost to agriculture annually. Of all such plantations, 85% are of pines, teak or *Eucalyptus*.

9.1.3   Ranching

In 1940, there were about $400000\,km^2$ of forest in Central America and this was reduced to about half by 1987, compared with perhaps $500000\,km^2$ standing at the time of the arrival of Europeans in the early sixteenth century (36, 37), when there was a large population: the Mayas with shifting cultivation and intensive terraced agriculture. With slavery and introduced disease, the population was reduced by 90% within two generations and, by 1550, there was a plantation economy exporting sugar, cocoa, indigo and cochineal but large amounts of the land originally cultivated reverted to forest. After independence in the 1820s, an expanding plantation economy removed the forest and became dominated by coffee and bananas, with a dangerous reliance on a few exports. With world price fluctuations in these commodities, diversification was sought by the 1940s, firstly cotton and then cattle.

Beef demand in the United States increased to a level equivalent to $60\,kg$ per head per annum by 1976, but increasing prices there encouraged commercial interest in Central American beef. However, such beef is lean as the animals are fed on grass rather than grain, so that it is fit only for the fast-food trade, one third of this meat goes into hamburgers. The trend was reversing by 1980 with American consumption falling and exports from Central America halved by 1985. Nevertheless, betwen 1950 and 1975 the amount of pasture doubled (4) and between 1966 and 1978, perhaps as much as $80000\,km^2$ of Brazil's forests disappeared in the formation of ranches. After the burning of the forest, pioneer trees appear and it is burnt again, and again the following year. The grass is then sown and it is periodically burnt thereafter to improve palatability.

With the addition of legumes and appropriate management, it is alleged that pasture productivity can be maintained but the grasslands are quickly invaded by inedible weeds such that by 1978, 20% of the pasture area in Amazonia was degraded. There are also major problems with compaction and the abandoned ranches become covered in a stunted tangle of indigenous pioneers and exotics.

9.1.4   Fuel

At present, only a small part of the fuel used in tropical countries is taken from tropical rain forest, little of it from 'primary' forest. Usually, it is gathered in scrub or secondary vegetation or local woodlots are utilized. Nevertheless, local supplies may become exhausted and, with an increase in the use of charcoal, particularly in urban areas, forests far away are beginning to be affected. Thus forests far from Bangkok are being converted into charcoal, as are those of northwest Kenya for the Nairobi market. The reduction of fuel availability leads to the use of dung, some 400 million tons of which are burnt in Africa and Asia alone each year. On the face of it, this seems good, but it must be remembered that each tonne used for fuel instead of as a fertilizer means a yield loss of some of 50 kg in grain production, for example.

9.2 The prospects

An attempt at modelling the future role of tropical forests in the world economy (39) had the export of tropical hardwoods doubling between 1986 and 2000 with the opening up of industries in other countries. This demand was extrapolated from the projected growth in such countries' gross national product. There was a projected shift of leading exporters from south-east Asia to tropical America as domestic consumption increased in the Asia-Pacific regions and as stocks became depleted. Because of the time needed to grow trees, plantations already established would not, before 2020, be capable of balancing that removed from natural forests, which would, in consequence, be under continued pressure. Imposed management is a necessary desideratum, as 11.3 million ha of mature forest land is being lost annually (40). Of this, 5.1 million is being converted to forest fallow, of which, in 1982, there was 409 million, with only 1 million ha of unforested land being taken over by natural regeneration.

Looking at different parts of the tropical world (41–43), it has been noted that by 1981 only 20% of all rain forests including tropical ones in Australia remained when compared with the total at European settlement (1788) and that there were only a few thousand ha left in northern Queensland, all accessible forest outside National Parks having been (or likely to be) logged, leaving an archipelago of 'refugia'. By 1983, Costa Rica had lost 78% of its wet lowland forest and all of its dry; by 1987 the Indonesian rain forests had been reduced from 65 million ha to 20 million and the Philippines from 16 million to one, though both countries had by then banned the export of unprocessed logs. Only 5% of the Madagascar rain forest presently survives and little remains of the Atlantic coastal forest of Brazil. Myers (44) has estimated that by the end of the century, at present rates, there will be only two relict blocks: western Brazilian Amazonia and Zaire, with smaller ones in New Guinea and the Guyana Shield of South

America, but that these are unlikely to last beyond 2050 because of cultivation as, for example, the population in Rondonia in south Brazilian Amazonia increased from 1100 in 1975 to well over a million by 1986, with an increase in cultivation from 1250 km^2 to 10000 in 1982 and 17000 in 1986. With such projections, then, over the next few decades, the world will lose the bulk of the rain forest remaining today. Many thousands of species of plants and animals will become extinct. Myers estimates a rate of loss of species from tropical rain forest at 10000 a year and compares this to a major 'evolutionary spasm', but unlike previous ones such as the end of the dinosaurs, the present phase has the plants as well as the animals in decline.

The explorers' accounts of the unending forests of the nineteenth century will seem as fabulous as mediaeval bestiaries. Politics and economics have heeded the advice of biologists and conservationists but little. Even where money could be made through tourism by conserving forests with their complement of birds and other animals to attract visitors, the animals have been shot and agriculture has encroached. This has led to the regazetting of boundaries and the diminishing of reserves. In this gloomy state of affairs, attention is being focused on the possibility of rehabilitation of habitats. This idea has borne some fruit in temperate countries, where spoil-tips, gravel-pits and other industrial eyesores have been returned to some sort of semblance of native vegetation with its associated fauna. What are the possibilities in the tropics?

9.2.1 Soils

In the Upper Rio Negro Region of southern Venezuela, a study on the recovery of *caatinga* forest was carried out (45). Two plots were cut and one of these was burnt after cutting. Shortly afterwards, an area nearby was bulldozed to make way for radar-tracking equipment. The sites were re-examined (Figure 9.3) 3 years after the initial disturbance, and recovery measured in terms of vegetation composition, biomass, nutrient accumulation, soil characteristics and nutrient leaching. The cut site was densely populated with many species of forest trees, the tallest of which had developed as coppice, while the burnt site had a loose canopy of pioneering *Cecropia* spp. 7m tall, and the bulldozed site had a thin layer of herbs, notably *Xyris* sp. The nutrient levels in the soil were higher, even 3 years after disturbance, in both the cut and cut-and-burnt sites, than in the surrounding untouched forest, probably because of steady transfer of nutrients from the forest ash to the soil. In the bulldozed plots they were

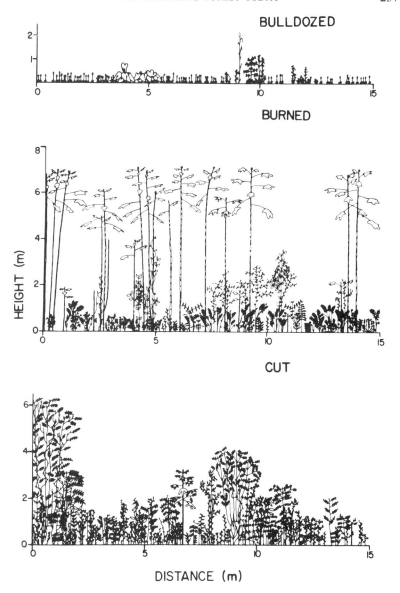

Figure 9.3 Profile diagrams representing the vegetation present in a 15 m × 1 m strip 3 years after the *caatinga* forest previously occupying the site had been cut (bottom), cut and burnt (middle), bulldozed clean (top). Reproduced with permission from C. Uhl *et al.* in *Oikos* **38** (1982) 315, Fig. 1.

lowest of all, because the topsoil had been removed. The above-ground biomass in the cut site was 1291 g/m², in the cut-and-burnt 879 g/m², while in the bulldozed only 77 g/m². From these figures, it was calculated that it would take some 100 years for the cut and cut-and-burnt sites to attain biomass levels of the original forest, while for the bulldozed forests, it is estimated that 1000 years would be needed.

Two forests in Costa Rica, one wet and one dry, also received cutting treatments (46). Biological features of the colonizing vegetation were seen to change with time and from the rate of change in these features a time for ecosystem recovery was calculated. Using mean seed weight, which increases with time, as a parameter, it was found that the dry site would take about 150 years to revert to the conditions obtaining in the original forest, whereas in the wet forest, the figure of 1000 years was obtained once again. This is far longer than in temperate ecosystems. On the Rio Negro in Amazonia it apparently takes 80 years to reach half the mature biomass (47, 48), suggesting that a return to forest is possible. Total stocks of calcium and potassium increase, apparently because of atmospheric input, phosphorus through mobilization of insolubles in the soil and nitrogen partly by biological fixation. Nitrification is higher under slash-and-burn regimes than under primary forest in southwest Venezuela and is apparently regulated by ammonium activity.

In northern Queensland (49, 50), soil under rain forest had markedly higher organic carbon, nitrogen, acid-extractable phosphorus and exchangeable calcium, magnesium and potassium than soil under a 41-year-old plantation of *Araucaria cunninghamii*. The loss of nutrients here is probably due to the original clearing of the forest and subsequent uptake by the crop. Soil-nutrients have been found to be at similar levels as under mature forest, at least so far as phosphorus, organic carbon and nitrogen are concerned, though the pH and cation concentrations are higher. Low-use sites were found to be depleted only in nitrogen though the heavily used ones were depleted of potassium and magnesium and possibly of calcium and phosphorus as well. The soils of the secondary fernlands of southwestern Sri Lanka (51) are dominated by *Dicranopteris linearis*. Rain-forest litter *in situ* is reduced by 63% in 1 year, but only by 41.7% in the fernland; fernland litter is reduced by 24.6% in the forest but only 15.7% *in situ*. Immobilization of nitrogen occurs in both types in both sites but phosphorus is immobilized only in the fern litter. Burning of the *Dicranopteris*, which acts as a nutrient-conservation system, leads to its shooting out from the rhizomes but trees cannot get established. For the growth of a forest, the fires have to be stopped and some enrichment planting of early successional species tried, e.g. *Canarium zeylanicum* (Burseraceae), *Artocarpus nobilis* (Moraceae) and *Vitex pinnata* (Verbenaceae), etc. to restore the tree cover.

Losses of soil through erosion vary greatly: under closed forest it may be 2.55 t per ha per annum in Trinidad, less than a tonne in Java or 0.41 in French Guyana, while under secondary forest in Madagascar it was nil, but up to 9 t when cleared and cultivated. Under *Eucalyptus* plantation it was 0.025 but, under other crops, up to 59 t (52). Loss of organic matter in the upper layers of the soil (35) was found to be at a rate of 9% per annum in the first 2 years in Trinidad, while after clear-felling and burning in Ghana, up to 13% was lost per annum and up to 33% of the cations. On volcanic soils in the Solomon Islands, even higher figures have been recorded. In 22 years, croplands in Costa Rica lost 60% of their calcium and 25% of their magnesium even though the fields in the first year were higher than the surrounding forest in these ions, through ash input.

In Colombia, a forest of known bioelement constitution was subjected to treatments leading to burnt and unburnt fallow. These treatments led to cation losses of 100–240 kg per ha per annum calcium and 30–80 of magnesium, for example, but it was estimated that such would be restored through rainfall input over 10–20 years of fallow; the 1300–1400 kg of nitrogen appears to be more easily replaced as fixation rates of 100–150 kg per ha per annum were recorded (53).

When forest is cut in Amazonia (54, 55), cations from the burn de-acidify the soil, releasing phosphorus from iron and aluminium compounds, but leaching rapidly removes the cations, only 18% of the standing crop potassium remaining after 3 years, while in crops only 3% of the nitrogen was incorporated (only 1% in the edible parts), 12% of the calcium (1% in edible parts) 24% potassium (4% in edible parts) and 5% of the phosphorus (1% in the edible parts). As the cations are leached, the acidity rises and the phosphorus becomes re-locked with the iron and aluminium, so that in some cases at least, phosphorus is the first limiting nutrient. In the Rio Negro region, there is a large increase in leachate of potassium, magnesium and nitrate in the first 2 years after burning, but, by the fifth year, it is not significantly different from those in undisturbed forest, when the general nutrient levels are similar to those in mature forest. Fires associated with cultivation volatilize 300–700 kg nitrogen per ha from vegetation and surface-litter and a further 500–2000 disappear from the surface soils in the next year or two. If overall losses are estimated at 1000 kg per ha per year, then the 20–25 million ha under shifting cultivation or permanent conversion from forest to other uses, generate more than the total nitrogen delivered by rivers to the oceans. At Turrialba in Costa Rica, the mineralization of nitrogen is higher in secondary forest there than in most forests so far measured elsewhere. With disturbance the rate is further increased by two or three times and then decreased to comparable levels after 6 months or so, the short-lived increase representing 300–400 kg nitrogen released to the site. The nitrogen level in the microbial biomass was only 50% of the original 1 year after clearing and remained low; little mineralized nitrogen was lost to the atmosphere by denitrification but much may be retained low in the soil profile. From a practical viewpoint, it has been found that *Eucalyptus deglupta* becomes established best in enrichment sites rather than after clearing or burning, apparently because of greater nitrogen availability in the unburnt system. Such reduces the need for supplementary fertilizers (57).

Amazon soils cannot provide nutrients from the parent material as can soils in temperate regions because the Basin has a Tertiary deposit of kaolinite derived from the sediments eroded from the surrounding highlands, but in some areas are patches of black soil, developed by the Indians for intensive cultivation by addition of organic matter. Such patches still retain high levels of fertility. Interest has therefore been taken in the effect of adding fertilizer in an attempt at rehabilitation. In a study, again in Costa Rica (58), the vegetation biomass, nutrient content and species composition were measured during the first year after clear-cutting of plots, some of which were treated with commercial inorganic fertilizer (70 kg N, 87.5 kg P and 166 kg K per hectare) while others were planted with the pioneering *Cecropia obtusifolia*, as seedlings. These latter plots produced the highest biomass and nutrient standing crop while neither of these features was enhanced in similar plots that had been fertilized. The plots allowed to regenerate naturally were dominated by shrubs and trees on the unfertilized ones, though by herbs on those that had been fertilized. These latter had the lowest biomass and the lowest nutrient standing crop. In short, fertilization enhanced neither the build-up of biological materials nor the capture of nutrients by it. It merely retarded the successional process by enhancing the competitive ability of the herbs, in this case *Phytolacca rivinoides* (Phytolaccaceae), which dominated. A similar retardation has been recorded from Fiji where the fertilizer applied was some 195 kg P per hectare (59).

Continuous agricultural crop production on tropical rain-forest lands hinges on the application of fertilizer and rather sophisticated advice and soil analyses year by year for success (60, 61). Tropical soils often have too little zinc for rice or citrus, or molybdenum for legumes while reclaimed peats have too little copper for many crops, while there are low levels of nitrogen and phosphorus almost everywhere. Some nitrogen-fixing trees are useful, e.g. *Inga jinicil* (Leguminosae), used as a coffee-shade in Mexico, fixes more than 40 kg nitrogen per ha per year, though *I. vera* fixes none. Nitrogen-fixing bacteria inhabit the rhizosphere of some tropical grasses but the amounts fixed for sugar-cane and rice, about 30 kg per ha per year at a maximum, are unlikely to satisfy crops, though they may be of significance in pasture and natural ecosystems. After clearing there seems to be a predictable cycle: firstly a phosphorus deficiency, then nitrogen, then phosphorus again. Nitrogen is abundant to begin with but there is an initial shortage of phosphorus because of its immobility, but nitrogen soon falls through leaching and uptake. Later, roots reach deeper phosphorus and this therefore is not limiting though nitrogen is, but later still all the phosphorus available has been taken up and is being recycled. Therefore in the early stages nitrogen and phosphorus fertilizer will improve growth, but the sooner the deep-rooted mycorrhizal species become established, the quicker there will be self-sufficiency in phosphorus, though phosphorus is likely to be limiting as the agroecosystem matures.

9.2.2 Successions after clearing

Africa Regrowth after cultivation in Nigeria is characterized by such tree colonists as the composite, *Vernonia conferta*, and *Trema orientalis* (Ulmaceae), with which the more slowly establishing *Musanga cecroptoides* germinates, to dominate the stand from year 3 to 15 or 20, when it is followed by the shade-bearers proper. In Ghana, the *Musanga* stage, which also includes *Macaranga* and *Ficus* spp., is taken over at about Year 10 by *Funtumia* (Apocynaceae), *Albizia* and *Chlorophora* (Moraceae) from which the high forest develops (62–64) but *Terminalia superba, Triplochiton scleroxylon* and *Chlorophora*, however, can regenerate on farmland without the intermediate *Musanga* stage.

Many observations on these successions have been made, as they have been in temperate regions, on abandoned cultivation. In the Ghana example, the pioneer tree stage is prefaced by coppice regrowth of the trees left in the agricultural system and a phase of herbaceous and sub-shrubby weeds and grasses, which is followed by a thicket of the lianes, such as *Adenia* (Passifloraceae), *Entada* and *Acacia* (Leguminosae). Where large gaps are formed in Ugandan rain forests, such light-demanders as the mahoganies (Meliaceae) become conspicuous, so that it has been suggested that the mahogany-rich forests of West Africa are truly secondary.

A 5-year succession on a stand of $800\,m^2$ in Ghana had short-lived pioneers, some with an annual increment in height of 4 m, which were unable to germinate in shade, and long-lived ones with similar germination but persisting into mature forest. Within 1 year, tree density was 2.5 per m^2, 95% of these being pioneers. Thereafter, the density declined exponentially, principally through death of pioneers, to about one tree per m^2 after 5 years. At that time about 90% of the pioneer species recorded there were present, as were some 60% of the 'primary' species, the number of pioneers declining with one or two primary ones being added each year, complete with about 10 non-tree species. The source of seeds was about equally divided between the seedbank and newly dispersed propagules. Within 3 years the species diversity was greater than in the surrounding forest and by 5 years some 25% greater.

America   A detailed study of regeneration after clearing in Mexican rain forests (65) took into account similar large-scale gaps in which five stages of regeneration could be recognized. The first was dominated by short-lived herbs with shrub and pioneer tree seedlings and could last for months or the succession could even remain in this state if pasturing ensued. The second stage was dominated by the shrubs that shaded out the herbs and saw the appearance of shade-tolerant species, which require lower temperatures and light levels for germination. This stage could last from 6 to 18 months and is one of rapid growth dominated by the short-lived shrubs in such genera as *Piper* and *Solanum*, reaching some 1.5 to 3m in height. The third stage, which lasted some 3–10 years, was dominated by the pioneering trees of low stature: *Trema*, *Miconia* (Melastomataceae), for example, yet also contained young examples of the taller pioneers such as *Cecropia, Didymopanax* and *Ochroma* (balsa), which when larger, dominated the next stage, which may last from 10–40 or more years. In their shade are saplings of the trees which eventually dominate the forest and reach some 25 m. The animal-dispersed pioneer species in this study have long fruiting seasons. Examination of the digestive tracts of birds showed that, out of a sample of 167 (37 different species), individuals of some six species had seeds of both *Cecropia* and *Trema*, individuals of 19 just the former and individuals of 24 just the latter.

In Suriname (66), it was found that an area deforested with heavy machinery and surrounded by *Pinus caribaea* plantations first bore a flush of grasses and sedges 1.5 m high after 9 months. By 16 months there was *Cecropia obtusa* to 4.5 m tall and many shrubby Solanaceae; by 30 months the *Cecropia* was 8–11 m tall and there was a distinctive shrubby layer, diminished grass cover but a 7% cover of lianes. After 5 years the *Cecropia* was at a maximum of 11–17 m but after a further 15 months started to decline so that by the end of 11 years some 45% were dead and all the Solanaceae had disappeared. Almost all the canopy trees at this stage had been recruited within the first few months after clearing. Of abandoned pastures in Amazonia (67, 68), the lightly used ones regenerated forest approximating to the young stages of mature forest after some 8 years; heavily used ones showed very poor regeneration and approaching 10% of them in Pará seem to be beyond restoration. Those with 'new' forest are sometimes very different, though this is not a new phenomenon in that the campinas—low open scleromorphic vegetation on leached white sands on the lower Rio Negro—are thought to have been cleared by Indians as they were occupied by the Guarito subculture in about AD 800. Having been abandoned, the land is recolonized very slowly after clearing because of the lack of nutrients and excessive drainage in the dry season.

Asia to Australia As discussed above, the fires in Borneo in 1983 had a devastating effect, unlogged forest being severely damaged, for example near Beaufort in Sabah where there was ground fire: in the logged forest there was not only severe ground fire but also crown fires killing most of the trees, the mortality of the larger trees being in inverse proportion to their girth. In 2 years, the logged forest was invaded by weedy lianes, herbs and ferns, the unlogged by pioneer trees such as *Macaranga* spp. After felling and burning in northern Queensland (69), it was found that of 82 tree species examined 23 months after the disturbance, 74 had coppiced, 10 produced root suckers and 34 had germinated from seed, these last having the highest growth rates.

To date, there have been few very long-term studies, though at Kepong, in Malaysia (70), a plot of some 0.3 ha isolated from forest seed-bearers was colonized by 21 woody species of which *Melastoma malabathricum* (Melastomataceae) was overwhelmingly dominant. As this declined, it was replaced by the rampant fern, *Dicranopteris linearis*, preventing further establishment of woody plants. Eventually, this was overcome by the canopy of the initial colonizers by year 14 and, by year 30, 51 species had established themselves but not until year 32 did the first dipterocarps colonize, although the nearest parent tree was only 180 m away.

9.2.3 Animals

If 10 000 ha of lowland forest in Sumatra is cleared, there is a loss (71) of some 30 000 squirrels, 5000 monkeys, 15 000 hornbills, 900 siamang, 600 gibbons, 20 tigers and 10 elephants. However, as was outlined above, logged forest can support some animal groups and will become more important as the unlogged ones dwindle. The habits of the animals may change, though, in logged forest in Pahang, Malaysia (72, 73), both the lar gibbon, *Hylobates lar*, and the banded leaf monkey, *Presbytis melalophos*, have made little change in their home ranges but spend less time feeding—perhaps food is more abundant—and, in consequence, spend more time resting. Even after the devastating Sabah fires, only a few months after the burn, four primate species and tree shrews were seen, though mammal densities were low and there was an almost complete absence of large frugivorous birds like hornbills and a decline in diurnal squirrels with fixed home ranges. Orang-utans turned to bark for food; long-tailed macaques, *Macaca fascicularis*, shifted to less preferred foods and exploited the increased insect population that accompanied the regeneration (74).

Conversely, some species survive in large tracts of logged forest but would be lost from isolated refuges of 'primary' forest, e.g. nectarivorous birds, hornbills and very large animals like Indian elephants that need to range widely. Other species can increase, even in the most transformed ecosystems, e.g. in oil-palm plantations in West Malaysia, the wood rat, *Rattus tiomanicus* can attain a population of 57 350 per ha., i.e. up to 5.7 t per km^2 (75). Generally, though, the overall diversity and species biomass of small mammals in plantations, as in northern Borneo, are much lower than in logged forest as canopy species, which make up a major element of the forest fauna, are almost absent (76).

In Malaysia, clearings 0.5–1.0 ha took about 10 years to acquire true shade-adapted birds (77). With selective logging, 'gaps' become the matrix and shade-birds suffer but in 20 to 25-year-old regrowth, where all forest species were present, they used it differently, with more itineracy, less territory holding and reduced breeding success. In southern New Guinea (78), the avifauna of secondary habitats is notable for the absence of some forest insectivores, while 15 or more resident and 11 occasional species are largely confined to secondary vegetation. Some 15% of the non-aquatic avifauna of Amazonia is restricted to riparian habitats, but 99 of the 169 species there rapidly invade anthropogenic secondary growth, rather as mangrove species do in Malesia (79). With major forest destruction, there are further ramifications as on Puerto Rico where the forest cover was reduced to 1% between 1508 and 1900, while the birds were persecuted by the inhabitants (80), so that 11.6% of the avifauna is extinct there. The extinction and restriction of the island's frugivore avian dispersers are perhaps responsible for the apparent failure of *Dacryodes excelsa* (Burseraceae) to regenerate.

As to insects, there are fewer euglossine visitors to baits in fragments of *terra firme* forest than in intact; a cleared area of only 100 m² is a barrier to at least four species. A failure of the insects to visit flowers could lead to reduced seed set so that it is clear for maintenance of euglossine-pollinated species, large intact, rather than fragmented, blocks of forest need to be conserved (81). After deforestation, termites (82) are reduced in species richness and certain generalists can proliferate to become pests of agriculture and forestry, such impoverished communities again resembling those found in hydro-edaphically limited sites, such as swamps and mangroves, rather like the avian examples above.

After burning in Borneo (83), there are no leeches to be seen but there are very large numbers of mosquitoes and spiders. Mosquitoes spread yellow fever in tropical America, where it is a lethal disease of monkeys and marmosets (84); when the trees are felled, human beings are exposed to clouds of mosquitoes and may be bitten, serving to spread the disease to urban areas should they be bitten when the virus is in the peripheral blood. In Africa, it is spread by primates visiting plantations where mosquitoes may bite them and then bite humans: the vector, *Aedes aegypti*, rapidly exploits domestic rubbish as well. The most important vector of malaria, *Anopheles gambiae*, breeds almost exclusively in earth-lined man-made pools, from footprints to bulldozer tracks, which are all increased during logging operations.

9.2.4 The 'new' forest

If it could be ensured that there would be no further interference from man during the re-establishement of tropical rain forests, would the resultant forest actually resemble that which has been lost? All the evidence suggests otherwise. Abandoning of logged-over forest leads to the establishment of trees that develop their mature crown characteristics at a height much lower than that at which trees in closed forest do (85–88). Ng has calculated that height reductions of 25 or even 50% can be expected. This leads to a vertical compression of forest structure and a reduction in the living space within the forest. This is already well known in temperate countries: trees established in parkland are shorter and with branches nearer the ground than in closed forest trees, and the 'bog-oaks' dug out of the fens of East Anglia show that the boles of the British oak forests of prehistory were much greater in size than any grown today. Even the great ecclesiastical buildings of the past incorporated such timbers taken from the relict forest surviving until the mediaeval period. Some documentary evidence survives in the form of paintings, notably by the Dutch and Flemish masters, who depicted the last of the spindly forest-grown giant trees of Europe as they disappeared.

Indeed, the modern clearing of the tropical forests is a continuation of the worldwide clear-felling initiated in the temperate regions. The removal of the Mediterranean forests in antiquity, much accelerated in recent times, and the clearing and desiccation of the region of the Tigris and Euphrates have had devastating effects on landscape and history. The cedars of Lebanon are a pathetic memorial to the lost vegetation. An example as recent as the late 1940s was the flora of the Mount Athos area in Greece, which was exposed to some 947 peasants and their 30 000 beasts for 2 years, in which time it was browsed off and twelve of the endemic plant species there have become extinct as a result.

In rain forest (86), 'large gap' species include many canopy or emergent tree species described as long-lived pioneers, e.g. *Cavanillesia platanifolia* (Bombacaceae), *Dipteryx panamensis* (Leguminosae), *Miconia argentea* (Melastomataceae), and these are often prominent after logging, though they are intolerant of continued disturbance. In the coastal rain forests of West Africa, one of the most common of the widespread taller trees is *Lophira alata* (Ochnaceae), which, when abundant, usually indicates abandoned cultivation; in the drier forests north and south of the African rain forest, *Triplochiton scleroxylon* and *Terminalia superba* sometimes occur gregariously and can regenerate well on old farmland and have thus greatly extended their range in recent time. In Cameroun, on the other hand, *T. superba* has penetrated deep into the forests and even reached the coast (87). Moreover, in Australian rain forests (88), secondary succession is often arrested by the absence of propagules of the later phases so that much of it in northern New South Wales, for example, appears to be held at an early secondary stage. Introductions can have a deflecting effect too: *Lantana camara* can block it at the pioneer phase: this South American verbenaceous shrub can form dense tangles several metres high penetrating into only slightly disturbed forests, as in parts of New Caledonia for example. Again, the entry of a particular species can change the disturbance regime, e.g. eucalyptus regeneration in Australia can increase fire effects.

As forest areas are reduced (89), the edge effects become greater in terms of immigration from anthropogenic communities of animals and plants, while the forest animals can forage out in the anthropogenic habitat. Paradoxically, when the forest is surrounded by agriculture, rather than disturbed forest vegetation, the effect is reduced. Introduced exotic organisms pose added problems, e.g. the American *Cecropia peltata* has been introduced to Africa, probably from the Caribbean, and now seems to be displacing the allied *Musanga cecropioides* in south-west Cameroun (90). Again, selected forms of the forest tree, *Melia azedarach* (Meliaceae), apparently domesticated independently in India and subtropical China, are now widely introduced as ornamentals and are running wild in New Caledonia, Madagascar and southern Africa, as well as in disturbed vegetation in the natural distribution of the 'wild' tree, e.g. in Australia (91).

Introduced animals (92, 93) are usually less obvious at first sight, but their effects can be far ranging, as in the case of 'Africanized' honey-bees in tropical America. Some 26 mated queens of *Apis mellifera scutellata* were introduced to southern Brazil in 1956 and from this breeding stock thousands of drones and some African/European hybrid queens and workers were released such that today there are approximately one trillion honeybees of African descent from northern Argentina to southern Mexico and they can consume 2 billion kg of pollen and 20 billion kg of nectar annually. The initial spread was through interbreeding with existing colonies of temperate origin or invading their nests: selection appears to have favoured the African genome over the others. They spread at a rate of 100 km a year and are aggressive to humans as well as to other pollinators. Between 1977 and 1982 in French Guyana for example, they began to oust native meliponine bees on *Mimosa pudica* (Leguminosae), the sensitive plant, such that being one of every 14 pollinators in 1977, they represented three out of four by 1982, apparently due to population increase and not actual aggression. Their use of nesting sites may be in competition with that of certain species of wasp and perhaps birds, bats and opossums. Certain parakeets excavate holes in the nests of arboreal termites: the bees move in when the birds have finished with them and they also occupy nests of terrestrial termites. As the bees use up to a quarter of the plant species, their presence upsets the relationships between these plants and their pollinators. Their large flight range may improve pollination in dioecious species. Moreover, their domination of certain entomofaunas may reduce the populations of, say, euglossines and thus affect the pollination success of plants they do not even visit; species with 'buzz-pollination', not effected by these bees, may also suffer.

The loss of much of the avifauna of Guam has been attributed to the brown tree-snake, *Boiga irregularis*, introduced about 1950 from Australia or Papuasia (94, 95). The whole forest avifauna is certainly very reduced and the one endemic bird is probably extinct, but the loss is perhaps not entirely due to the snake, because the island of Rota, Guam's snake-free neighbour, has also suffered contractions and losses in the 1970s in a pattern recalling the start of the Guam decline. Smaller organisms may be partly responsible too, as in the case of the native avifauna of Hawaii, which has halved in species numbers since 1778, the present native birds being restricted to high-mountain forest refuges. Although avian malaria was a theoretical possibility before humans arrived (because of migrant birds), there was no vector, as Hawaii was free of mosquitoes before Europeans landed. With the introduction of the night-flying avian malaria-carrier, *Culex pipiens*, from Mexico in 1826, came the spread of the disease and the collapse in the fauna. Moreover, hippoboscid flies, carriers of birdpox were probably introduced with domestic fowls. The honeycreepers (drepaniids), which were important pollinating agents, had never before been exposed to mosquitoes and are now restricted to mosquito-free sanctuaries above 600 m and have no appreciable immunogenetic capacity against either birdpox or avian malaria (96).

Increasing industrial and other human activity has affected the mangrove swamps near Rio de Janeiro, Brazil, such that the trees have more or less disappeared (97) to be replaced by a pollution-tolerant community of succulent *Sesuvium portulacastrum* (Aizoaceae) and *Iresine vermicularis* (Amaranthaceae), an association long believed to be a 'natural' one. Again, the distributions of certain tree species in the Guineo-Congolian region of Africa led to the recognition of a 'Sangha River Interval', separating Lower Guinea and Congolia, extending 14–18°E, but this interval, unlike the Dahomey Interval (see section 2.2), is not to be found in the distribution of, e.g. Dichapetalaceae, which may reflect the fact that plants in this family are tolerant of human activities like shifting cultivation, though many of the tree species in the original sample are not, i.e. the 'interval' may be another 'natural' result of human activity (98).

Rather more surreptitious conversion of forest must also be considered (99, 100). For example, *Shorea javanica* (Dipterocarpaceae) tapped for resin in forest, is left during clearing for shifting cultivation and planted after deforestation in Sumatra, leading to forest-plantation agroforestry systems with durian, langsat and other valuable trees (see section 8.5). A further stage is seen in West Africa, where the freshwater swamp forest that was rich in species has been reduced to a community dominated by *Raphia hookeri*, used for fruit and wine, and in which seedling enrichment is carried out. Under the cover of forest in southern India and Sri Lanka, cardamon (*Elettaria cardamomum*, Zingiberaceae) is planted, but when this is abandoned, species of *Ochlandra* bamboo take over and prevent regeneration, leading to *Ochlandra* 'breaks' (101). So the fabric of the forest may be changed.

Other human effects may be more subtle (102–104). For example, frugivorous bats in Costa Rica use paths as flyways and these include the paths used by man, so that bat observations made from such paths could well have exaggerated population sizes. Indeed, it is difficult to find any area where humans have not reduced the population of one or more frugivores and/or one of their principal predators to the point that it is unlikely that the remaining members still maintain the ecological relationships that have gone hand in hand with their evolution. With new introductions, there are new opportunities, for example, in the Jari scheme, the fallen fruits of the exotic *Gmelina arborea* (see section 9.1.1) are sought by the native *Mazama gouazoubira* deer. The African *Kigelia pinnata* (Bignoniaceae) and the Asiatic *Durio zibethinus* (Bombacaceae), the durian, with their 'bat flowers' introduced to Honduras are attractive to night-flying insects and some of these, notably large moths, can be effective pollinators (105). Indeed, many crop plants are pollinated effectively well beyond their native range (106), e.g. the Asiatic mango (*Mangifera indica*, Anacardiaceae) and the African akee (*Blighia sapida*, Sapindaceae) are visited by *Polistes* wasps in Jamaica. Many plants are flexible enough to cope with faunal change, an Australasian example being *Syzygium cormiflorum* (Myrtaceae), which is bat-pollinated but the flowers are actively visited by birds and insects; it is also capable of self-pollination. Its nectar flow seems to suit both nocturnal and diurnal visitors and possibly the species is evolving from a night to a day system but it may well be that there is selection for its being a generalist (107).

9.2.5 *Conservation*

There are many arguments for conservation of the biological heritage of the world; in view of the immediate plight of much of humanity, in a local context there may also be many humane arguments against it. These have been well put elsewhere and here only some ecological arguments, which may bear repetition, will be reiterated.

In the world, until recently, living with the possibility of instant Armageddon, generated by advances in the knowledge of atomic physics, the threat of devastation through slower biological processes was possibly heeded less than it might otherwise have been. Nevertheless, one possible consequence of the removal of rain forest is frequently brought to the fore, namely the effect on the carbon dioxide balance of the earth's atmosphere and, thence, the possible warming of the earth through the 'greenhouse effect' leading to the melting of the icecaps and glaciers, with devastation in the form of flooding. What is the evidence for this potential 'eco-disaster'?

The carbon cycle   There is more carbon in tropical moist forest than there is in all temperate forest and cultivated land put together. The tropical forests (108) affect the carbon balance of the world by fixing large amounts of carbon at a rapid rate, by releasing it as carbon dioxide to the atmosphere through respiration, and exporting carbon as organic matter to the sea, to peat, or as forest products, largely timber. Those who argue that the forests are potential sources of carbon, believe that rapid rates of forest removal will increase the levels of carbon dioxide in the air through burning and decomposition of the vegetation. Those who argue the reverse, citing the rapid rates of succession and high carbon dioxide utilization of forest, believe that regenerating forests are a sink for carbon dioxide. A balanced view is provided in Figure 9.4.

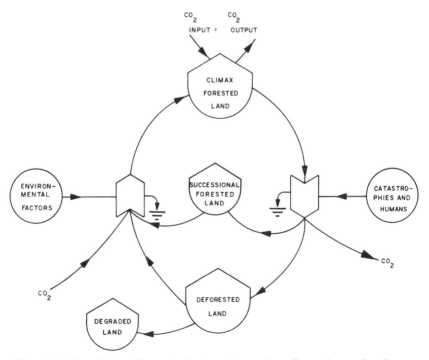

Figure 9.4   Flow diagram illustrating land-use changes that affect carbon cycling. Reproduced from *Unasylva* **32** n. 129 (1980) 11, courtesy FAO.

Carbon release to the atmosphere is deemed to increase with more drastic forms of land-use change, for the original forest is held to be in gaseous balance with the atmosphere. Some have argued that up to 95% of the carbon in the forests is given up to the atmosphere in this way. The successional stages are the net consumers of carbon dioxide. When human beings are completely dependent on solar energy for survival, there is a stable, steady-state with respect to atmospheric carbon, for then they are simple consumers using traditional shifting cultivation, hunting or gathering as a means of obtaining food. The activities of such human beings are not extensive enough to outweigh the effects of regeneration. This assumes that the mature forests are in equilibrium with the atmosphere. Evidence suggests that this is not always so, for such forests are not 'closed', and 'leak' small but measurable amounts of dissolved carbon to downstream vegetation. When extrapolated, such figures as 0.2×10^{14} g carbon per year exported from the Amazon are produced, though lately estimates for this figure have been raised fivefold. There are few such figures as yet and their significance is questionable, for it is not known how much of this carbon is lost irrevocably in the bottom of the seas and how much is respired back to the atmosphere by marine organisms.

It has been argued that those areas of the world relying on fossil fuels rather than on living forests and those where agriculture is less consumptive of land than formerly, as in Puerto Rico for example, may be net sinks of carbon, so that an assessment of the effect on carbon balance has to include an assessment of land use worldwide, not, then, a simple equation and, in short, we do not know the answer.

It is essential to be able to establish current and future land-use changes and to have an effective predictive model: neither exists (109). Models must take account of the canopy's importance in surface energy and water budgets but from current models a wide range of potential effects have been suggested—from overall global increases in temperature and decreases in rainfall to neither significant temperature change nor an increase in rainfall. In a simulated replacement of the Amazonian rain forest by impoverished grassland, it was suggested that there would be increased run-off due to less interception by the canopy and an increase in surface temperatures by $3-5°C$ and, although there would be no overall change in rainfall, the period of driest conditions would be extended.

So far as the world in general (110) is concerned (Figure 9.5), atmospheric CO_2 is increasing by 1–2% per annum, so that, at this rate, the level will have doubled by the middle of the next century, by which time the mean temperature will have increased by 1–5°C. At present, our atmosphere

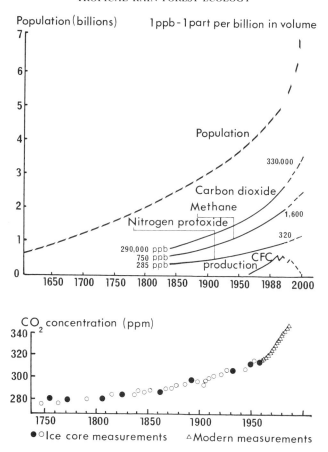

Figure 9.5   Population of the world and greenhouse gas concentrations (1 ppm = 1 part per million in volume). Redrawn from Financial Times 16 March 1990.

keeps us 33°C warmer than would be expected from our distance from the sun: Venus, with a thick layer of CO_2, is several hundred degrees higher than would be expected. The result of changes in modern CO_2 is not just the average change in climate but the frequency of 'extreme' events such as droughts. Twenty per cent of the excess carbon is alleged to be coming from fires in the Amazon, where satellite monitoring lately showed 178 000 fires of more than $1 km^2$ in area. A 50% reduction in present-day CO_2 would require the complete cessation of rain-forest destruction as well as the planting of two million km^2 of forest at a cost of $100 billion, though this is less than the cost of building sea walls around the United States for example.

Conservation principles   In a time of world recession, demand for tropical timbers has declined. Some lumber companies have ceased operations in the tropical rain forest and some have even gone bankrupt. Thus the definition of what is a 'commercial' species changes with markets, or indeed with fashion, but this is only a temporary respite in one sector of the pressures on rain forest. Forest people have come sufficiently into contact with the ideas of the developed world and its values to be changed irrevocably. The animals are gone from many forests, which are changed in other ways as a consequence. It seems that the frequency of cyclonic storms is increasing in the tropics. In the longer term, the weather will change as the next ice age arrives, and ultimately the continents will change their orientation with respect to one another. In this world of inevitable change, what is the rationale of conservation if it has a basis beyond sentiment? When reduced to the minimum, it must surely be an effort to put a brake on accelerating universal change arising from the direct action of man, so that resources may be saved for later consumption. Unless unchecked, at least in some places, we lose not only what is labelled our 'heritage' but we lose its variety, possibly before the value of this variety can be assessed, both in terms of diversity, valuable in itself, and the properties of the components of that diversity.

In Chapter 1, it was argued that there may have been a tropical origin for ecology; it could be said that conservation also has such roots in that the amateur and later professional naturalists employed in the colonial services, particularly in India from the end of the eighteenth century onwards, advocated conservation measures. Combined with the initially conservative silvicultural policies deriving from German practice, such apparent brakes on development were possibly accceptable to the distant European governments because they had the effect of preventing ecological changes that might have led to crop failures and consequently threaten the suzerainty of the colonial powers. The theme of this book is of change and diversity and no further remarks deliberately praising and encouraging conservation will be made here, for the readership of this book must be largely counted among the 'converted'. Given then that conservation is an essential part of the overall management of rain forests in the tropics, what should be top priority and how should it be conserved?

Myers (111, 112) has drawn up a list of top priority areas which include peninsular Malaysia, Madagascar and the relict coastal forest strip in

Atlantic Brazil. Choosing what to concentrate on relies on a knowledge that may not yet be sufficiently deep to make decisions, but representative areas of at least eight main phytogeographical zones in Amazonia, the major forest types of Indomalesia and the different formations of the Zaïre basin in Africa would be a minimum. A more recent survey based on levels of endemism in vascular plants is presented in Table 9.1 and a series of reports has been prepared by the International Union for the Conservation of Nature (IUCN), that dealing with Oceania, for example, proposing that 77 islands there should be given priority (113).

Some would argue that the Pleistocene refugia recognized in South America and posited elsewhere would serve as a basis, though this concept seems inapplicable to Indomalesia at least. The centres of diversity may not correspond with these supposed refugia and, even if they do, how big should preserved tracts be? The impossible task of estimating how large a population of rain-forest trees needs to be, to be self-sustaining has meant that the minimum size for such reserves is not known accurately. In earlier chapters, the prevalence of dioecy, and the graduated rather than 'all-or-nothing' flowering response were stressed: these and the high degree of endemism of moist forest plants, compared with those of seasonal forest suggest that larger and larger reserves are required (85). Many of the rain forests of Australia compose a relic archipelago of gallery and gully forests, some very small indeed, while many relict forests kept as water catchment areas or for scientific interest are believed to be declining because they are too small. Small reserves are known to have increased tree fall at the margins and isolated pockets of forest with trees out of synchrony with the surroundings are likely to be targets for insect predation. Loss of fauna and lowering of humidity will prevent germination of many species. In Brazil an effort has been made to find the appropriate size, and plots of 1, 10, 100, 1000 and 10 000 hectares have been set aside so that the decline in species numbers can be assessed more accurately (114).

Looking at Amazonia, then, Myers has estimated that the eight major phytogeographical regions and the 16 centres of species diversity would require some 185 000 km^2 of reserves including buffer zones around them. This represents an area equivalent to 80% of the land surface of Great Britain. The buffers (115) are essential in that even when reserve boundaries are clearly demarcated and there are guard patrols in the absence of such buffers, there tends to be agricultural encroachment, illegal hunting and gathering and timber felling because local people often see the reserves as government-imposed restrictions on their traditional rights. Of the 23 Biosphere Reserves established under the UNESCO Man and Biosphere

Table 9.1 Botanically rich areas in tropical forests. From Myers (112)

Area	Extent of forest (sq. km.) Original	Extent of forest (sq. km.) Present (primary)*	Vascular plant species in original forests	No. of endemics in original forests (and percentage)		Original endemics as proportion of earth's higher-plant total (%)	Present forest areas as proportion of earth's land surface (%)
Madagascar	62 000	10 000	6 000	4 900	82	1.96	0.00675
Atlantic-Coast Brazil	1 000 000	20 000	10 000	5 000	50	2.0	0.0135
Western Ecuador	27 000	2 500	10 000	2 500	25	1.0	0.0011/
Colombian Chocó	100 000	72 000	10 000	2 500	25	1.0	0.0486
Uplands of Western Amazonia	100 000	40 000	20 000	5 000	25	2.0	0.027
Rondonia/Acre	400 000	250 000	5 000	800	16	0.32	0.169
Tanzania/Kenya Montane Forests	20 000	10 000	2 400	750	31	0.3	0.00675
Eastern Himalayas	340 000	53 000	9 000	3 500	39	1.4	0.047
Sri Lanka	150	89	845	500	59	0.2	0.00006
Peninsular Malaysia	120 000	26 000	8 500	2 400	28	0.96	0.0175
Northwestern Borneo	120 000	60 000	9 000	3 300	37	1.32	0.04
Philippines	250 000	12 000	8 500	3 700	44	1.48	0.008
New Caledonia	15 000	1 500	1 580	1 400	89	0.56	0.001
	2 554 150	557 089	**	36 250		14.5	0.3868
For comparison:							
Hawaii	14 000	6 000	825	745	88	0.30	0.004
Queensland	13 000	6 300	1 165	435	37	0.17	0.004

* Some primary-forest species can survive in degraded forest.

**It is unrealistic to total these figures for vascular plant species because there is some overlap between adjacent regions.

Note: There is a great range of accuracy in these figures.

Programme (MAB) up to 1988, only six had such buffers, though four more were proposed. (Table 9.2). Across the tropical world, some 280 areas had some protection by 1985, making up over 39 million ha: Africa 44 (almost 9 million ha), Indomalesia 122 (over 5 million ha), Australasia 53 (almost 8 million ha) and Americas 61 (over 17 million ha); by 1990, Amazonia had 24 million ha in reserves and more were being designated for protection. In the

Table 9.2 Tropical moist forest biosphere reserves. From Oldfield, 1988 (115)

Country	Reserve	Total area (ha)	Major land user	Buffer zone (ha)
Bolivia	Pilon-Lajas	100 000	Human settlement	—
Cameroun	Reserve Forestière et de Faune du Dja	500 000	None	—
Central African Republic	Basse-Lobaye Forest	18 200	Human settlement	—
China	Dinghu Nature Reserve	1 200	Tourism	950
Congo	Odzala National Park	111 000	None	—
Gabon	Reserve naturelle integralle d'Ipassa-Makokou	15 000	Human settlement	5 000
Ghana	Bia National Park	7 770		22 800 Outside the park
Guinea	Monts Nimba	17 130	Engineering works	7 130
	Ziama Massif	116 170	Human settlement Forestry, agriculture	56 170
Côte d'Ivoire	Taï National Park	330 000	None	—
Indonesia	Gunung Gede-Pangrango	14 000	None	Proposed
	Lore Lindu	231 000	Settlement	Proposed
	Tanjung Puting	205 000	Logging, agriculture	None
	Gunung Leuser	946 000	Settlement	Proposed
	Siberut	56 000	Logging	Proposed
Mexico	Montes Azules	331 200	Agriculture	—
Nigeria	Omo Reserve	460	None	—
Panama	Parque Nacional Fronterizo Darien	597 000	Settlement	180 000
Peru	Manu Reserve	1 881 200	Human settlement Prospecting Canal construction	—
Philippines	Puerto Galera	23 545	Tourism	—
Sri Lanka	Sinharaja Forest Reserve	8 850	Forestry, agriculture	—
Zaire	La Luki Forest Reserve	33 000	None	—
	Reserve Floristique de Yangambi	250 000	Traditional agriculture	—

Tropical Forestry Action Plan, certain sums were set aside for conservation measures in the period 1987–1991: Africa 105 million dollars, Asia 148 and Americas 195, a total of 498 million dollars (116–118). There are plans for the conservation of some 625 000 km² of forest around the tropical world. This represents well over 7% of the whole biome but falls short of the 10–20% required by certain theorists. But, as Myers points out, those countries in the developing tropical world are being asked, for the good of mankind, to set aside a tenth of their land surface. How would such a request be received in a developed country?

At a local level, meandering streams and other unstable habitats favour the conservation of diversity, as seen in Manu National Park, Peru, but it is essential to conserve the 'keystone' resources, which include figs, palm fruits and nectar (119, 120). In Peru, for example, palm fruits are important for peccaries, capuchins, agoutis, squirrels and macaws, which account for up to 30% of the frugivore biomass there. Figs are a dry-season staple food for birds and mammals there and they sustain up to 60% of the frugivore biomass. Nectar sustains only about 10%, though this includes a wide range of mammals, birds and insects. At Cocha Cashu, less than 1% (i.e. 12 species out of 2000) sustain the entire frugivore community for 3 months of the year. Figs are also important in Kutai in Borneo, while some Myristicaceae and Meliaceae are critical for large avian herbivores—especially hornbills. Nectar and palm fruits are not so readily available there, so foliage and bark eating are more conspicuous but migration, e.g. of pigs, is a feature of Kutai, again with major implications for conservation.

An unexpected effect of the generalist nature of rain-forest pollinators has profound economic and conservational significance (121) in Malaysia. Bats there feed on a wide range of plants flowering throughout the year: species of *Arenga* (Palmae), *Duabanga* and *Sonneratia* (Sonneratiaceae) and *Musa* (Musaceae) among them. But one of the bats, *Eonycteris spelaea* is an important pollinator of the durian, a seasonally flowering commercial fruit, and turns to the other species when its 'favourites' are not in flower, such that the apparently 'useless' other trees actually maintain the pollinators out of the durian flowering season and therefore maintain the durians themselves. A completely different example but with similar effects is that of fishes in the Amazon, where the most important commercial species is *Colossoma macropomum* (tambaqui), which as adults feed almost exclusively on fruits and seeds from flooded forests. Other fish species are reliant on the flooded forest and as fish is the most important source of protein in the region (122), the supply of this is threatened not by overfishing but by threats to turn the forest into rice fields, which would not only remove the food but increase the flow of herbicides and insecticides into the river system.

Some examples Taking a much smaller area, there has been a very active move to conserve the rain forest left in Sri Lanka (123–127). Remains of Mesolithic Balangoda Man indicate the typical 'hunter-fisherman-gatherer' of the period and it is suggested that the forest-living Vaddhas of today derive from miscegenation between such ancient peoples and the metal-using Aryans, who moved in from India. There has thus been a very long contact between the forest and humans. Using the incidence of the vernacular names of trees in place names, it has been argued that rain-forest patches may have existed in the eastern areas of the central highlands, where they no longer survive. Under the Kandyan kings, shifting cultivation (*chena*) was under careful supervision of the village elders, such that damage to trees in the surrounding forest led to fines. But in 1824, extensive tracts of the highland forests were passed over to plantation agriculture for coffee under the aegis of a British administration. These were forests not only over which the local people had rights but which were the source of the waters for the elaborate irrigation schemes of their paddy-fields. As large accessible timber trees began to become scarce, the British imposed rules concerning tree felling: without trees, erosion was known to accelerate and make the land unsuitable for the establishment of commercial plantations. This system broke down and *chena* cultivation advanced uncontrolled with the result that the Director of Kew Gardens, then the centre of a colonial network of botanic gardens and with a very important role in the establishment of commercial plantation species, petitioned the Prime Minister, in 1873, in a report dealing with the effects on water supply and erosion in extraordinarily 'modern' terms.

At the end of the eighteenth century, 90% of the country's land area had been forested; between the years 1956 and 1983, the area under 'natural' high forest fell from 2.9 m ha, to 1.75 m ha, a rate of loss of 42 000 ha per annum and by the 1980s less than 1% and 3% of the land area in the wet lowlands and montane zone respectively, have 'relatively undisturbed' primary forest. It has been estimated that the island has 11 genera and 830 species of endemic angiosperms and that 90% of those that are trees with a bole diameter at breast height greater than 30 cm are either endangered, vulnerable or rare. The Sinharaja Biosphere Reserve harbours over 70% of the endemics but the rest are not adequately protected. For some of the 77 endemic lowland species in cultivation in the Peradeniya Botanical

Gardens near Kandy, the garden population (of some species a single individual) is the only one known.

All this has to be seen in the context of the island's population, which increased from some 817 000 in 1789 to about 400 000 in 1871, whereafter it increased annually by 1.4% until 1946; in 1988, it stood at 16.5 million. Inevitably there are pressures on those areas designated as reserves, often coming from the local people who have traditionally used the forest as a resource, particularly when such forests are a dwindling resource, a state of affairs hardly their fault but, in the main in this case, due to large-scale logging. Even when national parks are set up, such pressures can lead to their being 'degazetted' as in the Mount Kinabalu National Park in Borneo, where a copper mine, dairy farm, golf course, luxury houses and plantations have replaced the world's richest forest on the Pinosuk Plateau (128). On the other hand, reserves can be profitable: the Kuna Indians of north-eastern Panama have established their own 230 square mile forest park and botanical reserve attractive to birdwatchers and other scientific tourists. In Rwanda, until fairly recently, only poachers among the local people knew of the gorillas there. After an education programme, poaching was much reduced in the 6 years to 1984 and tourism based on viewing wild gorillas became the country's third highest foreign currency earner.

Reserves in countries, where, as in Europe, the original vegetation has been largely removed, have problems similar to those in the 'developed' world. The last population of the Javan rhinoceros, *Rhinoceros sondaicus* (129), is in Ujung Kulon Nature Reserve, preserved largely by its remoteness. However, the area does not in all its parts provide an optimal food supply, and experiments to enhance the growth of food plants have been initiated, notably the cutting of the palm cover to promote seedling growth. Such is, of course, a manipulation not unlike a silvicultural procedure but, in this case, seems even more 'unnatural' and on the road to a 'zoological garden' approach somewhere between *in situ* and *ex situ* conservation. A similar case is seen in populations of, say, deer in the tiny woods of central England or the avid 'gardening' needed to keep going the only British population of the annual buttercup, *Ranunculus ophioglossi-folius* in the world's smallest nature reserve (0.04 ha) at Badgeworth, Gloucestershire.

The rationale   The reasons for conservation of rain forest have been set out by Myers (111) as follows:

1. To allow evolutionary processes to continue
2. To safeguard the role of the moist forests in regulating the biosphere,

especially in maintaining climatic stability—local, regional and maybe even global

3. To provide a stock of plants and animals for pure and applied research
4. To provide undisturbed (*sic*) ecosystems for benchmark monitoring in comparison with which land-use strategies for the forest can be evaluated
5. To conserve gene pools of plants and animals for their future use to man and for maintaining ecosystem stability
6. To safeguard watersheds to prevent flooding and soil erosion and to maintain water supplies
7. To provide wildlands for recreation, for enjoyment and for education
8. To provide local income and foreign exchange through tourism.

Many of these have been touched on in the foregoing chapters, while others are beyond the scope of a book such as this. One which has not been mentioned and may prove to be of the greatest importance, is the significance of many 'minor' forest products, that is, those other than timber. It must be borne in mind that trade in these commodities long antedates that in timber, the commercial extraction of which in the tropics is a relatively modern development (130). Second to timber in commercial importance in SE Asia today is the rattan industry, based on the stems of climbing palms. Many other species, especially plants, have made thousands of contributions to modern agriculture, to medicine and to industry (111).

One pharmaceutical in four is derived from wild plants, by 1987 worth $35 billion annually (131). Such include vincristine, a drug that has quadrupled the chance of recovery for children suffering from leukaemia, anti-cancer drugs and compounds that combat heart disease, while many believe that the anti-fertility drugs used by forest people may have great potential as contraceptives.

Corynanthe johimbe (Rubiaceae) of tropical Africa which contains the alkaloid yohimbine, popularly considered to be an aphrodisiac, has lately been shown to be effective in increasing sexual motivation in rats and may be of value in the treatment of sexual libido problems in humans (132). Not all medically important plants have been discovered by local people, however, e.g. species of *Rauvolfia* (Apocynaceae) with useful hypotensive alkaloids and *Tabebuia* (Bignoniaceae), with a patented anti-cancer agent, lapachol. Indeed, few locally used ones have proved to be of commercial importance, though ipecacuanha (*Cephaelis ipecacuanha*, Rubiaceae) and *Quassia amara* (Simaroubaceae) from South America are exceptions, being used in the treatment of stomach disorders.

Some trees and other plants may produce enough alcohol or oils suitable for petroleum substitutes. Despite trading in gums and rubber and a hundred other forest products for centuries (e.g. Brazil gets its name from 'pau brasil' (*Caesalpinia echinata*, Leguminosae) because so much dye extracted from this plant was exported to Portugal), the costs involved mean that probably only 1% of rain-forest plant species have been screened for pure chemicals (133, 134); moreover, analysis is based on preconceived hopes of what the plants might contain. Some striking successes include *Astrocaryum* (Palmae) fruits which contain vitamin A in concentrations three times those in carrots; fruits of *Jessenia bataua* (Palmae) in Amazonia have 40% more protein than soya; from *Paullinia cupana* (*guarana*, Sapindaceae) is obtained a fizzy stimulating drink rich in caffeine—by 1979, 15 million bottles a day were being consumed in Brazil alone. Already, though, some forest plants, used in a 'minor' way, are endangered, like *Garcinia epunctata* and *G. kola* (Guttiferae), used as chewing sticks in W. Africa, and now becoming rare (124).

In the world as a whole, some 3000 species of plant have been grown at one time or another for food. Of these, some 150 have been grown on a commercial scale, though the world is now largely reliant on about 20. Some of these are genetically rather uniform and with few known wild relations. The tropics are experiencing increased levels of attack from pests and diseases while one in 10 of all plant species has recently become extinct or is in danger of becoming so. Many of these latter persist only as small numbers of mature individuals with little possibility of regenerating and multiplying *in situ*. For these, their evolutionary life in the rain forest is at an end: if they are to be saved, for whatever reason, seeds must be collected for arboreta and gardens. In the last century, a triumph was the spreading of commercially important tropical crop plants, such as rubber, through the agency of botanic gardens. Their role in an ever more quickly changing world becomes clear and it is heartening to see that such an organization as the Royal Botanic Gardens at Kew, with its old connections in the tropical world, has set up such a system of seedbanks and information. International agencies are now devoted to this aim too, e.g. the International Board for Plant Genetic Resources, which promotes an international network of genetic resource centres to further the collection, conservation, documentation and breeding of plant germplasm and thereby contribute to raising the standard of living and welfare of people throughout the world (135).

In such a broad context, Poore (136) has put the least disturbed rain forests in his simplified scheme of worldwide land-use categories as 'natural', compared with 'transformed', where the original vegetation has been largely removed and replaced by an artificial system, and 'modified' where the land is being changed by human use, but which still holds naturally occurring plants and animals, having some continuity with the original. The 'natural' ecosystems are now largely confined to remote or inaccessible regions, where neither agriculture nor commercial forestry is economically possible, though such places are now under other pressures—mineral exploitation or tourism, for example—while others are getting cut off as islands in seas of other types of land-use even if they are conserved. They may represent some 2.5% of the world's land area whereas 11% is under agriculture and less than 1% under plantation forestry. This leaves over 80% of the 'modified' lands, which receive little attention from the agriculturist because they are 'marginal', or from conservationists in the tropics, though it is, of course, such lands that are those being most vociferously fought over in the greatly modified landscapes of northern Europe, for example. Furthermore, mismanagement here will affect both the 'natural' and 'transformed'.

There are very strong pressures in many parts of the world to intensify the use and improve the efficiency of the transformed ecosystems: this receives universal acclaim, for these ecosystems contribute to satisfying the world's rising requirements for food and other raw materials. They are profitable, modern and provide outlets for products of technology and industry: they respond to the human instincts to be tidier, more up-to-date and more productive than one's neighbour. In other parts of the world with higher percentages of agricultural workers, however, the introduction of even modest improvements can lead to rural unemployment and migration to the urban slums, so that much of the developing world would not, contrary to much in aid programmes, welcome such innovation in the long run. Nevertheless, in many parts of the world, the intensive use of such systems may be the only relief of the pressure on other less transformed ones, such that it can be argued that research in agriculture should be stepped up as an aid to conservation (137). Indeed, well-intentioned pressures from without can have unfortunate consequences. Incensed by the 'hamburger connection', the developed world might boycott Costa Rican beef (138), but this would mean, among other things, that the tax income of the country would be reduced, reducing in turn conservation effort from within the country. The banning of all tropical hardwood imports would perhaps do the same, but pressures from consumer campaigns in this area are having an effect in reducing imports from sources other than 'sustainable' ones.

Conservation cannot be divorced from economics nor the aspirations of tropical peoples and it must be remembered that the populations of such territories are increasing as medical and nutritional conditions improve (139–143). The population of Asia is believed to have doubled in 36 years, tropical America in 30 and Africa in 23. If tropical rain forest is to be saved, then more emphasis has to be given to improving the income and quality of life of many of these who make up the 200 million subsistence farmers. Schemes to make use of 'minor' products may do this as well as take pressure off the forest itself in terms of timber. In the Philippines, bamboo has been promoted as a cash crop being harvestable after 3 years and annually thereafter; such would also take the pressure off rattans. Other alternative construction materials would protect forests if used fully: rubberwood is much used in Malaysia and Sri Lanka; in Thailand cassava wood has been mooted for chipboard and particleboard (of the 4.04 million tonnes available there in 1976, some 2.0 million were used). Myers alleges that all the world's timber needs could be got from secondary forests, leaving the primary for posterity; he argues that the indigenous animals could be farmed leading to more profit than the timber wood. He also points out that the funding of tropical biological study is not more than $30 million a year ($2 million from two American botanical gardens alone) and that this is equivalent to what the world spends every 10 minutes on armaments.

POSTSCRIPT

The one big lesson of the study of tropical ecology must surely be that concepts derived from purely temperate studies are often too rigid in definition, or at least, have too great an emphasis on one or a few particular aspects, to be of universal application. This is as true for morphology, as was first perceived by Alexander's army confronted with aerial roots in India, as it is for evolutionary and taxonomic studies in the light of rampant parallelism and ochlospecies, and ecology, which is now facing a relaxing of the concepts of pollination 'syndromes', tightly co-evolved dispersal mechanisms, the course of successional cycles, forest stratification and the nature of the coexistence of species. The relaxing of the straitjacket allows us to look afresh at temperate ecology in a world context as was advocated for the British flora long ago by A. H. Church (1).

Morphology and taxonomy were the bases of biology and were followed by physiology and biochemistry. The first two, in the light of tropical work, are advancing. As yet, the diversity of physiology and biochemistry of tropical plants is by comparison, little investigated. Very probably, though, a broadening of concepts will follow such investigations, as it has in cytology where the sweeping statements of chromosomal uniformity in tree groups seem absurd, now that variation as great as in temperate herbs has been found in tropical tree groups (2). In the area of genetics, even the most sacred of cows, the importance of cross-fertilization, has been called into question, as has the whole dogma of speciation involving populations rather than individuals (3), again, from a consideration of tropical organisms. To be honest, we know little about the biological world in all its diversity and this is a point that is rarely made in textbooks. There has been an effort, understandable enough, to draw up schemes and discover 'laws', possibly in an attempt to imitate essentially different physical sciences.

Ecology cannot be studied in isolation from evolution, biogeography or taxonomy, and the circumscription of the 'subject' is bound to be vague. To many scientists, then, ecology seems woolly and scarcely respectable: in many ways it is nearer economics than other sciences such as physics. When integrated with its sister disciplines, patterns do emerge, however, and some sense can be made of them (4). It is heartening to see that recent tropical work, for example, is increasingly imbued with a better taxonomic base than hitherto.

It will not have escaped the reader's notice that, in this book, many examples have been taken from the tree family Meliaceae: this is not merely because it has had the effect of loading the text with fewer plant names (nor just because the author happens to be studying them) but it demonstrates that within such a family, a great diversity of form and ecology can be found. The Meliaceae are not unique in this respect, for their architecture and relationships with animals are paralleled again and again in other groups. Indeed, the parallelism in the different groups of angiosperms in terms of pioneers or rheophytes are as striking as the far better-known zoological examples of marsupial and placental mammal parallelisms. This brings us back to a consideration of the importance of 'form' and the potential within particular groups for diversification. Why do, for example, the nutmegs and their allies, Myristicaceae, all have very similar seeds (arillate) and the same architectural model, whereas Euphorbiaceae are variable in almost all features and pervade temperate as well as tropical regions, unlike the Myristicaceae which are bound to the rain forest?

Parallelisms underline the importance of historical considerations, in that organisms are not evenly spread and completely mixed throughout the tropics. The paucity of species of bees in south-east Asia or of palms in Africa must have a historical explanation just as the importance of Wallace's Line and the isthmus of Panama have such explanations. Furthermore, it is unreasonable to suppose that species are not still extending their ranges, or indeed succumbing to those that are. Also, species are still in the making, and the frustrating variation patterns of certain tree species, for example, may mirror just this. Moreover, there must be caution, for the distributions may have been affected further by man as, for example, the bird fanciers of pre-Columbian America, or those transporting crops including trees, such as *Ceiba pentandra*. As we have seen, these may appear very much 'at home' and very aggressive in their new environments; so much so that weeds like *Lantana camara* (Verbenaceae), native to America, may completely arrest succession in the Old World.

The importance of time in terms of the asynchrony of the spread or extinction of associated organisms, such as dispersal agents in the case of certain Central American trees discussed in Chapter 6 and thus that of the concept of anachronistic traits in organisms, leads to a consideration of the wisdom or otherwise of viewing organisms in terms of their constituent features rather than viewing them as ragbags of compromise and opportunism. Again, with time, the disappearance of the large mammals from many tropical forests is certainly going to have a great effect, but with the exception of tree species that seem to be exclusively elephant-dispersed and similar examples, that effect may not be a specific, but a general, one. In Africa, the removal of these animals has accelerated in recent years such that the diversified mosaic landscapes maintained by what Kortlandt (5) has called the 'bulldozer herbivores', may be greatly modified. In turn, the great diversity of such faunas seems to argue that formerly rain forest areas may have been more diverse in their structure, which might well have an effect on the interpretation of vegetation history. Certainly these animals must have had some considerable effect through their foraging, debarking, uprooting, digging and trampling.

Ecology is concerned with the balance of a range of influences, and only rarely can one such influence or factor be recognized as having overriding importance. This has great implications for conservation. The type of forest that returns after complete clearance is different not only in terms of its height, but also its constituents. It has been estimated (6) that in Central Borneo, dipterocarps may spread at an average rate of 2 m per annum and that rivers may be a major obstacle to reinvasion. In the rehabilitation of forest, then, old trees should be left to act as seed sources (7). The dipterocarps may have invaded the Asiatic rain forests from more seasonal ones and may have been important forces in their modification, leading to mass flowering in the plants and migrations in the animals. Less successful, perhaps, has been the secondary invasion by humans, modifying the forest, though apparently never without a symbiotic relationship with other humans outside.To modern man, that modification seems not to be all for the good—the environmental consequences, though still not fully understood, seem unacceptable. Rather than seeing human activity as an extension of that of other animals, we see a need for conservation.

Much conservation effort has been based on the presence of conspicuous mammals or birds, so that areas rich in these tend to come high on lists of priority. This began in temperate regions and, in the case of primates, is encouraged by medical considerations, and such animals are used as the 'flagships' for international conservation bodies involving the tropics too. It is a way of capturing the imagination but it diverts attention from the importance of the integrated system of physical and biological factors, not least those of the world of micro-organisms and chemical cues, both not apparent to us through our enormous size and poorly developed olfactory senses. Mycorrhizae, nitrogen-fixers, decomposers, rumen bacteria, nectar, scents and pheromones are just part of the hidden workings of tropical rain forest. Despite the enormous increase in tropical rain forest studies in the last decades, we are only just beginning to understand some aspects of this dynamic and most diverse of the world's ecosystems.

Almost half a century ago, Corner (8) drew attention to what needed to be done in improving our knowledge of the forest before it disappeared. Today, the very topic of tropical rain forest ecology is endangered, in that the object of study is in retreat or radically changing. It is the duty of biologists (9) to draw attention to the possible fate of what is left, not least because it is in their own interests. For at least 90% of all tropical species the ecological future is either extinction or restriction to reserves totalling less than 10% of the tropical land surface. The force behind the subjection of the tropical ecosystems is rising populations with the apparent goal of converting the world into a sustaining and sustainable agroecosystem. Janzen has argued that this leaves three kinds of organism in the tropics: the 'living dead', ranging from species comprising adults with no reproductive future to cultigens unable to survive on their own; those 'exapted' to the new conditions and able to thrive in them; and thirdly the farmed organisms. Many wild breeding populations subsist on materials provided by the living dead, while many managed tropical habitats are rich in parasitoid-free insects, herbivore-free plants, sympatric carnivores with the same diet and trees without pollinators or dispersers. It seems almost inevitable that such things will be the materials for the tropical rain forest ecology of the future. But, in some senses, they always have been.

FURTHER READING

Cranbrook, Earl of (ed.) *Malaysia.* Pergamon, Oxford, 1988.

Crosby, A.W. *Ecological Imperialism.* Cambridge University Press, Cambridge, 1986.

Denslow, J.S. and Padoch, C. *People of the Tropical Rain Forest.* University of California Press, Berkeley, 1988.

Hallé, F., Oldeman, R.A.A. and Tomlinson, P.B. *Tropical Trees and Forests. An Architectural Analysis.* Springer, Berlin, 1978.

Janzen, D.H. *Ecology of Plants in the Tropics.* Arnold, London, 1975.

Kitching, R. (ed.) *The ecology of Australia's wet tropics. Proc. Ecol. Soc. Austr.* **15** (1988).

Leigh, E.G., Rand. A.S. and Windsor, D.M. (eds) *The Ecology of a Tropical Forest.* Oxford University Press, Oxford, 1983.

Longman, K.A. and Jenik, J. *Tropical Forest and Its Environment.* (2nd edn), Longman, London, 1987.

Meggers, B.J., Ayensu, E.S. and Duckworth, W.D. (eds) *Tropical Rain Forest Ecosystems in Africa and South America: A Comparative Review.* Smithsonian Institution, Washington, 1973.

Myers, N. *The Primary Source: Tropical Forests and Our Future.* Norton, New York and London, 1984.

Prance, G.T. and Lovejoy, T.E. (eds) *Amazonia.* Pergamon, Oxford, 1984.

Richards, P.W. *The Tropical Rain Forest.* Cambridge University Press, Cambridge, 1952 (with later reprints).

Riehl, H. *Climate and Weather in the Tropics.* Academic Press, London, 1979.

Rubeli, K. *Tropical Rain Forest in South-East Asia—A Pictorial Journey.* Tropical Press, Kuala Lumpur, 1986.

Sutton, S.L., Whitmore, T.C. and Chadwick, A.C. (eds) *Tropical Rain Forest: Ecology and Management.* Blackwell, Oxford, 1983.

Tomlinson, P.B. and Zimmermann, M.H. (eds) *Tropical Trees As Living Systems.* Cambridge University Press, Cambridge, 1978.

Tucker, R.P. and Richards, J.F. (eds) *Global Deforestation and The Nineteenth-Century World Economy.* Duke Press, Durham, North Carolina, 1983.

White, F. *The Vegetation of Africa.* UNESCO, Paris, 1983.

Whitmore, T.C. *Tropical Rain Forests of the Far East* (2nd edn) Clarendon, Oxford, 1984.

Whitten, A.J., Damanik, S.J., Anwar, J. and Hisyam, N. *The Ecology of Sumatra.* (2nd edn) Gadjah Mada University Press, Yogyakarta, 1987.

From the references that follow, the reader will see that to keep abreast of developments in the subject, it is valuable to read issues of certain journals, such as *Acta amazonica, Ambio, American Naturalist, Annual Review of Ecology and Systematics, Biotropica, Ecological Reviews, Ecology, Evolution, Journal of Animal Ecology, Journal of Ecology, Journal of Tropical Ecology, Oikos, Trends in Ecology and Evolution, Vegetatio.*

REFERENCES

Chapter 1 (pp. 1–16)
1. Stearn, W.T., *Gdns' Bull., Sing.* **29** (1977), 13–18.
2. Mabberley, D.J., *Jupiter Botanicus: Robert Brown of the British Museum.* Cramer, Braunschweig, 1985.
3. Goodland, R.J., *Oikos* **26** (1975), 240–245.
4. George, W. *J. Soc. Bibphy Nat. Hist.* **9** (1979), 503–514.
5. Ewel, J. (ed.), Tropical succession. *Biotropica* **12** (2) (Suppl.), 1980.
6. Myers, N., *The Conversion of Tropical Moist Forests.* Nat. Acad. Sci., Washington, 1980.
7. Lanly, J.F., *For. Abstr.* **44** (1983), 287–318.
8. Dawson, J.W., *Biotropica* **12** (1980), 159–160.
9. Diamond, J.M., *Nature* **315** (1985), 538–539.
10. Myers, N., *Env. Conserv.* **15** (1988), 293–298.
11. Keng, H., *J. trop. Geogr.* **31** (1970), 43–56.
12. Cranbrook, Earl of, *Malaysia.* Pergamon, Oxford, 1988, pp. 146–166.
13. Gentry, A.H. and Dodson, C., *Biotropica* **19** (1987), 149–156.
14. Whitmore, T.C. *et al. J. Trop. Ecol.* **1** (1985), 375–378.
15. Longman, K.A. and Jenik, J., *Tropical Forest and Its Environment,* (2nd edn) Longman, London, 1987.
16. Opler, P.A. *et al.* in J. Cairns *et al., Recovery and Restoration of Damaged Ecosystems.* University Press of Virginia, Charlottesville, 1977, pp. 379–421.
17. Stenseth, N.C., *Oikos* **43** (1984), 417–420.
18. Swaine, M.D. and Hall, J.B. in H. Synge (ed.), *The Biological Aspects of Rare Plant Conservation,* Wiley, Chichester, 1981, pp. 355–366.
19. Janzen, D.H., *Ecology of Plants in the Tropics.* Arnold, London, 1975.
20. Harriss, R.C. *et al., J. Geophys. Res.* **43** (1988), 1351–1360.
21. Brown, S. and Lugo, A.E., *Science* **223** (1984), 1290–1293.
22. Yamakura, T. *et al., Vegetatio* **68** (1986), 71–82.

Chapter 2 (pp. 17–30)
1. Howarth, M.K. and Adams, C.G. in L.R.M. Cocks (ed.), *The Evolving Earth,* British Museum (Natural History) and Cambridge University Press,1981, pp. 197–220, 221–236.
2. Coney, P.J., *Ann. Mo. Bot. Gdn* **69** (1982), 432–443.
3. Ashton, P.S., *Gdns' Bull., Sing.* **29** (1977), 19–23.
4. Hallé, F., Oldeman, R.A.A. and Tomlinson, P.B. *Tropical Trees and Forests. An Architectural Analysis,* Springer, Berlin, 1978.
5. Brown, T.M. *et al., J. Human Ecol.* **11** (1982), 603–632.
6. Hall, J.B. and Swaine, M.D., *Distribution and Ecology of Vascular Plants in a Tropical Rain Forest.* Junk, The Hague, 1981.
7. Morley, R.J., *J. Biogeogr.* **8** (1981), 383–404.
8. Walker, D. and Chen, Y. *Quat. Sci. Rev.* **6** (1987), 77–92.

9. Liu, K.-B. and Colinvaux, P.A., *Nature* **318** (1985), 556–557.
10. Haffer, J. in Whitmore, T.C. and Prance, G.T. (eds). *Biogeography and Quaternary History in Tropical America,* Clarendon, Oxford, 1987, pp. 1–18.
11. Mawson, R. and Williams, M.A.J. *Nature* **309** (1984), 49–51.
12. Whitmore, T.C. (ed.), *Wallace's Line and Plate Tectonics.* Clarendon, Oxford, 1981.
13. Barlow, B.A. in Calder, M. and Bernhardt, P., *The Biology of Mistletoes,* Academic Press, London, 1983, pp. 19–46.
14. Longman, K.A. and Jenik, J., *Tropical Forest and Its Environment,* (2nd edn), Longman, London, 1987.
15. White, F. *The Vegetation of Africa.* UNESCO, Paris, 1983, pp. 71–85.
16. Salati, E. in Prance, G.T. and Lovejoy, T.E. (eds), *Amazonia,* Pergamon, Oxford, 1985, pp. 18–48.
17. Manokaran, N., *Malays. For.* **42** (1979), 174–201.
18. Sugden, A.M., *J. Arnold Arb.* **62** (1982), 1–61.
19. Jordan, C.F. *et al., Acta Amaz.* **11** (1981), 87–92.
20. Osuji, G.E., *J. Environ. Manag.* **28** (1989), 227–233.
21. Lightbody, J.P., *Biotropica* **17** (1985), 339–342.
22. Dean, J.M. and Smith, A.P., *Biotropica* **10** (1978), 152–154.
23. Riehl, H. *Climate and Weather in the Tropics.* Academic Press, London, 1979.
24. Shaw, W.B., *Pacific Sci.* **37** (1983), 405–414.
25. Emmanuel, K.A., *Nature* **326** (1987), 483–485.
26. Harris, T.M., *J. Ecol.* **46** (1958), 447–453.
27. Henniker-Gotley, G.R., *Indian For.* **62** (1936), 422–423.
28. Malingreau, J.P., *Ambio* **14** (1988), 314–321.
29. Sanford, R.L. *et al., Science* **227** (1986), 53–55.
30. Uhl, C. *et al., Oikos* **53** (1988), 176–184.
31. Johns, R.J., *Blumea* **31** (1986), 341–371.
32. Garwood, N. *et al., Science* **205** (1979), 997–999.
33. Lee, D.W., *Biotropica* **19** (1987), 161–166.
34. Lee, D.W. *et al., Biotropica* **11** (1979), 70–77.
35. Bone, R.A. *et al., Appl. Optics* **24** (1985), 1408–1412.

Chapter 3 (pp. 31–51)
1. Harcombe, P.A. in Ewel, J. (ed.) Tropical succession. *Biotropica* **12**(2) (Suppl.), 1980.
2. Burnham, C.P. in Whitmore, T.C., *Tropical Rain Forests of the Far East,* (2nd edn.), Clarendon, Oxford, Chapter XI.
3. Lathwell, D.J. and Grove, T.L., *Ann. Rev. Ecol. Syst.* **17** (1986), 1–16.
4. Jordan, C.F. in Prance, G.T. and Lovejoy, T.E., *op. cit.* pp. 83–94.
5. van Steenis, C.G.G.J., *Blumea* **29** (1984), 395–397.
6. Anderson, A.B. *Biotropica,* **13** (1981), 199–201.
7. van Steenis, C.G.G.J., *Bot. J. Linn. Soc.* **89** (1984), 289–292.
8. Lieberman, M. *et al., J. Ecol.* **73** (1985), 505–516.
9. Crowther, J., *J. Biogeogr.* **9** (1982), 65–78.
10. Brooks, R.R., *Serpentine and Its Vegetation,* Dioscorides Press, Portland, 1988.
11. Proctor, J. *et al., J. Ecol.* **76** (1988), 320–340.
12. Lee, J. *et al., Bryol.* **80** (1977), 203–205.
13. Buckley, R.C. *et al., Biotropica* **12** (1980), 124–136.
14. van Steenis, C.G.G.J., *Bot. J. Linn. Soc.* **79** (1979), 97–178.
15. Sugden, A.M., *Bot. J. Linn. Soc.* **90** (1985), 231–241.
16. Golley, F.B. *et al., Biotropica* **10** (1978), 144–151.
17. Hardy, F., *Biotropica* **10** (1978), 71–72.

18. Baillie, I.C. *et al.*, *J. Trop. Ecol.* **3** (1987), 201–220.
19. Whitmore, T.C. in Proctor, J. (ed.), *Mineral Nutrients in Tropical Forest and Savanna Ecosystems.* Blackwell, Oxford, 1989, pp. 1–13.
20. Newberry, D.M. *et al.*, *Vegetatio* **65** (1986), 149–162.
21. Gartlan, J.S. *et al.*, *Ibid.* 130–148.
22. Jordan, C. *Ecology* **63** (1982), 647–654.
23. Vitousek, P.M. and Sanford, R.C., *Ann. Rev. Ecol. Syst.* **17** (1986), 131–167.
24. Jordan, C. *et al.*, *Biotropica* **12** (1980), 61–66.
25. Tukey, H.B. in Odum, H.T. and Pigeon, R.F. (eds), *A Tropical Rain Forest.* U.S. Atomic Energy Commission, Tennessee, 1970, Chapter H-11.
26. Witkamp, M., *Ibid.*, Chapter H-14.
27. Nadkarni, N.M., *Biotropica* **16** (1984), 249–256.
28. Williamson, G.B., *Biotropica* **13** (1981), 228–231.
29. Jordan, C. and Herrera, R., *Amer. Nat.* **117** (1981), 167–180.
30. Manokaran, N., *Malays. For.* **43** (1980), 266–289.
31. Spain, A.V., *J. Ecol.* **72** (1984), 947–961.
32. Proctor, J. *et al.*, *J. Ecol.* **71** (1983), 503–527.
33. Enright, N.J., *Malays. For.* **42** (1979), 202–207.
34. Anderson, J.M. *et al.*, *J. Ecol.* **71** (1983), 503–527.
35. Weigert, R.G. and Murphy, P. in Odum H.T. and Pigeon, R.F., *op. cit.*, Chapter H-5.
36. Lang, G.E. and Knight, D.H. *Biotropica* **11** (1979), 316–317.
37. Lowman, M.D., *J. Ecol.* **76** (1988), 451–465.
38. Bunvong Thaiutsa and Granger, O., *Unasylva* **31** (1979), 28–35.
39. Marrs, R.H. *et al.*, *J. Ecol.* **76** (1988), 466–482.
40. Lewis, W.M., *J. Ecol.* **67** (1986), 1275–1282.
41. Jordan, C. and Herrera, R., *Amer. Nat.* **117** (1981), 167–180.
42. Gower, S.T., *Biotropica* **19** (1987), 171–175.
43. Redhead, J.F. and Bowen, C.D. in Mikola, P. (ed.), *Tropical Mycorrhiza Research,* Clarendon, Oxford, 1980, Chapters 16, 21.
44. Janos, D.P. in Sutton, S.L. *et al.* (eds), *Tropical Rain Forest: Ecology and Management.* Blackwell, Oxford, 1983, pp. 327–345.
45. Alexander, I.J. and Högberg, P., *New Phytol.* **102** (1986), 541–549.
46. Allen, O.N. and Allen, E.K. *The Leguminosae.* Macmillan, London, 1981.
47. Janos, D.P. in Janzen, D.H. (ed.), *Costa Rican Natural History.* University of Chicago Press, 1987, pp. 340–344.
48. Lee, K.E. in Satchell, J.E. (ed.), *Earthworm Ecology.* Chapman and Hall, London, 1983, pp. 179–193.
49. Gould, E. and Andau, M., *Biotropica* **19** (1987), 370–372.
50. Collins, N.M. in Sutton, S.L. *et al.*, *op. cit.*, 381–412.
51. Collins, N.M. in Cranbrook, Earl of, *Malaysia.* Pergamon, Oxford, 1988, pp. 196–211.
52. Jordan, C.F. and Medina, E., *Ann. Mo. bot. Gdn.* **64** (1977), 737–745.
53. Proctor, J. *et al.*, *J. Ecol.* **71** (1983), 237–260.

Chapter 4 (pp. 52–79)

1. Drury, W.H. and Nisbet, I.C.T., *J. Arnold Arb.* **54** (1973), 331–368.
2. Hewetson, C.E., *Emp. For. Rev.* **35** (1956), 274–291.
3. Hallė, F., Oldeman, R.A.A. and Tomlinson, P.B., *Tropical Trees and Forests: An Architectural Analysis.* Springer, Berlin, 1978.
4. Mabberley, D.J. in Bramwell, D. (ed.), *Plants and Islands.* Academic Press, London, 1979, pp. 259–277.
5. Lorimer, C.G., *Ecology* **70** (1989), 565–567.

6. Taylor, A.R. in Goldie, R.H., *Lightning*. Academic Press, London, 1977, vol. 2, 831–849; Salo, J. *et al.*, *Nature* **322** (1986), 254–258.
7. Tho, Y.P., *Malays. For.* **45** (1982), 184–192.
8. Rijksen, H.D., *Meded. Landb. Wageningen* **78–82** (1978).
9. Poore, M.E.D., *J. Ecol.* **56** (1968), 143–196.
10. Whitten, A.J. *et al.*, *The Ecology of Sumatra*, (2nd edn), Gadja Mada University Press, 1987, p. 258.
11. Martínez-Ramos, M. *et al.*, *J. Ecol.* **76** (1988) 700–716.
12. Swaine, M.D. and Whitmore, T.C., *Vegetatio* **75** (1988), 81–86.
13. Hartshorn, G.S. in Tomlinson, P.B. and Zimmermann, M.H., *Tropical Trees as Living Systems*. Cambridge University Press, 1978, pp. 617–638.
14. Ewel, J. (ed.), Tropical succession. *Biotropica* **12** (2) (Suppl.), 1980.
15. Popma, J. *et al.*, *J. Trop. Ecol.* **4** (1988), 77–88.
16. Chiarello, N.R. *et al.*, *Funct. Ecol.* **1** (1987), 3–11.
17. Bazzaz, F.A. in Medina, E., Mooney, H.A. and Vázquez-Yánes, C., *Physiological Ecology of Plants in the Wet Tropics*. Junk, The Hague, 1984, pp. 233–243.
18. Aide, T.M., *Biotropica* **19** (1987), 284–285.
19. Putz, F.E. and Brokaw, N.V.L., *Ecology* **70** (1989), 508–512.
20. Putz, F., *Symposium—The Tropical Rain Forest*. Poster abstracts. Leeds, 1982.
21. Lieberman, D. *et al.*, *J. Ecol.* **73** (1985), 915–924.
22. Sanford, R. *et al.*, *J. Trop. Ecol.* **2** (1986), 277–282.
23. Brokaw, N.V.L., *Ecology* **66** (1985), 682–687.
24. Brokaw, N.V.L. and Scheiner, S.M., *Ecology* **70** (1989), 538–541.
25. Lawton, R.O. and Putz, F.E., *Ecology* **69** (1988), 764–777.
26. Uhl, C. *et al.*, *Ecology* **69** (1988), 751–763.
27. Lieberman, M. *et al.*, *Ecology* **70** (1989), 550–552.
28. Putz, F.E., *Ecology* **64** (1983), 1069–1074.
29. Whitmore, T.C., *For. Abstr.* **44** (1983), 767–779.
30. Augspurger, C.K. and Franson, S.E., *J. Trop. Ecol.* **4** (1988), 239–252.
31. Kennedy, D. and Swaine, M.D., *Trop. Biol. Newsl.* **53** (1989), 1–2.
32. Cheke, A.S. *et al.*, *Biotropica* **11** (1979), 88–95.
33. Ng, F.S.P. in Tomlinson, P.B. and Zimmermann, M.H., *op. cit.* pp. 129–162.
34. Whitmore, T.C. and Hartshorn, G.S. *Ibid.*, pp. 617–638, 639–655.
35. Finegan, B., *Nature* **312** (1984), 109–114.
36. Schupp, E.W. *et al.*, *Ecology* **70** (1989), 562–564.
37. Young, K.R. *et al.*, *Vegetatio* **71** (1987), 157–173.
38. Perez-Nasser, N. and Vásquez-Yánes, C., *Malays. For.* **49** (1986), 352–356.
39. Putz, F.E. and Appanah, S., *Biotropica* **19** (1987), 326–333.
40. Garwood, N.C., *Ecol. Monogr.* **53** (1983), 159–181.
41. Vásquez-Yánes, C. and Smith, H., *New Phytol.* **92** (1982), 477–485.
42. Uhl, C. and Clark, K., *Bot. Gaz.* **144** (1983), 419–425.
43. Marquis, R.J. *et al.*, *Biotropica* **11** (1986), 273–278.
44. Schupp, E.W. *et al.*, *Ecology* **70** (1989), 562–564.
45. Hopkins, M.S. and Graham, A.W., *Biotropica* **15** (1983), 90–99.
46. Foster, S.A., *Bot. Rev.* **52** (1986), 260–299.
47. Bongers, F. *et al.*, *Funct. Ecol.* **2** (1988), 379–390.
48. Brokaw, N., *Symposium—The Tropical Rain Forest*, Poster Abstracts, Leeds, 1982.
49. Yeaton, R., *Biotropica* **11** (1979), 155–158.
50. Kenworthy, J.B. and Riswan, S., *Symposium—The Tropical Rain Forest*, Poster Abstracts, Leeds, 1982.
51. Riswan, S. *et al.*, *J. Trop. Ecol.* **1** (1985), 171–182.
52. Raich, J.W., *Biotropica* **21** (1989), 299–302.
53. Primack, R.B., *Ecology* **66** (1985), 577–588.

54. Schemske, D.W. and Brokaw, N., *Ecology* **62** (1981), 938–945.
55. Shelly, T.E., *Biotropica* **20** (1988), 114–119.
56. Becker, P., *Malays. For.* **48** (1985), 263–265.
57. Levey, D.J., *Ecology* **69** (1988), 1076–1089.
58. Woolley, A. and Bishop, C., *New Scientist* **25**, viii (1983), 561–567.
59. Self, S. and Rampino, M.R., *Nature* **294** (1981), 699–704.
60. Thornton, I.W.B., *Ambio* **13** (1984), 216–225.
61. Whittaker, R.J. *et al.*, *Ecol. Monog.* **59** (1988), 59–123.
62. Tagawa, H. *et al.*, *Vegetatio* **60** (1988), 131–145.
63. Mabberley, D.J., *New Phytol.* **74** (1975), 365–374.
64. Sugden, A.M., *J. Arnold Arb.* **62** (1982), 1–61.
65. Sugden, A.M., *Biotropica* **14** (1982), 208–219.
66. Ball, E. and Glucksman, J., *Proc. Roy. Soc. London* (B) **190** (1975), 421–447.
67. Beard, J.S., *Vegetatio* **31** (1976), 69–77.

Chapter 5 (pp. 80–132)
1. Simberloff, D., *Synthèse* **43** (1980), 3–39.
2. Turkington. R. and Harper, J.L., *J Ecol.* **67** (1979), 245–254.
3. Thorne, R.F., in Meggers, B.J., Ayensu, E.S. and Duckworth, W.D. (eds) *Tropical Forest Ecosystems in Africa and South America: A Comparative Review*. Smithsonian Inst., Washington, 1973, pp. 27–47.
4. White, F., *The Vegetation of Africa*. UNESCO, Paris, 1983.
5. Koechlin, J., in Richard-Vindard, G. and Battistini, R. (eds) *Biogeography and Ecology in Madagascar*. Junk, The Hague, 1972, pp. 145–190.
6. Bourlière, F., in Meggers, B.J., *et al.*, *op. cit.* pp. 279–292.
7. Janzen, D.H., *Ann. Mo bot. Gdn* **64** (1977), 706–736.
8. Emmons, L.H. and Gentry, A.H., *Amer. Nat.* **121** (1983), 513–524.
9. Michener, C.D., *Ann. Mo bot. Gdn* **66** (1979), 277–347.
10. Dubost, G., *Afr. J. Ecol.* **17** (1979), 1–13.
11. Bourlière, F., *Ecol. Stud.* **69** (1989), 153–168.
12. Fleming, T.H. *et al.*, *Ann. Rev. Ecol. Syst.* **18** (1987), 91–109.
13. Karr, J.R., *Wilson Bull.* 88 (1976), 433–458.
14. Scott, N.J., *Biotropica* **8** (1976), 41–58.
15. Goulding, M., in Prance, G.T. and Lovejoy, T.E. (eds), *Amazonia* , Pergamon, Oxford, 1985, pp. 267–276.
16. Sutton, S.L. in Sutton, S.L. *et al.* (eds), *Tropical Rain Forest: Ecology and Management*. Blackwell, Oxford, 1983, pp. 77–91.
17. Collins, N.M. in Cranbrook, Earl of (ed.), *Malaysia*, Pergamon, Oxford, 1988, pp. 196–211.
18. Wilson, E.O., *Biotropica* **19** (1987), 245–251.
19. Dixon, A.F.G. *et al.*, *Amer. Nat.* **129** (1987), 588–592.
20. Hall, J.B. and Swaine, M.D., *Distribution and Ecology of Vascular Plants in a Tropical Rain Forest*. Junk, The Hague, 1983.
21. Putz, F.E. *et al.*, *Amer. Nat.* **112** (1984), 24–28.
22. Mitchell, A.W., *Reaching the Rain Forest Roof*. Leeds Philosophical Society, Leeds, 1982.
23. Hallé, F., Oldeman, R.A.A. and Tomlinson, P.B., *Tropical Trees and Forests: An Architectural Analysis*. Springer, Berlin, 1978.
24. Hallé, F. and Mabberley, D.J., *Gdns' Bull., Sing.* **29** (1977), 175–181.
25. Mabberley, D.J., *Gdns' Bull., Sing.* **29** (1977), 41–55.
26. Horn, H.S., *Scient. Amer.* **232**(5) (1975), 90–98.
27. Hamilton, C.W., *Amer. J. Bot.* **72** (1985), 1081–1088.
28. Tomlinson, P.B., *Ann. Rev. Ecol. Syst.* **18** (1987), 1–21.

29. Ashton, P.S., *Ann. Rev. Ecol. Syst.* **19** (1987), 347–370.
30. Holbrook, N.M. *et al., Principes* **29** (1985), 142–146.
31. Smith, A.P., *Biotropica* **11** (1979), 159–160.
32. Johnson, P.W., *Ghana J. Agric. Sci.* **5** (1972), 13–21.
33. Warren, J.D. *et al., Ecology* **69** (1988), 532–536.
34. Richter, W., *Ecology* **65** (1984), 1429–1435.
35. Lewis, A.R., *Biotropica* **20** (1988), 280–285.
36. Jeník, J., *Preslia* **45** (1973), 250–264.
37. Bodley, J.D. and Benson, F.C., *Biotropica* **12** (1980), 67–71.
38. Swaine, M.D., *Biotropica* **15** (1983), 240.
39. Nadkarni, N.M., *Science* **214** (1981), 1023–1024.
40. Sanford, R.L., *Science* **235** (1987), 1062–1064.
41. Putz, F.E. and Holbrook, N.M., *Selbyana* **9** (1986), 61–69.
42. Davison, G.W.H. (ed.), *Endau Rompin. A. Malaysian Heritage.* Malayan Nature Society, Kuala Lumpur, 1988, p. 117.
43. Lawton, R.O. and Putz, F.E., *Ecology* **69** (1988), 764–777.
44. Madison, M., *Selbyana* **2** (1979), 1–13.
45. Kress, W.J., *ibid.* **9** (1986), 2–11.
46. Huxley, C., *New Phytol.* **80** (1978), 231–268.
47. Benzing, D.H., *Ecol. Stud.* **76** (1989), 167–199.
48. Pócs, T., in Smith A.J.E., *Bryophyte Ecology.* Chapman and Hall, London, 1982, pp. 59–104.
49. Richards, P.W., in Schuster, R.M. (ed.) *New Manual of Bryology,* Nichinan, Hattori Labs, 1983, pp. 1233–1270.
50. Berrie, G.K. and Eze, J.M.O., *Ann. Bot.* **39** (1975), 955–963.
51. Kiew, R., in Cranbrook, Earl of, *Malaysia.* Pergamon, Oxford, 1988, pp. 56–76.
52. Putz, F.E., *Biotropica* **15** (1983), 185–189.
53. Putz, F.E. and Chai, P., *J. Ecol.* **75** (1987), 523–531.
54. Putz, F.E., *Ecology* **65** (1984), 1713–1734.
55. Peñalosa, J., *Biotropica* **16** (1981), 1–9.
56. Caballé, G., *Adansonia* **19** (1980), 467–476.
57. Stevens, G.C., *Ecology* **68** (1987), 77–81.
58. Juniper, B.E. *et al., The Carnivorous Plants.* Academic Press, London, 1989.
59. Corner, E.J.H., in Cranbrook, Earl of, *Malaysia.* Pergamon, Oxford, 1988, pp. 88–101.
60. Burtt, B.L., *Gdns' Bull., Sing.* **29** (1977), 73–80.
61. van Steenis, C.G.G.J., *Rheophytes of the World.* Sijthoff and Noordhoff, Alphen aan den Rijn, 1981.
62. Fink, S., *Amer. J. Bot.* **70** (1983), 532–542.
63. Mabberley, D.J., *Bull. Brit. Mus. (Nat. Hist.), Bot.* **6** (1979), 301–386.
64. Mabberley, D.J., *Taxon* **33** (1984), 77–79.
65. Chan, H.T., *Malays. For.* **45** (1982), 354–360.
66. Hamrick, J.L. and Loveless, M.D., in Bock, J.H. and Linhart, Y.B. (eds) *Evolutionary Ecology of Plants.* Westview Press, Boulder, Colorado, 1989, pp. 129–146.
67. Putz, F.E., *Malays. For.* **42** (1979), 1–24.
68. Corlett, R.T., *Biotropica* **19** (1987), 122–124.
69. Ng, F.S.P., *For. Res. Inst., Kepong, Res. Pamph.* **96** (1984).
70. Croat, T.B., *Flora of Barro Colorado Island.* Stanford University Press, Stanford, 1978.
71. Borchert, R., *Biotropica* **15** (1983), 81–89.
72. Augspurger, C.K., in Leigh, E.G. *et al. The Ecology of a Tropical Forest.* Oxford University Press, Oxford 1983, pp. 133–156.
73. Hopkins, M.S. and Graham, A.W., *Austral. J. Ecol.* **12** (1987), 25–29.
74. Voeks, R.A., *Biotropica* **20** (1988), 107–113.
75. Clark, D.A. and Clark, D.B., *J. Ecol.* **75** (1987), 135–149.
76. Yap, S.K., *Malays. For.* **45** (1982), 21–35.

77. Ng, F.S.P., *Malays. For.* **40** (1977), 126–137.
78. Ashton, P.S. *et al.*, *Amer. Nat.* **132** (1988), 44–66.
79. Dayandanan, S., *Reproductive Biology of Some Shorea spp. in Sri Lanka*, M. Phil. Thesis, University of Peradeniya, 1989.
80. Ng, F.S.P. and Loh, H.S., *Malays. For.* **37** (1974), 127–132.
81. Janzen D.H. in Burley, J. and Styles, B.T. (eds) *Tropical Trees: Variation, Breeding and Conservation.* Academic Press, London, 1976, pp. 179–188.
82. Bullock, S.H. and Bawa, K.S., *Ecology* **62** (1981), 1494–1504.
83. Bawa, K.S. in Jones, C.E. and Little, R.J., *Handbook of Experimental Pollination Biology.* Van Nostrand Reinhold, New York, 1983, pp. 394–410.
84. Bentley, B.L., *Ann. Bot.* **43** (1979), 119–121.
85. Ng, F.S.P. in Cranbrook, Earl of, *Malaysia.* Pergamon, Oxford 1988, pp. 102–125.
86. Corner, E.J.H., *Wayside Trees of Malaya*, 2 vols, (3rd edn) Malayan Nature Society, Kuala Lumpur, 1988.
87. Reich, P.B. and Borchert, R., *Biotropica* **20** (1988), 60–69.
88. Worbes, M. and Junk, W.J., *Ecology* **70** (1989), 502–507.
89. Hazlett, D.L., *Biotropica* **19** (1987), 357–360.
90. Primack, R.B., *Ecology* **66** (1988), 577–588.
91. Núñez-Forfán, J. and Dirzo, R., *Oikos* **51** (1988), 274–284.
92. Clark, D.B. and Clark, D.A., *Biotropica* **19** (1987), 236–244.
93. Martínez-Ramos, M. *et al.*, *Ecology* **70** (1989), 855–858.
94. Swaine, M.D. *et al.*, *J. Trop Ecol.* **3** (1987), 359–366.
95. Lieberman, D. *et al.*, *J. Trop. Ecol.* **1** (1985), 97–109.
96. Piñero, D. *et al.*, *J. Ecol.* **72** (1984), 977–991.
97. Barbault, R., *Oikos* **43** (1984), 77–87.
98. Caldecott, J. and S., *New Scientist*, **15**, viii (1985), 32–35.
99. Appanah, S., *J. Trop. Ecol.* **1** (1985), 225–240.
100. Pearson, D.L. and Derr, J.A., *Biotropica* **18** (1986), 244–256.
101. Brown, R.G. and Hodkinson, I.D., *Taxonomy and Ecology of the Jumping Plant Lice of Panama*, Brill, Leiden, 1988, pp. 287–294.
102. Wolda, H., *Ann. Rev. Ecol. Syst.* **19** (1988), 1–18.
103. Godfray, H.C.J. and Hassell, M.S., *J. Anim. Ecol.* **58** (1989), 153–174.
104. Wolda, H. and Foster, R., *Geo-Eco-Trop* **2** (1978), 443–454.
105. Chakrabarti, K. and Chaudhuri, A.B., *Sci. and Cult.* **38** (1972), 269–276.
106. Corner, E.J.H. in Cranbrook, Earl of, *Malaysia*, Pergamon, Oxford, 1988, pp. 88–101.
107. Levings, S.C. and Windsor, D.M., *J.Anim. Ecol.* **54** (1985), 61–69.
108. Diamond, A.W., *Symposium—The Tropical Rain Forest*, Poster Abstracts, Leeds, 1982.
109. Karr, J.A., *Amer. Nat.* **110** (1976), 973–994.
110. Fogden, M.P.L., *Ibis* **114** (1972), 307–343.
111. Bell, H.L., *Emu* **82** (1982), 65–74.
112. Bell, H.L., *Emu* **82** (1982), 143–162.
113. Crome, F.H.J., *Austr. Wildl. Res.* **2** (1975), 155–185.
114. Fleming, T.H. *et al.*, *Ann. Rev. Ecol. Syst.* **18** (1987), 91–109.
115. Greenberg, R., *Biotropica* **13** (1981), 241–251.
116. Brosset, A., *Rev. Ecol.* **35** (1981), 109–129.
117. Cranbrook, Earl of, *Malaysia*, Pergamon, Oxford, 1988, pp. 146–166.
118. Quris, R. *et al.*, *Rev. Ecol.* **35** (1981), 37–53.
119. Gittins, S.P., *Fol. Primatol.* **38** (1982), 39–71.

Chapter 6 (pp. 133–186)
1. Eastop, V.F., in Forey, P.L. (ed.) *The Evolving Biosphere.* British Museum (Natural History) and Cambridge University Press, Cambridge, 1981, pp. 179–190.
2. Southwood, T.R.E., *Oikos* **44** (1985), 5–11.

3. Vitousek, P.M. and Sanford, R.L., *Ann. Rev. Ecol. Syst.* **17** (1986), 137–167.
4. Lieberman, D. and Lieberman, M. *Symposium—The Tropical Rain Forest*, Poster Abstracts, Leeds, 1982.
5. Lowman, M.D., *Biotropica* **16** (1984), 264–268.
6. Coley, P.D., *Ecol. Monogr.* **53** (1983), 209–233.
7. Lemen, C., *Oikos* **36** (1981), 65–67.
8. Landsberg, J. and Ohmort, C., *Trends Ecol. Evol.* **4** (1989), 96–100.
9. Whitten, J.E.J. and A.J., *Biotropica* **19** (1987), 107–115.
10. Hendrix, S.D. and Marquis, R.J., *Biotropica* **15** (1983), 108–111.
11. Corner, E.J.H., in Cranbrook, Earl of, *Malaysia*, Pergamon, Oxford, 1988, pp. 88–101.
12. Pires, J.M., in Prance, G.T. and Lovejoy, T.E. (eds), *Amazonia*, Pergamon, Oxford, 1984, pp. 109–145.
13. Janzen, D.H. in Prance, G.T. and Lovejoy, T.E., *op. cit.*, pp. 207–217.
14. Barlow, H.S., in Cranbrook, Earl of, *op. cit.*, pp. 212–214.
15. Messer, A.C., *Malays. For.* **48** (1985), 266–267.
16. Baldwin, I.T. and Schultz, J.C., *Science* **221** (1983), 277–279.
17. Van Hoven, W., *Abstracts 2nd Int. Symp. The Tree*. Montpellier, 1990.
18. Braam, J. and Davis, R.W., *Cell* **60** (1990), 357–364.
19. Gartlan, J.S. *et al.*, *Biochem. Syst. Evol.* **8** (1980), 401–422.
20. Cooke, F.P. *et al.*, *Biotropica* **16** (1984), 257–263.
21. Waterman, P.G. *et al.*, *Biol. J. Linn. Soc.* **34** (1988), 1–32.
22. Davies, A.C. *et al.*, *Biol. J. Linn. Soc.* **34** (1988), 33–56.
23. McKey, D. *et al.*, *Science* **202** (1978), 61–64.
24. Marshall, A.G., *Zool. J. Linn. Soc.* **83** (1988), 351–389.
25. Montgomery, G.G. and Sunquist, M.E. in Golley, F.B. and Medina, E., *Tropical Ecological Systems*. Springer, Berlin, 1975, pp. 69–98.
26. Rogers, M.E. and Williamson, E.A., *Biotropica* **29** (1987), 278–281.
27. Short, J., *Mammalia* **45** (1981), 177–185.
28. Dransfield, J., in Cranbrook, Earl of, *op. cit.*, pp. 37–48.
29. Harborne, J.B., *Phytochemical Ecology*, (3rd edn.) Academic Press, London, 1988, esp. pp. 186–213.
30. Powell, R.J. and Stradling, D.J., *Symposium—Interactions Between Ants and Plants*, Abstracts. Oxford, 1989.
31. Stevens, G.C., in Janzen, D.H. (ed.) *Costa Rican Natural History*, University of Chicago, 1983, pp. 688–691.
32. Angeli-Paoa, J. and Eymé, J., *Ann. Sci. Nat., Bot. XIII*, **7** (1988), 103–129.
33. Cherrett, J.M., in Sutton, S.L. *et al.* (eds), *Tropical Rain Forest: Ecology and Management*. Blackwell, Oxford, 1983, pp. 253–263.
34. Fowler, H.G. *et al.*, in Lofgren, C.S. and Vander Meer, R.K. (eds) *Fire Ants and Leaf-cutting Ants: Biology and Management*. Westview, Boulder and London, 1986, pp. 123–145.
35. Huxley, C.R., in Juniper, B.E. and Southwood, T.R.E (eds) *Insects and the Plant Surface*. Arnold, London, 1986, pp. 257–282.
36. Elias, T.S. and Prance, G.T., *Brittonia* **30** (1978), 175–181.
37. Risch, S.J. *et al.*, *Amer. Midl. Nat.* **98** (1977), 433–444.
38. Rickson, F.R. and Risch S.J. *Amer. J. Bot.* **71** (1984), 1268–1274.
39. Letourneau, D.R., *Symposium–Interactions between Ants and Plants*, Abstracts, Oxford, 1989.
40. Janzen, D.H., *Ecology of Plants in the Tropics*. Arnold, London, 1975.
41. Davidson, D.W. *et al.*, *Ecology* **69** (1988), 801–808.
42. Page, C.N. and Brownsey, P.J., *J. Ecol.* **74** (1988), 787–796.
43. Janzen, D.H., *Ecology* **53** (1972), 885–892.
44. McKey, D., *Biotropica* **16** (1984), 81–99.

45. Rickson, F.R. and M.M., *Biotropica* **18** (1986), 337–343.
46. Benson, W.W., in Prance, G.T. and Lovejoy, T.E., *Amazonia*. Pergamon, Oxford, 1984, pp. 239–266.
47. Kiew, R., in Cranbrook, Earl of, *Malaysia*. Pergamon, Oxford, 1988, pp. 56–76.
48. Davidson, D.W., *Biotropica* **21** (1989), 64–73.
49. Fiala, B. *et al.*, *Symposium—Interactions Between Ants and Plants*, Abstracts. Oxford, 1989.
50. Pemberton, R.W. and Turner, C.E., *Amer. J. Bot.* **76** (1989), 105–112.
51. Turner, C.E. and Pemberton, R.W. in Bock, J.H. and Linhart, Y.B., *Evolutionary Ecology of Plants*, Westview, Boulder, Colorado, 1989, pp. 341–359.
52. Janzen, D.H., in Rosenthal, G.A. and Janzen, D.H. (eds), *Herbivores and Their Interaction with Secondary Plant Metabolites*. Academic Press, London, 1979, pp. 331–350.
53. Mackinnon, J., *Anim. Behav.* **22** (1974), 3–74.
54. Rijksen, H.D., *Meded. Landb. Wageningen* **78-2** (1978).
55. Owen, D.F., *Oikos* **35** (1980), 230–235.
56. Petelle, M., *Oikos* **38** (1982), 125–127.
57. Dickinson, T.A. and Tanner, E.V.J., *Biotropica* **10** (1978), 231–233.
58. Fleming, T.H. *et al.*, *Ann. Rev. Ecol. Syst.* **18** (1987), 91–109.
59. Janzen, D.H., *Amer. Nat.* **123** (1984), 338–353.
60. Janzen, D.H. in Futuyama, D.J. and Slatkin, M. (eds) *Coevolution*. Sinauer Assoc., Sunderland, Mass., 1983, pp. 232–262.
61. Wheelwright, N.T. and Orians, C.H., *Amer. Nat.* **119** (1982), 402–413.
62. Kubitzki, K. in Prance, G.T. and Lovejoy, T.E. (eds), *Amazonia*. Pergamon, Oxford, 1984 pp. 192–206.
63. Murray, K.G., *Ecol. Monogr.* **58** (1988), 271–298.
64. Izhaki, I. and Safriel, U.N., *Oikos* **54** (1989), 23–32.
65. Herrera, C.M., *Oikos* **42** (1984), 203–210.
66. Regal, P.J., *Science* **196** (1977) 622–629.
67. Snow, D.W., in Forey, P.L., *op. cit.*, pp. 169–178.
68. Levey, D.J. *Auk* **104** (1987), 173–179.
69. Greig-Smith, P.W., *Amer. Nat.* **127** (1986), 246–251.
70. Wheelwright, N.T., *Trends Ecol. Evol.* **3** (1988), 270–274.
71. Snow, D.W., *Oikos* **35** (1965), 274–280.
72. Wheelwright, N.T., *Oikos* **44** (1985), 465–477.
73. Wheelwright, N.T., *Biotropica* **16** (1984), 173–192 and *Ecol. Monogr.* **65** (1985), 808–818.
74. Wheelwright, N.T., in Estrada, A. and Fleming, T.H. (eds), *Frugivores and Seed Dispersal*. Junk, Dordrecht, 1986, pp. 19–35.
75. Howe, H.F. in Janzen, D.H. (ed.) *Costa Rican Natural History*. Chicago Univ. Press, 1983, pp. 603–604.
76. Howe, H.J. and de Steven, D., *Oecologia* **39** (1979), 185–196.
77. Crome, F.H.J., *Austral. Wildl. Res.* **2** (1975), 155–185.
78. Smythe, N., *Biotropica* **21** (1989), 50–58.
79. Stiles, E.W. and White, D.W. in Estrada, A. and Fleming, T.H. (eds), *op. cit.* pp. 45–54.
80. Herrera, C.M., *Oikos* **51** (1988), 383–386.
81. Leighton, M. and D.R. in Sutton, S.L. *et al.* (eds), *Tropical Rain Forest: Ecology and Management*, Blackwell, Oxford, 1983, pp. 181–196.
82. Croat, T.B., *Flora of Barro Colorado Island*, Stanford University Press, Stanford, 1978.
83. Janson, C.H., *Science* **219** (1983), 187–189.
84. Pannell, C.M. and Koziof, M.J., *Phil. Trans. Roy. Soc. London* B **316** (1987), 303–333.
85. Schupp, E.W., *Oikos* **51** (1988), 71–78.
86. Estrada, A. *et al.*, *Biotropica* **16** (1984), 315–318.
87. Vázquez-Yanes, C. and Orozco-Segovia, A. in Estrada, A. and Fleming, T.H., *op. cit.*, pp. 71–77.

88. Sazima, I. *et al.*, *Symposium—The Tropical Rain Forest*, Poster Abstracts, Leeds, 1982.
89. Cont, J.G.H., *Biotropica* **11** (1979), 122.
90. Bonaccorso, F.J. *et al.*, *Rev. Biol. Trop.* **28** (1980), 61–72.
91. Howe, H.F., *Ecology* **61** (1980), 944–959.
92. Ashton, P.S. in *Tropical Forest Ecosystems*. UNESCO, Paris, 1978, pp. 180–215.
93. Breitwitsch, R., *Biotropica* **15** (1983), 125–128.
94. Snow, D.W., *Biotropica* **13** (1981), 1–14.
95. Milton, K. *et al.*, *Ecology* **63** (1982), 752–762.
96. Morrison, D.W., *J. Mammal.* **59** (1978), 622–624.
97. Janzen, D.H., *Biotropica* **10** (1978), 121.
98. Marshall, A.G., *Zool. J. Linn. Soc.* **83** (1985), 351–369.
99. Terborgh, J. in Estrada, A. and Fleming, T.H. *op. cit.*, pp. 371–384.
100. Charles-Dominique, P. *et al.*, *Rev. Ecol.* **35** (1981), 341–436.
101. Terborgh, J. in Prance, G.T. and Lovejoy, T.E. *Amazonia*, Pergamon, Oxford, 1984, pp. 284–304.
102. Garber, P.A., *Amer. J. Primatol.* **10** (1986), 155–170.
103. Wrangham, R.W. and Waterman, P.G., *Biotropica* **15** (1983), 217–222.
104. Estrada, A. and Coates-Estrada, R., *Amer. J. Primatol.* **6** (1984), 77–91.
105. Goulding, M., *Sonderb. Naturwiss. Ver. Hamburg* **7** (1983), 271–283.
106. Pacini, E. and Grieco, L. *Symposium–Interactions Between Ants and Plants*, Poster Abstracts, Oxford, 1989.
107. Kaufman, S.F. *et al.*, *l.c.*
108. Roberts, J.T. and Heithaus, E., *Ecology* **67** (1986), 1046–1051.
109. Thomas, D.W., *Biotropica* **20** (1988), 49–53.
110. Madison, M., *Selbyana* **5** (1979), 107–115.
111. Davidson, D.W., *Ecology* **69** (1988), 1135–1152.
112. Lee, M.A.B., *Oikos* **45** (1985), 169–173.
113. Mason, R., *Nature* **191** (1961), 408–409.
114. Webber, M.L., *Malays. For.* **3** (1934), 18–19.
115. Gentry, A.H., *Sonderb. Naturwiss. Ver. Hamburg* **7** (1983), 303–314.
116. Gandawijaja, D. and Arditti, J., *Ann. Bot.* **52** (1983), 127–130.
117. Gradstein, S.R. *et al.*, *Acta bot. Hung.* **29** (1983), 127–171.
118. van Zanten, B.O. and Pócs, T. in Schulze-Motel, W. (ed.) *Advances in Bryology* **1**. Cramer, Vaduz, 1981, pp. 479–562.
119. Pirozynski, K.A. and Malloch, D.W. in Pirozynski, K.A. and Hawksworth, D.L. (eds), *Coevolution of Fungi with Plants and Animals*. Academic Press, 1988, pp. 227–246.
120. Lamont, B.B. *et al.*, *New Phytol.* **101** (1985), 651–656.
121. Hopkins, H.C. and M.J.G. in Sutton, S.L. *et al.* (eds), *Tropical Rain Forest: Ecology and Management*. Blackwell, Oxford, 1983, pp. 197–209.
122. Janzen, D.H. and Martin, P.S., *Science* **215** (1982), 19–27.
123. Stocker, G.C. and Irvine, A.K., *Biotropica* **15** (1983), 170–176.
124. Van Strien, N.J., *Meded. Landb. Wageningen* **74–16** (1974).
125. Dinerstein, E. and Wemmer, C.M., *Ecology* **69** (1988), 1768–1774.
126. Dawkins, R. *The Extended Phenotype*. Freeman, Oxford, 1982, pp. 35 *et seq.*
127. Mabberley, D.J., *New Phytol.* **75** (1975), 289–295.
128. Keeler, K.H., *Oikos* **44** (1985), 407–414.
129. Cox, P.A., *Oikos* **41** (1983), 195–199.
130. Sykes, *Bull. N.Z. Dept. Sci. Ind. Res.* **200** (1970), p. 140.
131. Lee, W.G. *et al.*, *Oikos* **53** (1988), 325–331.
132. Corner, E.J.H. *The Natural History of Palms*. Weidenfeld & Nicolson, London, 1966.
133. Alexandre, D.-Y., *La Terre et la Vie* **32** (1978), 47–62.
134. Short, J., *Mammalia* **1** (1981), 177–185.
135. Temple, S.A., *Science* **197** (1977), 885–886.

136. Sugden, A.M. in Sutton, S.L. *et al.* (eds), *Tropical Rain Forest: Ecology and Management.* Blackwell, Oxford, 1983, pp. 43–56.
137. Keeler-Wolf, T., *Biotropica* **20**, (1988), 38–48.
138. Bawa, K.S. *International Botanical Congress Abstracts*, Sydney, 1981.
139. Bawa, K.S. *et al.*, *Amer. J. Bot.* **72** (1985), 346–356.
140. Crepet, W.L., *BioScience* **29** (1979), 102–108.
141. Appanah, S., *Malays. For.* **44** (1981), 37–42.
142. Janzen, D.H. *Costa Rican Natural History.* Chicago University Press, Chicago, 1983, pp. 619–645.
143. Yap, S.K., *Malays. For.* **45** (1982), 21–35.
144. Bawa, K.S., *Amer. J. Bot.* **72** (1985), 331–345.
145. Baker, H.G. in Tomlinson, P.B. and Zimmermann, M.H. (eds) *Tropical Trees as Living Systems.* Cambridge University Press, Cambridge, 1978, pp. 57–82.
146. Baker, I. in Jones, C.E. and Little, R.J., *Handbook of Experimental Pollination.* Van Nostrand Reinhold, New York, 1983, pp. 117–141.
147. Bawa, K.S., *op. cit.*, pp. 394–410.
148. Boyden, T.C., *Evolution* **34** (1980), 135–136.
149. Guerrant, E.O. and Fiedler, P.L. *Biotropica* **13** Suppl. (1981), 25–33.
150. Beattie, A.J. *et al.*, *Amer. J. Bot.* **71** (1984), 421–426.
151. Prance, G.T. in Prance, G.T. and Lovejoy, T.E. (eds), *Amazonia*, Pergamon, Oxford, 1984, pp. 161–191.
152. Buchmann, S.L. in Jones, C.E. and Little, R.J., *op. cit.*, pp. 73–113.
153. Appanah, S. *et al.*, *Malay Nat. J.* **39** (1986), 177–191.
154. Johnson, L.K. in Janzen, D.H., *Costa Rican Natural History.* Chicago University Press, Chicago, 1983, pp. 770–772.
155. Steiner, K.E., *Biotropica* **17** (1985), 217–229.
156. Buchmann, S.L., *Ann. Rev. Ecol. Syst.* **18** (1987), 343–369.
157. Dressler, R.L., *Ann. Rev. Ecol. Syst.* **13** (1982), 373–394.
158. Ackerman, J.D. and Montalvo, A.M., *Biotropica* **17** (1985), 79–91.
159. Ackerman, J.D., *Biol. J. Linn. Soc.* **20** (1983), 301–314.
160. Ackerman, J.D., *Biotropica* **21** (1989), 340–347.
161. Roubik, D.W., *Ecology and Natural History of Tropical Bees*, Cambridge University Press, Cambridge, 1989.
162. Ashton, P.S., *Ann. Rev. Ecol. Syst.* **19** (1988), 347–370.
163. Burquez, A. *et al.*, *Bot. J. Linn. Soc.* **94** (1987), 407–419.
164. Anderson, A.B., *Biotropica* **20** (1988), 192–205.
165. Beaman, R.S. *et al.*, *Amer. J. Bot.* **75** (1988), 1148–1162.
166. Nilsson, L.A. *et al.*, *Biotropica* **19** (1987), 310–318.
167. Janzen, D.H., *Costa Rican Natural History.* Chicago University Press, Chicago, 1983 pp. 696–700.
168. Mabberley, D.J., *The Plant-Book*, repr. ed. Cambridge University Press, Cambridge, 1990, pp. 226–227.
169. Galil, J. *et al.*, *New Phytol.* **72** (1973), 1113–1127.
170. Bronstein, J.L., *Ecology* **69** (1988), 1298–1302.
171. Compton, S.G. and Robertson, H., *Ecology* **69** (1988), 1302–1305.
172. Bronstein, J.L., *Biotropica* **20** (1988), 215–219.
173. Bronstein, J.L., *Oikos* **48** (1987), 39–46.
174. Whittaker, R.H., Personal Communication, 1990.
175. Gardner, R.O., *Auckl. Bot. Soc. Newslett.* **38** (1983)
176. Kedric-Brown, A. *et al.*, *Ecology* **65** (1984), 1358–1368.
177. Stiles, F.G. in Janzen, D.H., *op. cit.* pp. 502–530.
178. Bolten, A.B. and Feinsinger, P., *Biotropica* **10** (1978), 307–309.
179. Pyke, G.H. and Waser, N.M., *Biotropica* **13** (1981) 260–270.

180. Hudson, P. and Sugden, A.M., *Symposium—The Tropical Rain Forest*, Poster Abstracts, Leeds, 1982.
181. Garber, P.A., *Biotropica* **20** (1988), 107–113.
182. Baker, H.G., *Aliso* **11** (1985), 213–229.
183. Ali, S.A., *J. Bombay Nat. Hist. Soc.* **35** (1931), 144–149.
184. Marshall, A.G., *Symposium—The Tropical Rain Forest*, Poster Abstracts. Leeds, 1982.
185. Hopkins, H., *J. Ecol.* **72** (1984), 1–13.
186. Ng, F.S.P. in Cranbrook, Earl of, *Malaysia*. Pergamon, Oxford, 1988, pp. 102–125.
187. Janson, C.H. *et al.*, *Biotropica* **13** suppl., (1981) 1–6.
188. Lumer, C. and Schoer, R.D., *Biotropica* **18** (1988), 363–364.
189. Webb, C.J., *Biotropica* **16** (1984), 37–42.
190. Hopkins, H.C. and M.J.G., *Brittonia* **34** (1982), 225–227.
191. Herrera, C.M., *Biol. J. Linn. Soc.* **35** (1988), 95–125.
192. Woodell, S.R.J., *Phil. Trans. Roy. Soc. London* **B 286** (1979), 99–108.
193. Davies, A.G. and Baillie, I.C., *Biotropica* **20** (1988), 252–258.
194. Oldroyd, H., *The Natural History of Flies*. Wiedenfeld & Nicolson, London, 1964, p. 84.
195. Whitten, A. *et al.*, *The Ecology of Sumatra*, ed. 2. Gadja Mada University Press, 1987, p. 303.
196. Juniper, B.E. *et al.*, *The Carnivorous Plants*. Academic Press, London, 1989, pp. 260–266.
197. Hoernicke, R., *Biotropica* **15** (1983), 237–239.
198. Fogden, S.C.L. and Proctor, J., *Biotropica* **17** (1985), 172–174.

Chapter 7 (pp. 187–204)
1. Rosen, B.R., in Forey, P.L. (ed.) *The Evolving Biosphere*. British Museum and Cambridge University Press, Cambridge, 1981, pp. 103–130.
2. Ha, C.O. *et al.*, *Bot. J. Linn. Soc.* **97** (1988), 317–331.
3. Pannell, C.M. and White, F., *Monogr. Syst. Bot.* **25** (1988), 639–659.
4. Mabberley, D.J., *Bull. Brit. Mus. (Nat. Hist.) Bot.* **6** (1979), 301–386.
5. Grey-Wilson, C., *Kew Bull.* **34** (1980) 661–668.
6. Linhart, Y.B. in Bock, J.H. and Linhart, Y.B. (eds) *Evolutionary Ecology of Plants*. Westview, Boulder, 1989, 393–430.
7. Gentry, A.H., *Ann. Miss. bot. Gdn.* **69** (1982), 557–593.
8. Fleming, T.H. *et al.*, *Ann. Rev. Ecol. Syst.* **18** (1987), 91–109.
9. Simpson, B.B. and Hoffer, J., *Ann. Rev. Ecol. Syst.* **9** (1978), 497–518.
10. Prance, G.T. in Whitmore, T.C. and Prance, G.T. (eds) *Biogeography and Quaternary History in Tropical America*. Clarendon, Oxford, 1987, pp. 46–55.
11. Brown, K.S., l.c., pp. 66–104.
12. Andersson, L., *Bot. Not.* **132** (1979), 185–189.
13. Diamond, A.W. and Hamilton, A.C., *J. Zool.* **191** (1980), 379–402.
14. Bourlière, F., *Ecol. Stud.* **69** (1989), 153–168. Emmons, L.H., *Ecol. Monogr.* **50** (1980), 31–54.
15. Mackinnon, J., *Primates* **18** (1977), 747–772.
16. Terborgh, J. in Prance, G.T. and Lovejoy, T.E., *Amazonia*. Pergamon, Oxford, 1984, pp. 284–304.
17. Crome, F.J.G. and Richards, G.C., *Ecology* **69** (1988), 1960–1969.
18. Bonaccorso, F.J., *Bull. Fla State Mus., Biol. Sci.* **24** (1979), 359–408.
19. Bell, H.L., *Emu* **84** (1983), 142–158.
20. Beehler, B., *Auk* **199** (1983), 1–12.
21. Ashton, P.S. in Grant, W.F. (ed.) *Plant Biosystematics*. Academic Press, Toronto and London, 1984, pp. 497–518.
22. Emmons, L.M., *Biotropica* **16** (1984), 210–222.
23. Keel, S.H.K. and Prance, G.T., *Acta Amaz.* **9** (1979), 645–655.

24. Pemadasa, M.A. and Gunatilleke, C.V.S., *J. Ecol.* **69** (1981), 117–124.
25. Huston, M., *J. Biogeogr.* **7** (1980), 147–157.
26. Ashton, P.S., in Holm-Nielsen, L.B. *et al.*, *Tropical Forests*. Academic Press, London, 1989, pp. 239–251.
27. Huston, M., *J. Biogeogr.* **7** (1980), 147–157.
28. Hubbell, S.P. in Sutton, S.L. *et al.*, *Tropical Rain Forest: Ecology and Management*. Blackwell, Oxford, 1983, pp. 25–41.
29. Benzing, D.H., *Selbyana* **5** (1979), 248–255.
30. Crome, F.J.H., *Amer. Wildl. Res.* **2** (1975), 155–185.
31. Appanah, S., *J. Trop. Ecol.* **1** (1985), 225–240.
32. Ashton, P.S., *Ann. Rev. Ecol. Syst.* **19** (1988), 347–370.
33. Snow, D.W., *Oikos* **15** (1965), 274–281.
34. Fleming, T.H. and Heithaus, E.R., *Biotropica* **13** (suppl.) (1981), 45–53.
35. Kiltie, R.A., *Biotropica* **13** (1981), 141–145.
36. Wright, S.J., *Ecology* **64** (1983), 1016–1021.
37. Howe, H.F. *et al.*, **66** (1985), 781–791.
38. Schupp, E.W., *Oikos* **51** (1988), 71–78.
39. Augspurger, C.K., *J. Ecol.* **71** (1985), 759–771.
40. Augspurger, C.K. and Kelly, C.K., *Oecologia* **61** (1984), 211–217.
41. De Steven, D. and Putz, F.E., *Oikos* **43** (1984), 207–216
42. Clark, D.B. and D.A., *Ecology* **66** (1983), 1884–1892.
43. Becker, P. and Wong, M., *Biotropica* **17** (1985), 230–237.
44. Lee, M.A.B., *Oikos* **45** (1985), 169–173.
45. Sterner, R.W. *et al.*, *J. Ecol.* **74** (1986), 621–633.
46. Connell, J.H. *et al.*, *Ecol. Monogr.* **54** (1984), 141–164.
47. Hay, M.E., *Amer. Nat.* **128** (1988), 617–641.
48. Gillett, J.B., *Syst. Assoc. Publ.* **4** (1962), 37–46.
49. Grubb, P.J., *Biol. Rev.* **52** (1977), 107–145.
50. Prance, G.T. and Mori, S.A., *Brittonia* **30** (1978), 21–33.
51. Richards, P. and Williamson, G.B., *Ecology* **56** (1975), 1226–1229.
52. Wheelwright, N.T., *Oikos* **44** (1985), 465–471.
53. Tanner, E.V.J., *Biol. J. Linn. Soc.* **18** (1982), 263–278.
54. Whitmore, T.C., *Commonw. For. Inst. Pap.* **46** (1974).
55. Colinvaux, P., *Quat. Sci. Rev.* **6** (1987), 93–114.
56. Colinvaux, P., *Nature* **340** (1989), 188–189.
57. Sugden, A.M. in Sutton, S.L., *et al.* (eds), *Tropical Rain Forest: Ecology and Management*. Blackwell, Oxford, 1983, pp. 43–56.
58. Karr, J.R. and Freemark K.E., *Ecology* **64** (1983), 1481–1494.
59. Silvertown, J.W. and Wilkin, F.R., *Biol. J. Linn. Soc.* **19** (1983), 1–8.
60. Hubbell, S.P. and Foster, R.B., in Diamond, J. and Case, T.J. (eds), *Community Ecology*. Harper and Row, New York, 1986, pp. 314–329.
61. Whitmore, T.C. *et al.*, *J. Trop. Ecol.* **1** (1985), 375–378.
62. Brünig, E.F. and Klinge, H., *Gdns' Bull., Sing.* **29** (1977), 81–101.

Chapter 8 (205–220)
1. Milton, K., *Amer. Anthropol.* **83** (1981), 534–548.
2. Hladik, L.M. in Montgomery, G.G. *The Ecology of Arboreal Folivores*. Smithsonian, Washington, 1978, pp. 373–393.
3. Repenning, C.A. and Fejfar, O., *Nature* **299** (1982), 344–347.
4. Church, A.H. in Mabberley, D.J. (ed.), *Revolutionary Botany*. Clarendon, Oxford, 1981, pp. 237–245.
5. Delson, E., *Nature* **318** (1985), 107–108.

6. Lieberman, M. and D., *Biotropica* **12** (1980), 316–317.
7. Mabberley, D.J. in Reynolds, E.R.C. and Thompson, F.B. (eds), *Forests, Climate and Hydrology: Regional Impacts*. United Nations University, 1988, pp. 6–15.
8. Ridley, H.N., *J. Straits Br. Roy. Asiat. Soc.* **25** (1984), 11–32.
9. House, C., *The Fieldbook of a Jungle-Wallah*. Witherby, London, 1929, pp. 117–118.
10. Corner, E.J.H., *Wayside Trees of Malaya* (3rd edn). Malayan Nature Society, Kuala Lumpur 1988, p. 712.
11. Corner, E.J.H. in *UNESCO, Symposium on the Impact of Man on the Humid Tropics Vegetation*. Canberra, 1962, pp. 28–41.
12. Jones, J.S. and Rouhani, S., *Nature* **319** (1986), 449–450.
13. Bray, W., *Nature* **321** (1986), 726.
14. Rijksen, H.D., *Meded. Landb. Wageningen* **78-2** (1978).
15. Whitmore, T.C., *Tropical Rain Forests of the Far East*. Clarendon, Oxford, 1975.
16. Kershaw, A.P., *Nature* **322** (1986), 47–49.
17. Lynch, T.F. in Jennings, J.D. (ed.) *Ancient South Americans*. Oxford University Press, Oxford, 1988, pp. 87–137.
18. Whitten, A.J. *et al.*, *The Ecology of Sumatra*, (2nd edn) Gadjah Mada University Press, 1987, pp. 65–67.
19. Brain, C.K. and Sillen, A., *Nature* **336** (1988), 464–466.
20. Maloney, B.K., *J. Biogeogr.* **12** (1985), 537–558.
21. Bush, M.B. *et al.*, *Nature* **340** (1989), 313–315.
22. Lowenstein, F.W. in Meggers, B.J. *et al.* (eds), *Tropical Rain Forest Ecosystems in Africa and South America: A Comparative Review*. Smithsonian, Washington, 1973, pp. 293–310.
23. Powell, J.M., *Monogr. Biol.* **42** (1982), 207–227.
24. Harris, D.R., *Amer. Scient.* **60** (1973), 180–193.
25. Headland, T.N. *Human Ecol.* **15** (1987), 463–491.
26. Bailey, R.C. *et al.*, *Amer. Anthropol.* **91** (1989), 59–82.
27. Headland, T.N. and Reid, L.A., *Curr. Anthropol.* **30** (1989), 43–66.
28. Headland, T.N., *Abstr. 88th Meeting Amer. Anthropol. Assoc., Washington DC, 15–19 Nov. 1989*.
29. Crosby, A.W., *Ecological Imperialism*, Cambridge University Press, Cambridge, 1986.
30. Purseglove, J. *Tropical Crops*, Vol. 1., Longman, London, 1968, p. 12.
31. Jett, S.C. in Jennings, J.D. *op. cit.*, pp. 337–393.
32. Dunn, F.L., *Monogr. Malays. Br. Roy. Asiat. Soc.* **5** (1975).
33. Mabberley, D.J., *The Plant-Book*, repr. ed. Cambridge University Press, Cambridge, 1990.
34. Boxer, C.R. in Livermore, H.V. (ed.), *Portugal and Brazil*, Oxford University Press, Oxford, 1953, p. 217 *et seq.*
35. Harries, H.C., *Turrialba* **27** (1977), 227–231.
36. Headland, T.N., *Trop. Ecol.* **29** (1988), 121–135.
37. May, R.M., *Nature* **312** (1984), 19–20.
38. Smith, N.J.H., *Ann. Assoc. Amer. Geogr.* **70** (1980), 553–566.
39. Meggers, B.J. in Meggers, B.J. *et al.*, *op. cit.*, pp. 311–314.
40. Carmichael, E. *et al.*, *The Hidden Peoples of the Amazon*. British Museum Pub., London, 1985, pp. 78–93.
41. Meggers, B.J., in Whitmore, T.C. and Prance, G.T., *Biogeography and Quaternary History in Tropical America*. Clarendon, Oxford, 1987, pp. 151–174.
42. Weinstock, J.A., *Econ. Bot.* **37** (1983), 58–68.
43. Dransfield, J., *Malay. For. Rec.* **29** (1979).
44. Carey, I., *Orang Asli*. Oxford University Press, Kuala Lumpur, 1976.
45. Rambo, A.T. in Cranbrook, Earl of, *Malaysia* Pergamon, Oxford, 1988, pp. 273–288.
46. Dransfield, J., in *op. cit.*, pp. 37–48.
47. Primack, R.B., *Sarawak Mus. J.* **33** (54 n.s.) (1984), 69–74.

48. Day, G.M., *Ecology* **34** (1953), 329–347.
49. Kühn, F. and Hammer, K., *Kulturpfl.* **27** (1979), 145–173.
50. Morren, G.E.B. and Hyndman, D.C., *Human Ecol.* **15** (1987), 301–315.
51. Myers, N., *Conversion of Moist Tropical Forests.* National Academy of Science, Washington, 1980.
52. Prance, G.T. *et al.*, *Econ. Bot.* **31** (1977), 129–139.
53. Meggers, B.J. in Sioli, H. (ed.), *The Amazon.* Junk, Dordrecht, 1984, pp. 627–648.
54. Posey, D.E., *Cienc. and Cult.* **35** (1983), 877–894.
55. Posey, D.E., *Agrifor. Syst.* **3** (1985), 139–158.
56. Denevan, W.M. *et al.*, in Hemming, J. (ed.), *Man's Impact on Forests and Rivers.* Manchester University Press, Manchester, 1985, pp. 137–155.
57. Barrau, J., *J. Soc. Océan.* **21** (1965), 55–78.
58. Hewetson, C.E., *Emp. For. Rev.* **35** (1956), 274–291.
59. Herrera, C.M., *Oikos* **36** (1981), 51–58.
60. Corlett, R.J., *Sing. J. Trop. Geogr.* **5** (1984), 102–111.
61. Clayton, W.D. and Renvoize, S.A., *Genera graminum.* HMSO, London, 1986, p. 356.
62. Cone, G. *Nature* **276** (1978), 704.
63. Haemig, P.D., *Biotropica* **10** (1978), 11–17.
64. Haemig, P.D., *Biotropica* **11** (1979), 81–87.

Chapter 9 (pp. 221–263)
1. Jordan, C.F., in Prance, G.T. and Lovejoy, T.E., *Amazonia*, Pergamon, Oxford, 1984, pp. 410–426.
2. Hemming, J.H. *Amazon Frontier. The Defeat of the Brazilian Indians.* Macmillan, London, 1987, Ch. 24.
3. Poore, D., *Unasylva* **28** (1976), 127–146.
4. Myers, N., *The Conversion of Moist Tropical Forests.* Nat. Acad. Sci., Washington, 1980.
5. Evans, J., *Plantation Forestry in the Tropics.* Clarendon, Oxford, 1982.
6. Dove, M.R., *Agrifor. Syst.* **1** (1983), 85–99.
7. Uhl, C. and Buschbacher, R., *Biotropica* **17** (1985), 265–268.
8. Wilkie, D.S. and Finn, J.T., *Biotropica* **22** (1990), 90–99.
9. Brewbaker, J.L., *Econ. Bot.* **33** (1979), 101–118.
10. Abrams, E.M. *Amer. Antiq.* **52** (1987), 485–499.
11. Whitten, A.J. *et al.*, *The Ecology of Sulawesi.* Gadjah Mada University Press, Yogyakarta, 1987, p. 631.
12. Janzen, D.H., *The Ecology of Plants in the Tropics.* Arnold, London, 1975.
13. Goodland, R.J.A. and Irwin, R.S., *Landscape Planning* **1** (1975), 123–254.
14. Anon, *Ecol. Stud.* **60** (1987), 58–75; *Newsweek* **25**, i (1982).
15. Fearnside, P.M. and Rankin, J.M., *Interciencia* **7** (1982), 329–339.
16. Rudel, T.K., *Human Ecol.* **11** (1983), 385–403.
17. Hay, A., *Nature* **302** (1983), 208–209.
18. Poore, D., *J. Roy. Soc. Arts, Proc. Jan., 1985*, pp. 136–149.
19. Myers, N., *The Primary Source: Tropical Rain Forests and our Future.* Norton, New York and London, 1984.
20. Laarman, J.G., *Columbia J. World Business* **21** (1977), 77–82.
21. Mabberley, D.J., *The Plant-Book*, repr. ed. Cambridge University Press, Cambridge, 1990, p. 443.
22. Keay, R.W.J., *Biologist* **37** (3) (1990), 73–77.
23. Salleh, M.S. and Ho, K.S., in Srivastava, P.B.L. *et al.* (eds), *Tropical Forests—Source of Energy through Optimization and Diversification.* Penerbit University, Pertania, Malaysia, 1982, pp. 101–106.

24. Longman, K.A. and Jeník, J., *Tropical Rain Forest and Its Environment*, (2nd edn.) Longman, London.
25. Woods, P., *Biotropica* **21** (1989), 290–298.
26. Whitmore, T.C., *Tropical Rain Forests of the Far East*, (2nd edn.) Clarendon, Oxford, 1984.
27. Johns, A.D., *Malays. Appl. Biol.* **10** (1981), 221–226.
28. Dean, W. in Tucker, R.P. and Richards, J.F. (eds) *Global Deforestation and the Nineteenth-Century World Economy*. Duke Press, Durham, North Carolina, 1983, pp. 50–93.
29. Frisk, T., *Unasylva* **30**, n. 122 (1978), 14–24.
30. Lamb, D., *Exploiting the Tropical Rain Forest*. UNESCO and Parthenon, 1990.
31. Poore, D. *et al.*, *No Timber Without Trees: Sustainability in the Tropical Forest*. Earthscan, London, 1989.
32. Dawkins, H.C., in McDermott, M.J. (ed.), *The Future of the Tropical Rain Forest*. Oxford Forestry Institute, Oxford, 1988, pp. 4–8.
33. Melville, R., *Bull. Misc. Inf. Kew* **1936** (1936), 193–210.
34. Melville, R., *Bull. Misc. Inf. Kew* **1937** (1937), 274–276.
35. Ewel, J. (ed.) Tropical succession. *Biotropica* **12** (2) (suppl.) 1980.
36. Grainger, A. *The Ecologist* **10** (1980), 1–54.
37. Myers, N. and Tucker, R., *Envir. Rev.* **11** (1987), 55–71.
38. Fearnside, P.M. in Prance, G.T. and Lovejoy, T.E., *Amazonia*, Pergamon, Oxford, 1984, pp. 393–418.
39. Grainger, A., *The Future Role of the Tropical Rain Forest in the World Economy*. D. Phil. Thesis, University of Oxford, Oxford, 1986.
40. Lugo, A.E. in Wilson, E.O., *Biodiversity*. National Academy Press, Washington, 1988, pp. 58–70.
41. Tracey, J.G. in Synge, H. (ed.), *The Biological Aspects of Rare Plant Conservation*, Wiley, Chichester, 1981, pp. 165–171.
42. Sader, S.A. and Joyce, A.T., *Biotropica* **20** (1988), 11–19.
43. Johnstone, B., *New Scientist* **2**, iv (1987), 18.
44. Myers, N. in Wilson, E.O., *op. cit.*, pp. 28–35.
45. Uhl, C. *et al.*, *Oikos* **38** (1982), 313–320.
46. Opler, P. *et al.* in Cairns, J. *et al.* (eds), *Recovery and Restoration of Damaged Ecosystems*. University Press of Virginia, Charlottesville, 1977, pp. 379–421.
47. Saldarriaga, J.G., *Ecol. Stud.* **60** (1987), 24–33.
48. Montagnini, I. and Buschbacher, R., *Biotropica* **21** (1989), 9–14.
49. Holt, J.A. and Spain, A.V., *J. Appl. Ecol.* **23** (1986), 227–237.
50. Buschbacher, R. *et al.*, *J. Ecol.* **76** (1988), 682–699.
51. Maheshwaran, J. and Gunatilleke, I.U.C.N., *Biotropica* **20** (1988), 90–99.
52. Fournier, P. in *Tropical Forest Ecosystems*, UNESCO, Paris, 1978, pp. 256–269.
53. De la Salas, G. and Fölster, H., *Turrialba* **26** (1976), 179–186.
54. Jordan, C.F., in Prance, G.T. and Lovejoy, T.E. (eds), *op. cit.*, pp. 83–94.
55. Uhl, C. and Jordan, C.F., *Ecology* **65** (1984), 1476–1490.
56. Matson, P.A. *et al.*, *Ecology* **68** (1987), 491–502.
57. Lamb, D., *Papua New Guinea Trop. Res. Note* SR **30** (n.d.).
58. Harcombe, P., *Ecology* **58** (1977), 1375–1383.
59. Nicholson, S.A., *Biotropica* **15** (1981), 110–116.
60. Sanchez, P.A. *et al.*, *Science* **216** (1982), 821–827.
61. Ewel, J.J., *Ann. Rev. Ecol. Syst.* **17** (1986), 245–271.
62. White, F., *The Vegetation of Africa*, UNESCO, Paris, 1983.
63. Hall, J.B. and swaine, M.D. *Distribution and Ecology of Vascular Plants in a Tropical Rain Forest*. Junk, The Hague, 1981.
64. Swaine, M.D. and Hall, J.B., *J. Ecol.* **71** (1983), 601–627.

65. Gómez-Pompa, A. and Vázquez-Yanes, C. in West, D.C. *et al.* (eds), *Forest Succession. Concepts and Applications.* Springer, New York, 1981, pp. 246–266.
66. Zwetsloot, H., *Turrialba* **31** (1981), 369–379.
67. Uhl, C. *et al., J. Ecol.* **76** (1988), 663–681.
68. Prance, G.T. and Schubart, H.O.R., *Brittonia* **30** (1978), 60–63.
69. Stocker, G.C., *Biotropica* **13** (1981), 86–92.
70. Kochummen, K.M. and Ng, F.S.P., *Malays. For.* **40** (1977), 61–78.
71. Whitten, A.J. *et al., The Ecology of Sumatra,* (2nd edn.) Gadja Mada University Press, Yogyakarta, 1987, p. 487.
72. Johns, A.D., *Biol. Conserv.* **31** (1985), 355–375.
73. Johns, A.D. *Ecology* **67** (1986), 684–694.
74. Berenstein, L., *Biotropica* **18** (1986), 257–262.
75. Cranbrook, Earl of, *Malaysia,* Pergamon, Oxford, 1988, pp. 147–166.
76. Stuebing, R.S. and Gasis, J., *J. Trop. Ecol.* **5** (1989), 203–214.
77. Wells, D.A. in Cranbrook, Earl of, *op. cit.*, pp. 167–195.
78. Bell, H.L., *Emu* **82** (1982), 217–224.
79. Remson, J.V. and Parker, T.A., *Biotropica* **15** (1983), 223–231.
80. Brash, A.R., *Biol. Conserv.* **39** (1987), 97–111.
81. Powell, A.H. and G.V.H., *Biotropica* **19** (1987), 176–179.
82. Collins, N.M. in Cranbrook, Earl of, *op. cit.*, pp. 196–211.
83. Speeden, M., *Straits Times* 3 May 1984, sect. 2, p. 1.
84. Copper, J.I. and Tinsley, T.W., *La Terre et la Vie* **32** (1971), 221–240.
85. Ng, F.S.P. in Sutton, S.L. *et al., Tropical Rain Forest. Ecology and Management.* Blackwell, Oxford, 1983, pp. 359–375.
86. Denslow, J.S., *Ann. Rev. Ecol. Syst.* **18** (1987), 431–451.
87. Hopkins, M.S. in White, F., *op. cit.*, p. 77.
88. Gillison, A.N. and Anderson, D.J. (eds), *Vegetation Classification in Australia.* CSIRO, 1981, pp. 42–52.
89. Janzen, D.H., *Oikos* **41** (1983), 402–410.
90. McKey, D., *Biotropica* **20** (1988), 262–264.
91. Mabberley, D.J., *Gdns' Bull., Sing.* **37** (1984), 49–64.
92. Prance, G.T. in Prance, G.T. and Lovejoy, T.E., *Amazonia.* Pergamon, Oxford, 1984, pp. 166–191.
93. Roubik, D.W., *Ecology and Natural History of Tropical Bees.* Cambridge University Press, Cambridge, 1989.
94. Savidge, J.A., *Ecology* **68** (1987), 660–668.
95. Pratt, H.D. *et al., A Field Guide to the Birds of Hawaii and the Tropical Pacific.* Princeton University Press, Princeton, 1987, pp. 42–43.
96. Warner, R.E., *Condor* **70** (1968), 101–120.
97. de Lacerda, L.D. and Hay, J.D., *Biotropica* **14** (1982), 238–239.
98. Breteler, F.J., *Garcia de Orta Ser. Bot.* **6** (1984), 111–118.
99. Torquebiau, E., *Agrofor. Syst.* **2** (1984), 103–127.
100. Profizi, J.-P., *Human Ecol.* **16** (1988), 87–94.
101. Pascal, J.P., *Trav. Sect. Sci. Tech. Inst. Franc. Pondichéry* **20** *bis* (1988), 284.
102. Palmeirim, J. and Etheridge, K., *Biotropica* **17** (1985), 82–83.
103. Smythe, N., *Ann. Rev. Ecol. Syst.* **17** (1988), 169–188.
104. Stallings, J.R., *Biotropica* **16** (1984), 155–157.
105. Marshall, A.G., *Zool. J. Linn. Soc.* **83** (1985), 351–369.
106. Free, J.B. and Williams, I.H., *Trop. Agric.* **53** (1976), 125–139.
107. Crome, F.J.H. and Irvine, A.K., *Biotropica* **18** (1986), 115–125.
108. Lugo, A.E. and Brown, S., *Unasylva* **32**, n. 129 (1980), 8–13.
109. Henderson-Stellers, A. in McDermott, M.J. (ed.), *The Future of the Tropical Rain Forest.* Oxford Forestry Institute, Oxford, 1988, pp. 13–14.

110. Dobson, A. *et al.*, *Trends Ecol. Evol.* **4** (1989), 64–68.
111. Myers, N. in Synge, H. (ed.), *The Biological Aspects of Rare Plant Conservation*. Wiley, London, 1981, pp. 141–154.
112. Myers, N., *Nature conservation at global level*. Inaugural lecture, Rijksuniversiteit Utrecht, Netherlands.
113. IUCN Commission on National Parks and Protected Areas, *Review of the Protected Areas System in Oceania*. IUCN, Cambridge and Gland, 1986.
114. Lovejoy, T.E. *et al.*, in Sutton, S.L. *et al.* (eds), *Tropical Rain Forest: Ecology and Management*. Blackwell, Oxford, 1983, pp. 377–384.
115. Oldfield, S., *Buffer Zone Management in Tropical Moist Forests. Case Studies and Guidelines*. IUCN, Cambridge and Gland, 1988.
116. MacKinnon, J. *et al.*, *Managing Protected Areas in the Tropics*. IUCN, Gland, 1986.
117. Eden, M.J., *Ecology and Land Management in Amazonia*. Belhaven, London, 1990.
118. Poore, D. and Sayer, J., *The Management of Tropical Moist Forest Lands: Ecological Guidelines*. IUCN, Cambridge and Gland, 1987.
119. Terborgh, J. in Soulé, M.E. (ed.), *Conservation Biology*. Sinauer Associates, Sunderland, Mass., 1986.
120. Howe, H., *Biol. Conserv.* **30** (1984), 261–281.
121. Ng, F.S.P. in Cranbrook, Earl of, *Malaysia*. Pergamon, Oxford, 1988, pp. 102–125.
122. Goulding, M. in Prance, G.T. and Lovejoy, T.E., *Amazonia* Pergamon, Oxford, 1984, pp. 267–276.
123. Deraniyagala, P.E.P., *Ancient Ceylon* **3** (1979), 45–66, 9tt.
124. Werner, W.L. in Schweinfurth, U. (ed.), *Forschungen auf Ceylon* **3** (1989), 43–72.
125. Karunaratna, N., *Forest Conservation in Sri Lanka from British Colonial Times*. Trumpet, Colombo, 1987.
126. Gunatilleke, C.V.S. *et al.*, in Bramwell, D. *et al.* (eds), *Botanic Gardens and the World Conservation Strategy*. Academic Press, London, 1987.
127. Gunatilleke, N., *Loris* **17** (4) (1986).
128. Sheldon, F., *Ibis* **128** (1986), 174–175; Myers, N., *Int. Wildl.* **17**, 4 (1987), 18–24.
129. Schenkel, R. *et al.*, *Malay. Nat. J.* **31** (1978), 253–275.
130. Jacobs, M., *Fl. Males. Bull.* **35** (1982), 3768–3782.
131. Caldecott, J., *Brit. Med. J.* **295** (1987), 229–230.
132. Clark, J.T. *et al.*, *Science* **225** (1984), 847–849.
133. Gottlieb, O.R. in Prance, G.T. and Lovejoy, T.E., *op. cit.*, pp. 218–231.
134. Ballick, M.J., *loc. cit.*, pp. 339–368.
135. International Board for Plant Genetic Resources Regional Committee for South-East Asia, *5-year plan 1985–1989*, Rome, 1982.
136. Poore, M.E.D., *New Phytol.* **90** (1982), 404–416.
137. Harley, J.L., *Proc. Roy. Soc. Lond. B* **197** (1977), 3–10.
138. Janzen, D.H., *Oikos* **51** (1988), 257–258.
139. Poore, D. *et al.*, *No timber Without Trees. Sustainability in the Tropical Forest*. Earthscan, London, 1989.
140. Spears, J.S., *Unasylva* **32** (1980), 2–12.
141. Madeley, J., *Financial Times* 23 February 90, p. 36.
142. *For. Prod. Abstr.* **61** (1983), 406.
143. Myers, N., *The Primary Source: Tropical Forests and our Future*. Norton, New York and London, 1984.

Postscript (pp. 264–267)

1. Church, A.H. *Oxf. bot. Mem.* **13** (1922), repr. in D.J. Mabberley (ed.), *Revolutionary Botany*, Clarendon Press, Oxford, 1981, pp. 133–235.
2. Styles, B.T. and Vosa, C.G. *Taxon* **20** (1971), 485–499.

3. van Steenis, C.G.G.J. *Rheophytes of the World*, Sijthoff & Noordhoff, Alphen aan den Rijn, 1981, pp. 98–142.
4. White, F. *Mitt. bot. Staatsamml. München* **10** (1971), 91–112.
5. Kortlandt, A. *Symposium–The Tropical Rain Forest*, Poster Abstract. Leeds, 1982.
6. Ashton, P.S. in *Tropical Forest Ecosystems*. UNESCO, Paris, 1978, pp. 180–215.
7. Ng, F.S.P. *Malays. For.* **40** (1977) 126–137.
8. Corner, E.J.H., *New Phytol.* **45** (1946), 185–192.
9. Janzen, D.H., *Ann. Rev. Ecol. Syst.* **17** (1986), 305–324.

Index

Abies grandis 94
Acacia 241
  nigrescens 137
Acanthaceae 111, 112
Acioa edulis 7
Acromyrmex 142
Acrostichum 79
Adenanthera pavonina 212
Aedes aegypti 245
aerial roots 104
Aerobryopsis longissima 39
Agaonidae 178
agaric fungi 129
Agaricus 1
Aglaia 72, 162, 197
  yzermannii 162
agouti 84, 156, 166, 196
agriculture 36, 37, 208, 254, 262
agrisilviculture 233
agroforestry 248
Albizia 28, 62, 117
alcohol 261
Aldabra 185
Aleurites moluccana 208
Alfisols 33
algae 49
  blue-green 108
Alisma 1
alkaloids 71, 137, 138, 139, 148, 156, 260
allelopathy 40, 54, 198
Allomerus demerarae 145
allophane 34
alluvial
  soils 36
  forest 45
alluvium 33
Amaryllidaceae 12
Amazon 7, 16, 25, 41, 47, 90, 164, 201,
  225, 240, 252, 257
Amazonia 8, 15, 22, 25, 160, 162, 164, 174,
  177, 189, 194, 200, 204, 213, 214, 218, 226,
  238, 239, 243
America 242

Amherstia 123
Amorphophallus 112
amphibians 85, 90, 134
anachronisms 165
Andes 7, 189, 194, 225
andosols 34, 36
Andropadus latirostris 131
Angylocalys oligophyllus 21
animal dispersal 66, 72, 75, 157, 243
annatto 218
Annonaceae 100
Anoectochilus 111
Anthocephalus chinensis 62
anthocyanins 30
Anthurium 164
ant plants 107
ants 62, 66, 92, 107,145, 146,162,173, 179,
  185
Aphanamixis 143
aphid 92
Apis 127, 176
  dorsata 170, 171
Apocynaceae 13, 72, 85, 136
aquatics 12
Araceae 106, 111
Araliaceae 4, 12, 13
Araucaria 18, 20
  cunninghamii 238
arboreal animals 229
architectural models 96, 97
Areca catechu 208
Argostemma 111
arils 153, 163
Arisaema triphyllum 118
Aristolochia 177
armadilloes 20, 66
arrowroot 218
arthopods 7, 74, 129, 135, 147
Artocarpus 125, 219
Asclepiadaceae 82, 136
Aspalathus 13
Astragalus 13
Astrocaryum 261

jauary 163
mexicanum 58, 125, 126
standleyanum 156
Atta cephalotes 142
Attalea funifera 118
Attamyces bromatificus 142
Attim's model 100
Australia 45, 243, 246, 254
avian malaria 248
Avicennia germinans 94
avocado pear 101, 156
Azadirachta indica 142
Azteca 186

Baccaurea motleyana 206
Bactris 172
Balanophoraceae 111
balsa 64, 69
bamboo 121, 217, 219, 263
Bambusa wrayi 217
banana 15, 20, 45, 98, 212, 218, 233, 234
banded leaf monkey 244
barbets 87
bark 101, 136, 141
Barringtonia 70
Barro Colorado Island 63, 64, 69, 72, 109, 116, 117, 123, 128, 157, 158, 161, 174, 175, 192, 195, 197, 200, 204
Barteria nigritana 145
bat-pollinated 182, 249
bats 7, 8, 86, 87, 88, 89, 90, 131, 140, 170, 181, 189, 192, 193, 196, 249, 257
bauxite 226
beans 233
bee-pollination 176
beeches 99, 101
beef 234
bees 91, 137, 172, 173, 184
beetle-pollination 176
beetles 133, 172, 176
Begonia 30
Begoniaceae 111
Bertholletia excelsa 54, 62, 94, 174
betel nut 208, 219
Bignonaceous lianes 173
Bignoniaceae 12, 15, 72, 109, 144
biomass 42
biotic factors 195
birds 75, 87, 88, 89, 90, 130, 150, 152, 192, 193, 202, 244
    dispersal 152
    dispersed seeds 152
    pollinated 75
    of paradise 152,193

birdpox 248
Bixa orellana 144, 218
blue-green algae 108
Bocconia 12
Boletus 110
Bombacaceae 72
Borneo 87, 120, 138, 177, 194, 215, 245, 257
botanic gardens 261
bovids 21
Brachypodium pinnatum 197
bracken 135
Brazil 40, 50, 54, 62, 221, 261
Brazil nuts 15, 54, 94, 174
breadfruit 73, 103
Breynia cernua 123
broadbills 88, 131
Bromeliaceae 106
bromeliads 44, 90, 107, 181, 186
Brosimum alicastrum 11
Broussonetia papyrifera 219
brown-tree snake 247
Brownea rosa-de-monte 181
bruchid beetle 182
Bruguiera 20
Bryonia 13
bryophytes 8, 52, 108, 165
bud dormancy 117
buffalo 86
buffer zones 254
butterflies 136, 170, 176
buttresses 52, 101, 103

caatinga 37, 38, 69, 236
Cactaceae 106
cacti 82, 155
Caesalpina echinata 261
Caesalpinioideae 40, 48
Caesearia praecox 122
caffeine 261
Calathea 163, 176
Camellia sinensis 124
Cameroun 40, 140, 145, 246
camphor 142, 212
Canarium 208
cannon-ball tree 114
canopy 94, 251
    height 94
Caprifoliaceae 4
Capsella 1
capuchin monkey 141, 160
capybara 84
Carapa 125, 229
carbon cycle 250

carbon dioxide 249
Carboniferous 133, 134, 170
Carex 13, 163
Caribbean 7
Carica papaya 48, 62
Cariniana 173
carnivores 8, 90
carnivorous plants 133
carpenter bees 127
carps 90
carrion flies 177
cassava 98, 101, 141, 209, 233
cassava wood 263
cassowaries 88, 89, 168
Castanea dentata 217
Casuarina 219
Casuarinaceae 23
caterpillar 58
catfish 90, 163
Catopsis berteroniana 107
cattle 226, 234
cauliflory 2, 9, 52, 114
Caytoniales 18
Cebidae 86
Cebus capucinus 141
Cecropia 48, 61, 62, 64, 141, 146
  obtusifolia 69, 73, 125
Cedrela odorata 229
Ceiba pentandra 62, 81, 140, 165
cellulose 134
Centrospermae 13
Cephaelis ipecacuanha 260
Cerberiopsis candelabra 121
Cestrum 61
chablis 55
chalk grassland 197, 202
chameleons 85
chaos equations 80
charcoal 235
chasmophytes 35
chestnuts 217
chewing sticks 261
chicle 11
chimpanzees 87, 161
Chinese Elm 135
chiropterophily 181
Chisocheton 114, 124, 143
  lasiocarpus 188
Chlamydomonas 1
Chrysobalanaceae 81
Chrysobalanus icaco 81
chrysomelid beetles 177
cinnamon 141
Citrus 94

clay 32, 33, 38, 41
clear-felling 246
Clematis vitalba 4
climax 53, 54, 55, 60
cloud forest 6, 36, 37, 52, 64, 77, 106, 169
clouds 25
cloves 141, 198
clubmosses 133
Clusia 8, 105
co-evolution 133, 151
coal-swamp forests 17
coccids 147
Cocha Cashu 160, 257
cochineal 234
cockroaches 133
cocoa 11, 101, 198, 233, 234
coconut 117
Codiaeum variegatum 219
coffee 101, 106, 141, 234, 258
Coleoptera 127, 129
Colobus 139
colobus monkeys 140
Colombia 39, 239
Combretum fruticosum 183
Commelinaceae 111
Compositae 12, 48
conifers 18, 123
conservation 249, 253
continental drift 17
contraceptives 215, 260
Copernicia tectorum 105
copper 226
coppice 37, 79, 100
coral reefs 22, 194
Cortinarius 110
Corvidae 156
Corypha 96, 121
Costa Rica 8, 29, 38, 47, 58, 62, 64, 66,
  135, 170, 172, 180, 204, 238, 249
cotingas 89, 150, 152, 153, 189
cotton 234
Couratari 165, 174
Couroupita 173
  guianensis 114
Coussapoa 105
Crescentia 144
  alata 166
Cretaceous 17, 18, 20, 133
cross-pollination 171
Croton 54, 219
cucumber 208
Cucurbitales 13
Culex pipiens 248
cultigens 209, 218

cyanide 149
cycads 18, 118
Cyclanthaceae 106
cyclones 27, 118, 194, 253
Cynometra alexandri 61
Cyrtandra mirabilis 112
Czechoslovakia 217

Dahomey Gap 24
Dalbergia
  decipularis 229
  parviflora 212
Danum Valley 66
day length 29
deciduousness 123
decomposers 74
deforestation 15
Delonix regia 123
Dendrobium crumenatum 118
denitrification 240
Denmark 8
deserts 13, 21, 24
Devonian 17, 133
dew 25
Diaphanopterida 133
Dicerorhinus sumatrensis 74, 166
Dicranopteris linearis 238, 243
Dictyoneurida 133
Didymocarpus 164
  platypus 122
Didymopanax pittierii 107
Dillenia 61
Dinizia excelsa 54
dioecy 254
Dioncophyllaceae 83
Dioscoreaceae 13
Diospyros 101, 187
Dipterocarpaceae 9, 15, 61, 81
dipterocarps 20, 48, 49, 71, 81, 89, 93, 94,
  100, 110, 115, 120, 121, 123, 126, 127,
  138, 139, 165, 171, 176, 188, 194, 243, 266
Dipteryx panamensis 125, 158, 197
dispersal 8, 75, 78, 148
  agents 116, 151, 156, 184
  mechanisms 199
diversity 80, 194
dodo 168
domestication 219
dorsata bees 129
doves 89
dragonflies 133
drip-tips 44, 45
Dromornithidae 166
Drosera 37

droughts 28, 194
drugs 15
Dryobalanops aromatica 94, 123
durians 70, 113, 116, 249, 257
Durio 70, 113
  zibethinus 116, 249
dye 261
dynastid beetle 177
angustifolium 162
Dysoxylum 157
  angustifolium 162
  parasiticum 2

earthquakes 28, 56, 58
earthworms 49
ebonies 101, 187
ectomycorrhizae 49
  fungi 165
Ecuador 8, 22, 200, 201, 228
edge effects 246
Egypt 20
El Nino phenomenon 120
elephants 21, 86, 168
elm leaf beetle 135
elms 12, 101
Endiandra palmerstonii 70
Endogonaceae 48
Endospermum formicarum 147
Entada 241
Entandrophragma 61
Enterolobium cyclocarpum 166
entomophily 169
Eocene 20
Eonycteris spelaea 182, 257
Epacridaceae 23
Eperua purpurea 38, 105
Epicharis 174
epiphyllae 39, 44, 108
epiphytes 4, 6, 8, 25, 36, 39, 44, 52, 56, 75,
  82, 105, 107, 195
equids 20, 166
Erica 13
Ericaceae 23
erosion 239
Erythrina 181
Escalloniaceae 23
Eschweilera 174, 191
Eucalyptus 79, 227, 233, 239
euglossine bees 174, 175
Eulaema 175, 176
  ignita 176
  moscaryi 176
Euphorbiaceae 12, 103, 115
European colonizations 221

evaporation 25, 26
evapotranspiration 15, 25
extended family 213, 214
extinction 12

Fagaceae 20, 126, 165
Fagraea 105
Fagus 99
Fanning Island 164, 197
farming 222, 225
fernlands 238
ferns 82, 106, 108, 133, 135, 146
fertilizer 235
Festuca 13
Ficus 8, 46, 77, 109, 123
  carica 178
  microcarpa 163
  pubinervis 179
  religiosa 178
  sur 179
figs 12, 73, 89, 114, 157-159, 161, 186
fig-wasps 159, 178
Fiji 8, 55, 56, 58
fire 194, 223, 229, 243, 252
fish 162, 257
Fistulina 110
flies 133, 176, 177, 184
flooded forests 257
flower piercers 181
flushing 123
flycatcher 152
fog 25
food bodies 144
forest 16, 40, 45, 49
forest clearing 208
forest people 208
Formicariidae 129
fossil 206
frangipani 98, 117
French Guiana 160, 247
freshwater fish 90
freshwater swamp forest 36, 37, 248
Freycinetia 172
  arborea 167
frogs 74, 85, 90
frugivores 75, 87, 88, 89, 148, 159, 189
frugivorous birds 152
fruit doves 131
fruit pigeons 131, 157
fruit trees 118
fruiting plants 75
fruits 149
Fucus 1
fuel 234

Funaria 1
fungi 79, 136
Funtumia elastica 231
fynbos 8

Gabon 85, 132
Galeola altissima 110
gall formation 178
gallery 254
gapfillers 232
gaps 55, 58, 59, 64, 73
  area 64, 65
  environment 62
gap phase 18, 66
  regeneration 63
Garcinia 188
  epunctata 261
  kola 261
  mangostana 116
genetic erosion 228
genetic resource centres 261
germination 29, 65, 68, 69, 70, 151, 199
Gesneriaceae 111, 112, 164
Ghana 26
giant armadillo 84
giant clam 22
giant groundsels 98
gibbons 162, 192
Gilbertiodendron dewevrei 52
ginger 141
glaciers 21
global carbon cycle 15
global interest 14
Glochidion tetrapteron 67
Gmelina arborea 226, 249
gold 226
gomphotheres 166
Gondwanaland 23, 168
Gonystylus bancanus 15, 37, 162
gorillas 141
Gramineae 12
grasses 13
Great Barrier Reef 209
great-tailed grackle 220
Greece 246
green algae 108
greenheart 228, 233
greenhouse effect 15, 250
greenhouse gas 252
ground sloths 166
Guam 247
guanacaste 166
guarana 261
Guarea 148

INDEX

glabra 155
guidonia 162
rhopalocarpa 121
Guatemala 11
guava 117
gums 15, 141
gymnosperms 18, 20, 111

Hakea 79, 198
halophytes 37
Hawaii 248
hawkmoths 170, 177
heath forest 9, 34, 37, 40, 45, 49, 64
Hedera helix 4, 155
Helianthus 1
Helichrysum 13
hemi-epiphytes 105, 107, 110
hemi-parasites 106
Hemiptera 128, 129
herbivores 44, 90, 266
herbivory 8, 133, 134, 135, 139
herbs 30, 111, 164
Heritiera macrophylla 123
Hevea brasiliensis 198
Hieracium 13
Hildegardia barteri 118
hippoboscid flies 248
Homalium grandiflorum 118
Honduras 224
honeycreepers 172, 248
honeyeaters 172, 179
Hopea 28, 196
Hoplestigmataceae 83
hornbills 87, 131
howler monkeys 140, 161
Huaceae 83
human fossils 207
hummingbirds 170, 179, 180
humus 47, 54
hunting 217
hurricanes 27
Hybanthus prunifolius 118
Hydnophytum 107
Hydrostachyaceae 112
hyperparasites 106
hyraxes 86

icecaps 21
Idiospoermaceae 83
igapo 37, 38, 43, 194 ,228
illipe 127
Impatiens hawkeri 188
Impatiens mirabilis 112
Imperata cylindrica 224

Imperata savanna 227
India 9
indigo 234
Indonesia 228
insects 9, 74, 90, 245
  attack 44, 196
  parasites 128
  pollination 169
  predation 254
insectivores 12, 86, 90, 130
introduced animals 247
introduced exotics 246
Intsia 28, 70, 223
invertebrates 78, 127
Iridomyrex cordatus 146, 147
island hopping 20
Isoglossa 112
isoprene 16
IUCN 254
Ivory Coast 29
ivy 13, 155

jak fruit 114
Jari scheme 226, 249
Java 123, 259
jays 156
Jessenia bataua 261
jicaro 166
Jurassic 27, 133

Kalimantan 27, 28
Kampuchea 11
kaolinite 32, 33, 34, 41, 240
kapok 62
Kayapo 218
kechapi 116
kerengas 37, 138
Kew 261
kingfisher 186
kinkajous 85, 183
knee-roots 103, 104
Koompassia 110
  excelsa 94, 102, 216
Krakatoa 55

La Selva 61, 118, 122
Labisia longistylis 150
Lagenaria siceraria 208
Lagerstroemia speciosa 117
land crab 164
landslips 55-58
Lannea welwitschii 141
lanseh 72, 116
Lansium domesticum 72, 116

laterite 33, 34
latosols 31, 34
Lauraceae 71, 100, 130, 150, 152, 156, 161
leaching 32, 33, 41, 45, 239, 239
leaf monkeys 132
leaf properties 138
leaf-cutting ants 142, 143, 186
leaf-monkey 139
leaf-retention 100
Lecythidaceae 72, 166, 199
leeches 245
legumes 47, 81, 164, 234
Leguminosae 12, 13, 20, 40, 48, 72
lemur 140
Lepidoptera 129, 170, 177
Lesser Sunda Islands 9
lianes 4, 8, 13, 20, 26, 36, 52, 56, 63, 67, 75, 85, 109, 165, 229
lichens 108
life-spans 126
light hardwoods 61
lightning 56
lignin 134
Liliaceae 12
limbfall 63
limestone 38, 45
limestone forest 46, 49
litter 45, 71
liverworts 165
lizards 90
llamas 20
Lobelia 12
lobelioids 166
Lodoicea maldivica 167
logging 227
London Clay Flora 20
long-distance dispersal 201
long-tailed macaques 244
Lonicera periclymenum 4
Lophopetalum multinervium 104
Loranthaceae 23, 106, 151
Lua 217
Lucilia papuensis 177
Lythrum 12

MAB 256
Macaranga 61, 66, 77, 147, 243
Machaerium arboreum 26
Macrotermitinae 49, 50
Madagascar 26, 87, 177
Madhuca longifolia 219
Magnoliaceae 23
mahogany 11, 15, 72, 101, 228, 233, 241
maize 11, 218, 233

maize mosaic virus 224
malaria 245
Malay Peninsula 7, 138, 212, 21
Malaysia 9, 28, 185, 206, 228
Mallotus 66
  oppositifolius 171
Malpighiaceae 72
mammals 84, 85, 90, 131, 134, 139, 192
manakins 87, 89, 150, 153, 189
mandrills 87
Mangifera 101
mango 101, 209
mangosteens 116
mangroves 6, 13, 20, 36, 37, 56, 79, 89, 92, 100, 164, 248
Manihot esculenta 98
Manilkara zapota 11
Manu National Park 257
mapping 40
Marantaceae 111
Maranthes 81
Marsdenia laxiflora 109
marsupials 23, 89, 160, 183
mass-flowering 118, 121, 122
mastodons 20
Mauritius 168
Maxburretia rupicola 35
Maya 11
Medusandraceae 83
Melastoma 66, 158
Melastoma malabathricum 243
Melastomataceae 49, 66
Melia 83, 94
  azedarach 247
  parasitica 2
Meliaceae 72, 94, 142, 241, 257, 265
meliponid bees 170, 176, 247
Menispermaceae 85
meranti 15, 61, 71
Metrosideros 6, 105
Metroxylon 36, 37
Mexico 11, 191
mice 85
Miconia 196, 242
  argentea 64
Microcerotermes dubius 58
micro-organisms 62
Middle Jurassic 17
midges 90, 177
migratory birds 155
Mimosa pudica 69, 247
Mimulopsis 112
minor forest products 212, 260

Miocene 20
mistletoe 23, 106, 155, 173
moist forests 38, 41
molluscs 49, 134
monarch butterfly 173
monkeys 85, 86, 87, 139, 206
monocotyledons 12
monsoon forest 4, 27
Monstera deliciosa 104
montane forest 25, 39, 43, 45, 52, 89
Montrichardia arborescens 164
mor 31
Mora 94
Moraceae 12, 20, 73, 103, 161
morphological diversity 92
mortality rates 64
mosquitoes 74, 90, 185, 245
moths 176
Motmot Island 78
Mount Athos 246
movements, thigmonastic, nyctinastic 26
mull 31
Muntingia calabura 117, 122, 158, 183
Musanga 61, 103
  cecropioides 69, 73, 103, 247
mycorrhizae 40, 47, 48, 62, 70, 79
mycotroph 110
Myristicaceae 71, 103, 257
Myrmecodia 107, 146
myrmecophytes 146

nanophanerophytes 60
Neckeraceae 108
Neckia 112
nectar 172, 179
Nematoda 134
nematodes 147, 178
Neobalanocarpus heimii 123
Nepal 166
Nepenthes 37, 110, 186
Nephelium lappaceum 206
Nephrodium 1
nettles 12
Neuroptera 128
Neurospora 1
New Guinea 28, 46, 62, 87, 130, 193
New South Wales 46
nickel 226
Nigeria 25, 145
nitrogen fixation 43
nitrogen-fixers 225
  bacteria 225, 241
  trees 241
nitrous oxide 16

noctuid moths 170
nocturnal mammals 131
nomads 59
non-flying mammal 86, 183, 194
nutcrackers 156
nutmeg 71, 141
nutrient cycling 44
nutrients 41, 43, 73

oaks 99
obeche 101
ochlospecies 115, 187, 188
Ochroma lagopus 64, 69
Ocotea 232
  cymbarum 229
  rodiaei 228
  skutchii 195
Octomeles 62
oilbirds 150
oleoresins 139
olive baboons 206
omnivorous birds 150
opossums 85, 183
orang asli 216
orang-utans 58, 87, 148, 162, 192, 207, 244
Orchidaceae 12, 106
orchids 13, 82, 107, 110, 165, 176, 178, 230
Ordovician 18
ornithophily 179
oxalates 149
oxisols 33, 34, 37, 43, 44, 45, 69
ozone levels 16

paca 84
pachycaul 18, 98
Pachysima aethiops 145
pagoda trees 98
painted jay 220
Palmae 152
palms 20, 36, 46, 72, 83, 85, 98, 103, 125,
  126, 136, 156, 167, 171, 177, 233
Panama 28, 128, 155, 159, 259
Panamanian isthmus 20
Panda oleosa 168
Pandanus 103, 209, 219
Pangaea 18
pangolin 84, 85, 185
papaya 62, 117
papyrus 13
parallelism 265
parasites 12, 44, 110

Paratecoma peroba 15
Parkia 165, 166, 182
  speciosa 217
parrots 131, 155, 156, 165
Pasoh 25, 50, 68, 115, 119
passerines 150, 181, 191
Passifloraceae 173
Paullinia cupana 261
peat-swamp forest 9, 15, 27, 28, 36, 56, 58
peatswamp 90
peccaries 196
pectins 161
Pentaclethra 63
  macroloba 125, 126, 197
Pentaphragma 108
  horsfieldii 122
periwinkles 13
Permian 18, 133
peroba 15
Persea americana 156
  theobromifolia 228
Peru 127, 145, 218, 225
phenolics 40, 71, 138
phenology 118, 160
Philippines 9, 228
Phyllanthus 115
phytochrome 29, 120
Phytolacca rivinoides 240
pigeon orchid 118
pigeons 88, 89
pigs 21, 22, 74, 86, 127
pines 48, 198, 226
pinnate leaves 52
Pinus caribaea 183
pioneers 59, 60, 61, 62, 66, 68, 122, 123,
  125, 138
Piper 13, 54
  aduncum 158
  auritum 62, 69
Piptadeniastrum africanum 123
pisolites 233
Pistacia lentiscus 154
pitcher-plants 110
Pittosporaceae 23
placentals 23
plant cycles 116
Plantaginaceae 13
Plantago 96, 149
plantains 13, 96
plantations 51, 233, 258
Platypodium elegans 197

Pleistocene 22, 33, 34, 166, 254
Pliocene 21
Plumbaginaceae 13
Plumeria alba 98
  plywood 15
pneumatophores 103, 104
Podocarpaceae 167
Podocarpus 20
  barteri 126
Podostemaceae 112
podzols 32, 33, 34, 36
pollination 8, 75, 150, 169, 184
Polygalaceae 171
Polyscias kikuyuensis 99
ponerines 163
poplars 100, 137
population 263
porcupines 85
Posoqueria grandiflora 121
Poulsenia armata 140
precipitation 24, 43
predation 71, 193
predator satiation 121, 131, 157
prehensile-tailed vertebrates 85
Presbytis 139
  melalophos 139
  rubicunda 185
prickly pears 198
primary forest 1, 27
primary successions 54, 76, 79
primates 20, 86, 87, 88, 89, 160, 205
Proteaceae 23
protein 261
protein/fibre ratios 139
protoangiosperms 18
provenances 115
Prunus nigra 217
Psidium 83
  guayava 117
Pterocarpus 28
  officinalis 103
Ptilinopus magnificus 130
Puerto Rico 45, 46, 251
pulp 226, 231
pygmies 210
pygmy marmoset 183
Pyrrhalta luteola 135
Pyrrosia nummulariifolia 106

Quararibea asterolepis 102, 123
  cordata 183

Quassia amara 260
Quaternary 21, 22, 34
Queensland 130, 238
Queensland walnut 70
Quercus petraea 124
quinine 142
Quiscalus mexicanus 220

radiation 15, 29
Rafflesia 111
  pricei 177
raindrop size 25
raining trees 148
rambutans 206
ramiflory 9
ramin 37
ramon 11
Ramphastos swainsonii 154
ranching 234
Ranunculus 1
  ophioglossifolius 259
Raphus cucullatus 168
rat dispersal 164
rattans 85, 145, 212, 260, 263
Rauvolfia 260
Ravenala madagascariensis 68, 98
red leaf-monkey 185
redwoods 93
refuge theory 201
refugia 189, 190, 201
regeneration 63, 65
regeneration niche 199
reiteration 100
Renealmia 123
reptiles 85, 90, 134
reserves 254
resins 137
resistance 136
resource partitioning 192
Restionaceae 23
rheophytes 112, 162
rhinoceros 22, 58, 74
Rhipsalis 82, 155
Rhizophoraceae 20, 103
Rhus typhina 98
rice 208, 218, 224
Ricinus communis 98
riparian forests 4
rodents 8, 85, 86, 183
root climbers 110
root feeders 135
root hairs 48

root-parasites 111
rootplate 58
roots 47
root suckers 100
rubber 101, 124, 136, 198, 221, 233
rubberwood 263
Rubiaceae 12, 66, 82, 111
ruminants 89
run-off 25, 56
Rutaceae 142
Rwanda 259

Sabah 39
Sacoglottis trichogyna 70
sago palm 218
Salix martiana 165
Sambucus nigra 153
Sandoricum koetjape 116, 124
Sapindaceae 72
saplings 72
Sapotaceae 103
saprolite 32
Saraca 123
Sarawak 58, 125, 216
sarcotesta 153
savanna 6, 22, 24, 79, 201
savoka 83, 98
Saxifragaceae 23
scarabs 176
Scarrone's model 100
Schefflera 105
sclerophylly 37, 39
Scrophulariaceae 12
Scytopetaceae 83
sea-grasses 20
sea-pinks 13
seasonal climates 33
seasonal forests 38
seasonality 116, 193
  patterns 127
seaweeds 198
secondary forest 9, 11, 77
secondary successions 54
sedges 13, 78, 79
seed dormancy 66, 68
seedbanks 65-68
seed-ferns 18, 133
seedlings 70, 71
seed predation 121, 155
seed-rain 65, 67, 68
seeds 69, 70, 71, 149
seed shadow 159
seed, size 70
Semecarpus 49

sesquioxides 33, 34
Sesuvium portulacastrum 248
sex 116
Seychelles 167
shade-bearers 55, 59, 61, 63
shifting cultivation 209, 211, 215, 216, 222, 217
Shorea 15, 58, 61, 138, 180, 196
  albida 56
  curtisii 62, 71
  javanica 248
  leprosula 75, 93, 115
  ovalis 123, 159
  parvifolia 117
  platyclados 117
short-lived shrubs 242
shrews 86, 192
shrubs 8
siamang 161, 192
Siberut Island 135
Sierra Leone 7
Silene 13
silviculture 227, 232
Singapore 117
sloths 20, 86, 140
snails 22
snakes 9, 85
Socratea durissima 72, 197
  exorrhiza 103
soil erosion 25
soil temperature 69
soil types 32, 236
Solanaceae 174
Solanum 13, 98
Solomon Islands 200
Sonerila 111, 112
South America 3
Spathelia 96
speciation 12, 187
species diversity 12, 86, 192
species patterns 195
species richness 195
spices 212
spider-monkeys 140, 160
spiders 78, 186, 245
Spirogyra 1
spittle-bugs 148
Spondias 105
squirrels 192
Sri Lanka 120, 194, 238, 258
starlings 88, 131
Stemmadenia donnel-smithii 158
Sterculiaceae 72, 177
stilt-roots 18, 103

storms 55, 56, 59, 194
strangling figs 105
Strobilanthes 112
strychnine 217
Stylosanthes humilis 220
succession 3, 53, 54, 56, 70, 75
sugar 234
sugar maples 137
sugarbirds 179
sugarcane 45
Sulawesi 23
sumach 98, 101
Sumatra 58
sunbirds 170, 172, 179
sunflecks 29, 94
Surinam 224, 242
swamp-cypress 103
Swietenia 72, 94, 228
  macrophylla 11, 229
Swiss cheese plant 104

Tabebuia 260
  rosea 123
Talbotiella gentii 13
talipot 96
tamarin monkeys 181
Tamarindus indica 219
tanagers 88, 89, 131, 189
tannins 38, 71, 134, 138, 139, 141, 148, 161
Tanysiptera galatea 130
tapirs 20
taproots 101
Taraxacum 13
Taxodium distichium 103
tea 124, 141
teak 123
Tectona grandis 123
temperate forests 45, 49
temperature 29
Terminalia 208
  catappa 123
  ivorensis 69
termites 32, 49, 50, 58, 91, 92, 137, 245
terpenoids 143
terra firme 6, 37, 92
Tertiary 18, 20
Tetragastris panamensis 158
Thailand 46, 66, 217
thaumatin 142
Three Kings Island 167
thrips 127, 171, 176, 180, 196
ticks 74
Tilia cordata 10
Tillandsia medusae 107

timber 15, 260
Timonius 105
  compressicaulis 78
tin 226
toad 74
Tobago 169
Toona ciliata 123
topography 195
toucans 87, 89, 184, 189
toxins 71, 136, 149
transmigration 225
trans-Saharan seaway 18
tree architecture 63, 117
tree form 92
Trema 48, 66, 117, 242
  micrantha 64
Trigona 127
  fulviventris 174
  pallens 174
trigonid bees 174, 182
Trinidad 169 ,196
triterpenoids 142
Tropical Forestry Action Plan 257
tropical climate 23
tropical moist forest 4, 250
tualang 216
Turraea 72
typhoons 27

Uganda 159
Ulex 163
Ulmaceae 12
Ulmus parvifolia 135
ultisols 32, 33, 43, 44, 45
ultrabasic 36
ultramafic 38, 39
Umbelliferae 12
UNESCO 254
ungulates 86, 92
Upper Baram River 127
Upper Carboniferous 17
Upper Cretaceous 19
Urtica 12
Urticaceae 12, 112

vanillin 141
varzea 37, 43, 44, 228

Venezuela 11, 28, 38, 69, 200
Veracruz 125, 220
Verbena 12
Vernonia 13
vertebrates 83, 87, 129
Vinca 13
vincristine 260
vine forest 6
Viola 163
Virola surinamensis 71
Viscaceae 23, 106, 151
Viscum album 106, 155
volcanic deposits 48
volcanic soils 34, 239
vulcanoseres 55

Wallace's line 23
warfare 214, 227
wasps 133, 170
water-dispersed palms 164
weaver ant 164
weevils 3, 71, 121
white-faced monkeys 158
Wightia 8, 105
willows 100
wind dispersal 72, 75, 157, 164, 165
wind-dispersed seeds 66, 67, 152
wind pollination 75, 170, 171, 182
windthrow 65
Winteraceae 23
wood 62, 227

xeromorphic leaves 39
Xerospermum intermedium 118
  noronhianum 118, 171
Xylocarpus 103, 129
Xylocopa 127, 170, 171, 174

yams 13, 209
yellow fever 245
yohimbine 260

Zaire 24, 50, 224
Zingiberaceae 108, 111, 112
Zosterops japonica 167
Zunacetha annulata 129
Zygomycetes 48